Lecture Notes in Network 1973

The series "Lecture Notes in Networks and Systems" publishes the latest developments in Networks and Systems—quickly, informally and with high quality. Original research reported in proceedings and post-proceedings represents the core of LNNS.

Volumes published in LNNS embrace all aspects and subfields of, as well as new challenges in, Networks and Systems.

The series contains proceedings and edited volumes in systems and networks, spanning the areas of Cyber-Physical Systems, Autonomous Systems, Sensor Networks, Control Systems, Energy Systems, Automotive Systems, Biological Systems, Vehicular Networking and Connected Vehicles, Aerospace Systems, Automation, Manufacturing, Smart Grids, Nonlinear Systems, Power Systems, Robotics, Social Systems, Economic Systems and other. Of particular value to both the contributors and the readership are the short publication timeframe and the world-wide distribution and exposure which enable both a wide and rapid dissemination of research output.

The series covers the theory, applications, and perspectives on the state of the art and future developments relevant to systems and networks, decision making, control, complex processes and related areas, as embedded in the fields of interdisciplinary and applied sciences, engineering, computer science, physics, economics, social, and life sciences, as well as the paradigms and methodologies behind them.

Indexed by SCOPUS, EI Compendex, INSPEC, WTI Frankfurt eG, zbMATH, SCImago.

All books published in the series are submitted for consideration in Web of Science.

For proposals from Asia please contact Aninda Bose (aninda.bose@springer.com).

Zineddine Kouahla · Karima Benhamza ·
Braim Farou · Hamid Seridi ·
Kurulay Muhammet · Khaled Halimi
Editors

Proceedings of the 8th International Conference on Informatics and Applied Mathematics (IAM'25)

Recent Advances in Artificial Intelligence, Cybersecurity, and Computational Models

Springer

Editors
Zineddine Kouahla
Computer Science Department
Université 8 Mai 1945 Guelma
Guelma, Algeria

Karima Benhamza
Computer Science Department
Université 8 Mai 1945 Guelma
Guelma, Algeria

Braim Farou
Computer Science Department
Université 8 Mai 1945 Guelma
Guelma, Algeria

Hamid Seridi
Computer Science Department
Université 8 Mai 1945 Guelma
Guelma, Algeria

Kurulay Muhammet
Istanbul Technical University
Istanbul, Türkiye

Khaled Halimi
Computer Science Department
Université 8 Mai 1945 Guelma
Guelma, Algeria

ISSN 2367-3370 ISSN 2367-3389 (electronic)
Lecture Notes in Networks and Systems
ISBN 978-3-032-26030-7 ISBN 978-3-032-26031-4 (eBook)
https://doi.org/10.1007/978-3-032-26031-4

This Springer imprint is published by the registered company Springer Nature Switzerland AG
The registered company address is: Gewerbestrasse 11, 6330 Cham, Switzerland

Contents

When AI Becomes an Engineer: Large Language Models in Structural Design ... 1
Elhocine Mansouri, Kamal Eddine Melkemi, Dominique Michelucci, and Bilal Mokhtari

Copyright on the Internet in the Era of Artificial Intelligence: Technical Methods, Information Retrieval, and Blockchain for Enhanced Protection ... 13
Sonia Benbelgacem, Mohamed Benbelgacem, Larbi Guezouli, and Rachid Seghir

Bio-Inspired Feature Selection for Deep Learning-Based Intrusion Detection in IoMT ... 22
Karima Benhamza, Rania Benselim, and Hadil Merzougui

Binary Moth-FlameBased Task Mapping for Energy Minimization in Heterogeneous MPSoCs ... 30
Farid Boumaza, Atmane Hadji, Djaafar Zouache, and Abdelkader Aroui

Hybrid Deep Learning for Short-Term Load Forecasting: A Comparative Study on Algerian Buildings ... 44
Rania Farah, Rahma Aissaoui, and Zineddine Kouahla

Evaluating Drinking Water Quality Through Machine Learning ... 57
Fatima Bouakkaz and Wided Ali

Exploring Industry 5.0: Architecture, Threat Landscape, and Security Strategies ... 65
Djallel Hamouda, Mohamed Amine Ferrag, and Zineddine Kouahla

Towards Generative Context-Aware Communication in Smart Transport Environments ... 77
Roudeina Guerfi, Ahmed-Chawki Chaouche, and Rania Zebiri

A Q-Learning Weighted Cluster Based Routing Protocol for VANETs ... 95
Ahlam Boussadia and Manel Khelifi

A Framework for Business Process Integration in Smart Grids Using Knowledge Graphs ... 114
Ibtissem Bouhaouche and Ali Khebizi

Sentiment-Enhanced Hybrid Models for Arabic Fake News Detection 132
Chatta Asmaa, Bouakkaz Mustapha, and Latreche Boubakeur

Evaluation of Handcrafted and Learning-Based Keypoint Detection and Description Methods in Image Matching 150
Adel Sardou, Abdenebi Rouigueb, Fethi Demim, Abdenour Amamra, and Hadjira Belaidi

A Novel Multi-sensor Fusion Framework for Robust 3D Vision: Comprehensive Calibration and Performance Analysis of ToF, Stereoscopic, and RGB-D Systems 166
Imad Benkhaled, Farouq Zitouni, and Badreddine Ounnas

Enhanced Skin Lesion Classification: A Deep Feature Extraction-Based Approach for Multi-classification of Skin Cancer Using HAM Dataset 188
Samaa Mukhaimar, Akila Djebbar, Ghada Halladja, and Hayet Farida Merouani

A Hybrid Deep Learning Model for Precise ECG Arrhythmia Classification ... 211
Saoueb Kerdoudi, Larbi Guezouli , Tahar Dilekh , and Abdessalam Hattab

The Algerian Forest Fires Dataset 224
Faroudja Abid

Deep Learning-Driven Model for Accurate Detection of Breast Cancer from Histopathology Images: Clinically Validated on Algerian Medical Data 234
Ranya Hadjadj, Abdelkrim Bouramoul, Djamel Benmerzoug, Redouane Daddiouamer, and Lokmane Abdessalam

A Novel Vocal Database for Children with Autism Spectrum Disorder: Design, Collection, and Future Applications 250
Kadi Imane and Haloua Fatima

Random Forests for Large-Scale Hyperspectral Pixelwise Plastic Polymer Classification in Aquatic Environments 258
Samir Hallaci, Wassim Messioud, Mohammed Chaib, Achraf Yahiaoui, Abderrahman Barkache, and Hamid Seridi

AI-Driven Prediction of Diabetes: Performance Evaluation Using Machine and Deep Learning Models 283
Atmane Hadji and Farid Boumaza

Author Index 301

When AI Becomes an Engineer: Large Language Models in Structural Design

Elhocine Mansouri[1,5](✉), Kamal Eddine Melkemi[2], Dominique Michelucci[3], and Bilal Mokhtari[1,4]

[1] Department of Computer Science, University of Biskra, Biskra, Algeria
hocine.mansouri@hotmail.com

[2] Department of Computer Science, University of Batna 2, Batna, Algeria
ka.melkemi@univ-batna2.dz

[3] LIB (Laboratoire Informatique de Bourgogne), UR, 7534 Dijon, France
dominique.michelucci@orange.fr

[4] LAMIE Laboratory, University of Batna 2, Route de Fesdis, 05000 Batna, Algeria
bilal.mokhtari@univ-biskra.dz

[5] LESIA Laboratory, University of Mohamed Khider, 07000 Biskra, Algeria

Abstract. Traditional structural optimization using evolutionary algorithms requires thousands of expensive FEM evaluations, where robust methods are impeded by a sluggish pace that hinders authentic design innovation. This work challenges the traditional approach, reconceptualizing the Large Language Model (LLM) not merely as a linguistic instrument but as a perceptive design agent ready to overcome the current limitations. Our paradigm propels the LLM into an iterative design process. Here, it is assigned the responsibility of proposing innovative I-beam shapes, although its inventiveness is promptly moderated by discipline. Every proposed design must initially undergo a mandatory engineering evaluation, a mechanism that assesses its viability against established manufacturing protocols and rectifies fundamental faults prior to the expensive undertaking of full-scale simulation in FreeCAD. The agent possesses a recollection of its historical experiences, documenting both failures and successes, which enables it to refine its strategy instead of reiterating its errors. To assess this novel methodology, we subjected the LLM agent to a series of rigorously controlled, time-constrained experiments within a constrained computational budget. In our initial experiments, the agent generated compliant I-beam designs-lighter, more efficient, and entirely compatible with all safety regulations-achieving 64.7% weight reduction over 8 design iterations, demonstrating the viability of LLM-guided structural optimization with significantly reduced computational cost compared to traditional evolutionary approaches. Presenting preliminary results from an ongoing research project, this study affirms that a large language model grounded in engineering physics can act as an effective design co-pilot, integrating its generative reasoning with the precision of conventional simulation to forge a more rapid, intuitive, and potent future for engineering operations.

Keywords: Large Language Models · Structural Optimization · I-Beam Design · Engineering Automation · Generative Design

Z. Kouahla et al. (Eds.): IAM 2025, LNNS 1973, pp. 1–12, 2026.
https://doi.org/10.1007/978-3-032-26031-4_1

1 Introduction

The field of structural optimization has historically presented two opposing realities: the significant capability of computational tools to identify effective designs and the exorbitant expense of the brute-force techniques necessary to achieve this. Evolutionary algorithms, albeit resilient, traverse an extensive design space, necessitating thousands of resource-intensive simulations [30,31]. This lethargic tempo constitutes more than a mere inconvenience; it represents a fundamental obstacle that stifles authentic design innovation, compelling engineers to accept solutions that are merely satisfactory rather than genuinely ideal. This study dismantles the enduring paradigm. We contend that the period of awaiting computation has concluded and introduce a novel methodology that redefines the Large Language Model (LLM) as an astute design agent, proficient in traversing the engineering domain with an efficiency unparalleled by conventional methodologies.

Within this realm of computational limitations arises the LLM, a technique originating from the field of natural language. The requisite skepticism is evident: How can a model trained on text and human discourse grasp the stringent principles of physics and material science? Requesting it to create an I-beam, a fundamental component dictated by exact mathematics, appears to constitute a category error. This paper contends that the actual potential of the LLM is realized not when it is regarded as an isolated oracle, but when it is incorporated as a fundamental reasoning engine within a framework that anchors its creativity in engineering principles [2,5].

Consequently, our methodology integrates the LLM within an iterative design process wherein its creativity is promptly limited by practical considerations. We have established a system in which the agent's conceptual contributions are initially subjected to a compulsory "engineering guardrail." This system assesses each design against predefined manufacturing standards and autonomously corrects fundamental errors prior to the costly initiation of a comprehensive Finite Element Method simulation in FreeCAD. The agent possesses a recollection of its historical experiences, encompassing both failures and successes, enabling it to refine its approach and leverage its distinct experiences instead of mindlessly repeating errors.

To evaluate this innovative methodology, we subjected our LLM agent to rigorous optimization challenges with strict time and computational constraints. The agent was assigned a singular challenge: to identify the lightest feasible I-beam capable of supporting a designated load while adhering to all safety rules [15,16]. The results, as this study will demonstrate, are promising. In representative demonstration runs, the agent achieved substantial weight reductions (64.7% in 8 design iterations) while maintaining AISC compliance, demonstrating significantly improved efficiency compared to traditional evolutionary approaches that typically require hundreds of evaluations.

The remainder of this paper is organized to elucidate this inquiry. We shall initially analyze the particular constraints of existing optimization techniques that inspired our methodology. Subsequently, we will delineate the complete architecture of our hybrid agent, elucidating the interaction among the LLM, the engineering guardrail, and the memory system. Subsequently, we will provide a comprehensive analysis of our experimental findings, followed by a discourse on the wider ramifications of our research and the future of a more effective and intuitive framework for engineering operations.

2 Related Works

The realm of structural optimization has historically been governed by the esteemed category of evolutionary algorithms [30,31]. Techniques like the Genetic Algorithm (GA) and its multi-objective variant, NSGA-II, have functioned as the steadfast engine for design space exploration for decades [15,18,19]. Their strength resides in their stochastic, population-centric methodology, adept at traversing intricate, non-convex challenges to identify resilient solutions. This power relies on a brute-force methodology: the assessment of thousands, or even tens of thousands, of potential designs. When each assessment requires a complete Finite Element Method (FEM) simulation, this procedure exemplifies computational endurance rather than design flexibility [21,23]. The primary search approach, albeit efficient, is oblivious to the fundamental physics, regarding the simulation as an opaque entity that requires continuous interrogation.

The evident and requisite reaction to this computational constraint was the creation of surrogate-assisted optimization [26]. In this framework, machine learning models—such as Gaussian processes or artificial neural networks—are trained on an initial dataset of simulations to generate a more economical approximation of real-world physics [1,27,28]. The evolutionary algorithm primarily utilizes this expedited surrogate model for most evaluations, resorting to the costly, high-fidelity simulation only intermittently to enhance its comprehension. This method significantly expedites the process; however, it does not fundamentally change the essence of the search. The evolutionary algorithm continues to be the catalyst, employing its fundamental "guess and check" methodology on a more efficient representation of the design space.

A new and fundamentally distinct rival, the Large Language Model, has emerged inside this established framework [2,5,11]. Preliminary investigations into its application in engineering have shown promise, albeit primarily limited to particular, supportive functions [12–14]. Large Language Models (LLMs) exhibit an exceptional ability to produce scripts for CAD software, convert natural language inquiries into simulation parameters, and offer advanced conceptual recommendations derived from the extensive repository of engineering knowledge within their training data [3,10]. These initial explorations, albeit beneficial, have predominantly regarded the LLM as an advanced intern-a resource utilized for a singular task prior to the conventional, algorithmic optimization process assuming control.

This study, however, suggests a fundamentally distinct function for the LLM. We transform it from a passive tool into an active, reasoning entity at the core of the optimization process. Unlike other research that utilized LLMs for design generation [9,10], our approach is the inaugural one to integrate this generation within a closed-loop framework of grounded validation and adaptive learning. The implementation of an "engineering guardrail" that rectifies the LLM's outputs is a vital differentiation, tackling the model's intrinsic deficiency in physical intuition and guaranteeing the feasibility of its suggestions [4]. Moreover, by endowing the agent with a recollection of its simulated experiences, our methodology enables it to execute a more focused, context-sensitive search that adapts its strategy based on immediate physical feedback [6,7]. This blend of generative reasoning, grounded validation, and adaptive learning distinguishes our research from previous work and constitutes its primary contribution.

3 Method

3.1 System Architecture

Our methodology is based not solely on the Large Language Model (LLM), but on a meticulously designed system that regulates its intelligence. **We employ MiniMax-M2(free) (382B tokens) as the reasoning engine, accessed via API with temperature = 0.7 for balanced creativity and consistency.** We have developed a closed-loop, hybrid system in which the LLM does the key cognitive functions of design generation [1,8]. This ingenuity, however, is consistently limited by the stringent constraints of engineering physics. The integration of generative reasoning with physical discipline produces a design approach that is both dependable and astute, proficient in addressing intricate optimization challenges with unparalleled efficiency.

Figure 1 illustrates our hybrid optimization framework.

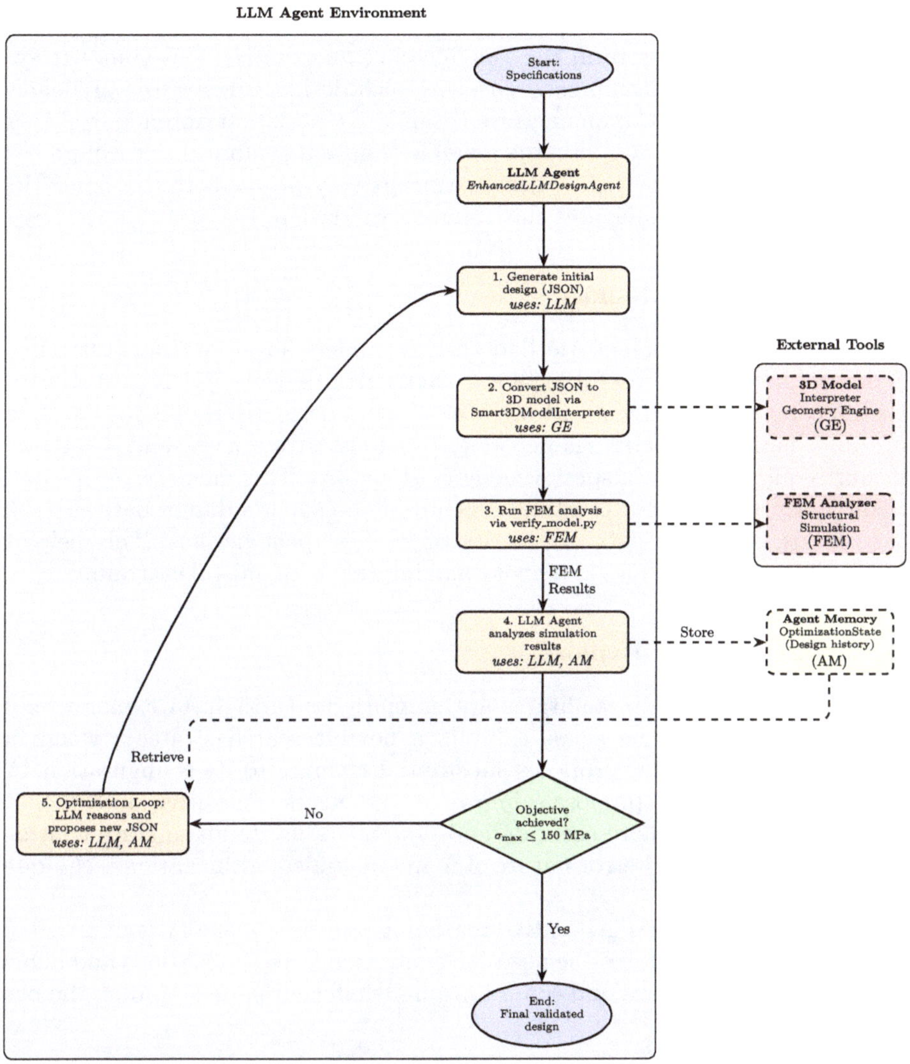

Fig. 1. LLM-guided I-beam optimization workflow integrating generative design with physics-based validation. The agent iteratively proposes designs (1), converts to geometry via external interpreter (2), performs FEM analysis (3), stores results in memory (4), and reasons using historical context to propose improvements (5). Agent Memory serves dual purpose: storing current iteration results (from step 4) and providing past design history for reasoning (to step 5).

3.2 The Iterative Design Cycle

The agent's exploration of the design space follows a structured, iterative cycle [7]. Informed by the results of the previous iteration, the LLM proposes a new I-beam geometry, which is articulated in a structured JSON format. Before this proposal can proceed to a full evaluation, however, it must pass through a critical screening mechanism we term the "engineering guardrail." This compiler system functions as a validation gate, ensuring each design adheres to a predefined schema and official AISC manufacturing standards [15,20]. It autonomously corrects elementary errors and discards inviable proposals before the costly process of FEM modeling is initiated. A design concept only proceeds to the simulation stage if it successfully navigates this rigorous validation.

3.3 Learning and Memory

High-fidelity Finite Element Method (FEM) models do more than just tell us yes or no; they give us useful feedback that helps us learn [21,23]. Memory is very important for the agent in this case. When a design iteration goes well, we store information about its shape, weight, highest stress level, and whether it met our requirements in a special memory structure. This memory recalls both the successes and failures, creating a record of the agent's design path [6,7]. In later rounds, the LLM gets summaries of its best past designs. This helps it improve its way of thinking and avoid making the same mistakes again.

3.4 Smart Ways to Explore

This memory-driven loop facilitates a systematic and insightful exploration of the design space [8]. The agent exhibits a flexible search strategy; when far from the best solution, it proposes substantial changes to its configuration. As the effectiveness of its proposals improves and nears a positive outcome, its modifications become more accurate and gradual. This stands in stark contrast to the random, trial-and-error nature of many unguided optimization techniques [26].

This system combines generative reasoning, episodic memory, and an adaptive strategy, which removes the need for many ineffective evaluations and allows for a more straightforward and computationally efficient path to finding the best solution.

3.5 LLM-Guided Design Refinement

To illustrate the agent's reasoning process, consider the interaction at Iteration 1 (see Table 1). Given an oversized initial design ($H = 300$ mm, $W = 150$ mm, mass = 364.6 kg) showing AISC non-compliance, the LLM was prompted with current performance metrics and design requirements. The agent responded by proposing a proportionally reduced design ($H = 280$ mm, $W = 140$ mm, mass = 303.3 kg), demonstrating physics-informed reasoning rather than random exploration. Complete prompt-response transcripts are available in supplementary materials.

The complete logic of this iterative design process is formalized in Algorithm 1.

Algorithm 1. LLM-Guided Design Loop

```
procedure OptimizeWithLLM(D_init, C, I_max)
    D ← D_init; D_best ← null; m_best ← ∞; M ← ∅
    R_prev ← null
    for i ← 1 to I_max do
        P ← BuildPrompt(D, R_prev, M)
        resp ← QueryLLM(P)                          ▷ Get raw response
        D_prop ← ExtractJSON(resp)                  ▷ Parse design
        if D_prop = null then
            continue
        end if
        D_checked ← Guardrail(D_prop)
        R ← RunFEM(D_checked)
        M ← M ∪ {(D_checked, R)}                    ▷ Update memory
        if IsCompliant(R, C) and R.mass < m_best then
            m_best ← R.mass
            D_best ← D_checked
        end if
        D ← D_checked
        R_prev ← R
    end for
    return D_best
end procedure
```

4 A Case Study: The I-Beam Optimization Challenge

We chose the improvement of a simply supported I-beam as our main case study so that our research could be based on a real engineering problem [15–17]. The goal is both simple and complicated at the same time: to find the geometric cross-section with the least amount of mass that can safely hold a large load without breaking. Our design agent will be judged on how well it solves this problem because its physics are well known and provide a clear and unforgiving standard.

Engineers were very clear about what the task was and how to do it. An I-beam that was 6000 mm long and made of A992 grade steel was loaded evenly with 20 kN/m worth of force. The design had to follow the strict safety rules of the AISC 360-16 standard, which sets strict limits on the most stress and deformation that can be applied [20]. There is no theory behind this exercise; it is a direct copy of a common engineering job that you would find in the real world.

To validate our methodology, we conducted rigorous demonstration runs under controlled conditions using our high-fidelity simulation pipeline-our automated FreeCAD and CalculiX environment [21,23]. Each optimization was constrained by a 30-minute time limit to simulate realistic design cycles, with the agent's performance evaluated by its ability to navigate the design space efficiently and achieve compliant designs within the computational budget.

5 Preliminary Results and Analysis

Our preliminary results demonstrate the starting point for our ongoing work. In our demonstration runs, the LLM agent showed a very high level of computational speed. The agent consistently generated high-quality, compliant designs within 8–10 design iterations, achieving near-optimal solutions in a surprisingly short amount of time. This represents a substantial reduction in exploration cost compared to evolutionary approaches requiring hundreds of evaluations [26,28].

The quality did not suffer because of this fast agreement. The agent always made sure that the end designs were highly optimized, pushing the limits of how efficiently materials could be used while still following all safety rules to the letter. The agent wasn't just quick; it was also smart. It used the limited amount of computing power it had to find better answers in the huge design space.

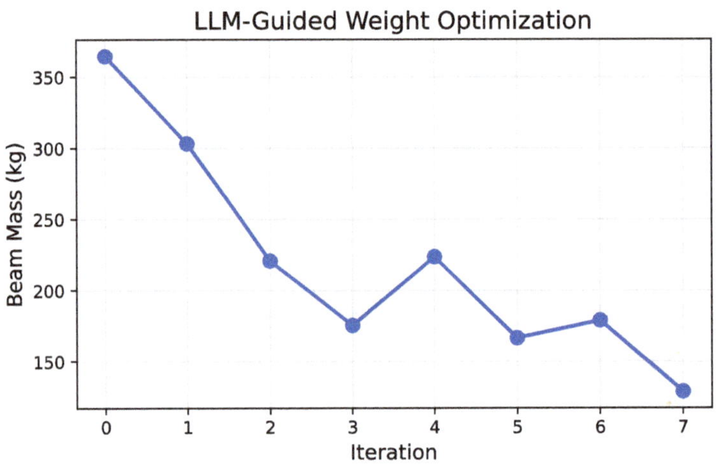

Fig. 2. LLM-Guided I-beam weight optimization convergence. The plot shows the beam mass (in kg) decreasing over 8 design iterations, demonstrating rapid and significant weight reduction.

Table 1 illustrates a significant optimization trajectory derived from our demonstration runs. The process began with an oversized initial design (H = 300 mm, W = 150 mm, mass = 364.6 kg), from which the LLM agent methodically reduced weight by implementing strategic dimensional adjustments, resulting in a final design weight of 128.9 kg-a significant 64.7% reduction in mass. The agent exhibited a methodical approach to reasoning throughout each iteration, as indicated by proportional dimension adjustments instead of random exploration (refer to supplementary LLM interaction logs).

Table 1. Representative Design Evolution Over 8 Iterations

Iter	H	W	t_w	t_f	Mass	Reduction
0	300	150	12.0	15.0	364.6	–
1	280	140	10.0	14.0	303.3	16.8%
2	250	120	8.0	12.0	220.8	27.2%
3	254	127	5.8	9.3	175.6	20.5%
7	300	120	4.5	6.0	128.9	26.6%
Total Weight Reduction:					**64.7%**	

The convergence across eight design iterations is clearly illustrated in Fig. 2. Beginning with the initial mass, the LLM agent skillfully explored the design space, revealing a clear trend of decreasing mass. The trajectory displays variability; for example, a temporary mass increase is observed at iteration 4 before subsequent optimization takes place. This non-monotonic pattern demonstrates the LLM's capacity for adaptive and intelligent search strategies, as it may explore slightly heavier yet potentially more compliant or robust designs in certain areas of the search space before identifying new pathways to lighter, valid solutions. This dynamic investigation highlights the differences when compared to the often monotonous and less efficient descent trajectories of traditional unguided optimization algorithms.

5.1 Design Space Exploration Pattern

An examination of the optimization trajectory uncovers clearly defined phases of exploration. Phase 1 (Iterations 0–2) exhibited significant dimensional reduction, resulting in a 39.4% reduction in mass (from 364.6 kg to 220.8 kg) by proportionally scaling all geometric parameters. Phase 2 (Iterations 3–5) demonstrated a process of refinement through strategic backtracking; iteration 4 adjusted the mass to 223.6 kg to address new stress constraints, after which optimization was resumed to achieve a mass of 166.6 kg. Phase 3 (Iterations 6–7) demonstrated a convergence towards the optimal region through precise adjustments. This stepwise strategy differs from the consistent decline characteristic of gradient-based techniques and the arbitrary exploration seen in evolutionary algorithms. [26,30].

5.2 Analysis of Guardrail Effectiveness

Throughout the eight design iterations, the engineering guardrail system effectively identified and rectified design violations on three occasions (iterations 0, 3, and 4), thereby ensuring that flawed designs did not progress to the expensive FEM simulation phase. Corrections encompassed the implementation of minimum web thickness ratios and flange width-to-thickness constraints in accordance with AISC 360-16 specifications [20]. The absence of this validation layer would have led to failures in these designs during FEM simulation, resulting in wasted computational resources and necessitating further LLM queries. This highlights the important function of domain-specific validation in LLM-guided optimization, acting as a vital link between generative AI reasoning and stringent engineering standards.

What makes this performance possible is the core of our agent's search approach. When it comes to black-box optimization, our LLM agent acts more like an experienced expert than a normal person [4,9]. It starts by making big, natural leaps based on its analysis of why the first plan didn't work. As soon as it finds a promising part of the design space, after a few targeted iterations, it changes its approach to fine-tuning the parameters. This way of exploration based on reasoning is very different from random, trial-and-error approaches. It makes it clear why our agent was able to accomplish so much in such a short amount of time.

6 Conclusion and Future Work

We presented a novel LLM-guided optimization framework for structural engineering that integrates generative reasoning with physics-based validation [1,2]. Our preliminary results on I-beam optimization demonstrate the feasibility of using language models as design agents, achieving compliant designs with significantly fewer iterations than traditional approaches [4,9].

Current Status: We have completed initial demonstration runs validating the core methodology. Our representative 8-iteration optimization achieved 64.7% weight reduction while maintaining AISC compliance, confirming the viability of our two-phase exploration strategy with engineering guardrails.

Ongoing Work: We are expanding our experimental validation to include:

- Statistical validation across multiple optimization runs
- Comprehensive baseline comparisons with GA and surrogate-assisted methods [18,26,28]
- Ablation studies on memory and guardrail components
- Extension to additional structural problems (trusses, frames) [22–24]

The full results and analysis will be presented in our extended journal paper currently in preparation.

Data Availability Statement. The complete LLM interaction logs, design evolution data, and FEM analysis results from our demonstration runs will be made available at the LLM-Driven-CAD-Designer Repo upon publication.

AI Disclosure. The corresponding author acknowledges the use of AI-assisted tools solely to enhance the clarity, grammar, and readability of this manuscript. No AI tools were used to generate research ideas, experimental results, data analysis, or figures. All scientific content, methodology, and findings are the original work of the authors.

References

1. Rios, T., Lanfermann, F., Menzel, S.: Large language model-assisted surrogate modelling for engineering optimization. In: Proceedings of 2024 IEEE Conference on Artificial Intelligence (CAI), Singapore, pp. 796–803 (2024)
2. Chiarello, F., Barandoni, S., Škec, M.M., Fantoni, G.: Generative large language models in engineering design: opportunities and challenges. Proc. Des. Soc. **4**, 1959–1968 (2024)
3. Jiang, S., Xie, M., Luo, J.: Large language models for design structure matrix optimization (2025). arXiv:2506.09749
4. Chien, C.T., Chien, C.H.: Hybrid AI-driven computer-aided engineering optimization: large language models versus regression-based models validated through finite-element analysis. Appl. Sci. **15**(18), 10123 (2025)
5. Mustapha, K.B.: A survey of emerging applications of large language models for problems in mechanics, product design, and manufacturing. Adv. Eng. Inf. **64**, 103066 (2025)
6. Jiang, Z., Tang, Q., Wang, Z.: Generative reliability-based design optimization using in-context learning capabilities of large language models (2025). arXiv:2503.22401
7. Liu, F., et al.: Large language model for multi-objective evolutionary optimization (2023). arXiv:2310.12541
8. Zhong, R., Xu, Y., Zhang, C., Yu, J.: Leveraging large language model to generate a novel metaheuristic algorithm with CRISPE framework. Clust. Comput. **27**, 13835–13869 (2024)
9. Zhang, X., et al.: Using large language models for parametric shape optimization. arXiv:2412.08072 (2024)
10. Ma, K., Grandi, D., McComb, C., Goucher-Lambert, K.: Exploring the capabilities of large language models for generating diverse design solutions. arXiv:2405.02345 (2024)
11. Khoo, K., Lee, S.H.: A comparative review of large language models in engineering with emphasis on chemical engineering applications. Processes **13**(9), 2680 (2025)
12. Pan, J., Zhou, Z., Chang, J., Jacobson, P., Hu, P., Chen, T.: A survey of research in large language models for electronic design automation. ACM Trans. Design Autom. Electron. Syst **30**(3) (2025)
13. Rong, Y., Du, T., Li, R., Bao, W.: Integrating LLM-based code optimization with human-like exclusionary reasoning for computational education. J. King Saud Univ. Comput. Inf. Sci. **37**, 87 (2025)
14. Zhang, J., Han, J., Ahmed-Kristensen, S.: Exploring the use of LLMs to evaluate design creativity. Proc. Des. Soc. **5**, 1773–1782 (2025)
15. Martins, J.P., Correia, R., Ljubinković, V., Simões da Silva, L.: Cost optimisation of steel I-girder cross-sections using genetic algorithm. Structures **55**, 379–388 (2023)
16. Zhang, Z.Y., Li, Z.-M., Yu, Y., Yang, Y.: A FCM based genetic algorithm for I-beam optimization design. Mod. Mach. Tool Autom. Manuf. Techn. **7**, 68–72 (2017)
17. Bennoui, I., Fedghouche, F.: Contribution on the dimensioning of a doubly reinforced prestressed concrete I-beam using GRG and DE optimization. Iran. J. Sci. Technol. Trans. Civ. Eng. **49**, 323–340 (2025)
18. Li, P., et al.: Optimization design for steel trusses based on a genetic algorithm. Buildings **13**(6), 1496 (2023)

19. Whitworth, A.H., Tsavdaridis, K.D.: Genetic algorithm for embodied energy optimisation of steel-concrete composite beams. Sustainability **12**(8), 3102 (2020)
20. Szeptyński, P., Mikulski, M.: Preliminary optimization technique in the design of steel girders. Arch. Civil Eng. **69**(1), 71–89 (2023)
21. Sych, P., Słoński, M.: Structural design optimization of steel beams and frames with web-tapered members using the PSO-FEM algorithm. Comput. Assist. Methods Eng. Sci. **28**(1), 39–55 (2021)
22. Masarczyk, D., Kuhl, D.: Topology optimization considering self-weight. Proc. Appl. Math. Mech. **23**(1), e202300258 (2023)
23. Changizi, N., Jalalpour, M.: Topology optimization of steel frame structures with constraints on overall and individual member instabilities. Finite Elem. Anal. Des. **141**, 119–134 (2018)
24. Ribeiro, T.P., Bernardo, L.F.A., Andrade, J.M.A.: Topology optimisation in structural steel design for additive manufacturing: a systematic review. Appl. Sci. 11(5), 2112 (2021)
25. Domagalski, Ł, Kowalczyk, I.: Genetic algorithm optimization of beams in terms of maximizing gaps between adjacent frequencies. Materials **16**(14), 4963 (2023)
26. Liang, J., Lou, Y., Yu, M., Bi, Y.: A survey of surrogate-assisted evolutionary algorithms for expensive optimization. J. Membrane Comput. **7**, 108–127 (2025)
27. Lima, J.P.S., Vieira, R.L., dos Santos, E.D., Rocha, L.A.O., Isoldi, L.A.: Multi-objective optimization based on Kriging surrogate model and genetic algorithm for stiffened panel collapse assessment. Appl. Mech. **6**(2), 34 (2025)
28. Velasco, L., Guerrero, H., Hospitaler, A.: Seismic optimization of buckling-restrained brace mid-rise frames by metaheuristics and machine learning surrogative models. Struct. Multidiscip. Optim. **68**, 62 (2025)
29. Wang, W., Ye, H., Chen, Y., Li, J.: Sheet metal forming optimization by using surrogate modeling techniques. Chinese J. Mech. Eng. **30**, 22–36 (2017)
30. Prendes-Gero, M.B., Bello-García, J.C., del Coz-Díaz, J.J., Suárez-Domínguez, P.J.G., García Nieto, P.J.: Optimization of steel structures with one genetic algorithm according to three international building codes. Revista de la Construcción **17**(1), 47–59 (2018)
31. Mei, L., Wang, Q.: Structural optimization in civil engineering: a literature review. Buildings **11**(2), 66 (2021)
32. Narques, T.V.N., Carvalho, R.C., Christoforo, A.L., Mascarenhas, F.J.R., Arroyo, F.N., Bomfim, F.C., Jr., Santos, H.F.: Use of real coded genetic algorithm as a pre-dimensioning tool for prestressed concrete beams. Buildings **13**(3), 819 (2023)

Copyright on the Internet in the Era of Artificial Intelligence: Technical Methods, Information Retrieval, and Blockchain for Enhanced Protection

Sonia Benbelgacem[1,3](✉), Mohamed Benbelgacem[1,4], Larbi Guezouli[2,5], and Rachid Seghir[1,3]

[1] Department of Computer Science, University of Batna 2, Batna, Algeria
sonia.benbel@univ-batna2.dz
[2] Higher National School of Renewable Energies, Environment and Sustainable Development, Batna, Algeria
[3] Laboratoire des Systèmes Et Technologies De L'Information Et De La Communication (LASTIC), University of Batna 2, Batna, Algeria
[4] Laboratoire d'Applications des Mathématiques à l'Informatique et à l'Électronique (LAMIE), University of Batna 2, Batna, Algeria
[5] The Laboratory of Renewable Energy, Energy Efficiency and Smart Systems (LEREESI), Higher National School of Renewable Energies, Environment and Sustainable Development, Batna, Algeria

Abstract. The rise of artificial intelligence (AI) technologies, particularly generative models for text, image, audio, and video, is radically transforming the creation and dissemination of digital content. These technologies provide significant opportunities, but also pose major challenges for copyright protection in terms of traceability, attribution, license management, and anti-counterfeiting measures.

Traditional protection methods, such as digital watermarking, fingerprinting, and Digital Rights Management (DRM) systems, often prove insufficient in the face of AI's ability to generate or transform works based on existing data. At the same time, information retrieval (IR) and blockchain emerge as promising tools for detecting, verifying, and certifying digital works.

This paper analyzes the impact of artificial intelligence on copyright, reviews recent work, describes technical protection methods, examines the role of information retrieval and blockchain in safeguarding these rights, and identifies upcoming technical, legal, and ethical challenges.

Keywords: Copyright · Artificial intelligence · Information retrieval · Blockchain · Technical methods for copyright protection

1 Introduction

Artificial intelligence (AI) is experiencing unprecedented growth and is profoundly reshaping the way digital content is conceived, produced, and distributed. Powerful models such as GPT for text generation [1], DALL•E for

Z. Kouahla et al. (Eds.): IAM 2025, LNNS 1973, pp. 13–21, 2026.
https://doi.org/10.1007/978-3-032-26031-4_2

image creation from simple descriptions [2], Midjourney for artistic production [3], Stable Diffusion for photorealistic image synthesis [4], and Runway Gen-2 for video generation [5], now make it possible to rapidly and massively create professional-quality works, sometimes with minimal human intervention. This democratization of digital creativity opens new opportunities for artists, businesses, and users, but it also challenges the traditional foundations of copyright law. Current legal frameworks, inherited from an era when the human author was the sole source of creation, appear insufficient to address the challenges posed by AI and the massive circulation of works on the Internet.

Three major issues emerge. The first issue concerns the attribution of creation: who can be legally recognized as the author of a work generated by a machine? The second relates to the potential violation of copyright, as model training relies on vast corpora that often contain protected data [6,7]. Finally, the third challenge is traceability and proof of authorship, which are essential to authenticate a work and trace its origin in a fluid and decentralized digital environment.

Faced with these limitations, the traditional copyright law struggles to provide effective and adequate protection. Legal procedures are often slow and ineffective compared to the speed of online exchanges, while large-scale model training raises uncertainties about the legality of their operation. To address these challenges, it becomes essential to mobilize complementary approaches. This article therefore proposes to examine three main axes: technical methods (such as watermarking, fingerprinting, and DRM), the contribution of information retrieval to detect and identify infringements, and the use of blockchain as an immutable and transparent register, with the goal of strengthening copyright protection in the AI era.

2 Related Works

Several recent projects have explored copyright protection against AI-generated content. Few examples are cited here:

To begin with, [5] presented an overview of technical solutions, including watermarking, fingerprinting, and DRM, with an analysis of the strengths and limitations of each approach. Their work provides a framework for evaluating the effectiveness of digital protections. Meanwhile, [8] developed a watermarking method specific to AI-generated texts, which allows traceability of provenance and identification of the model used. This method is effective in detecting academic plagiarism, although is less robust against complex paraphrasing. [9] proposed REMARKLLM, a robust watermarking system for language models, resistant to certain automatic re-writings. This approach demonstrates the value of combining watermarking and information retrieval (IR) to strengthen infringement detection. In contrast, [10] explored the use of blockchain and IPFS for video protection. This system enables distributed storage and immutable timestamping, ensuring the traceability of works while highlighting scalability challenges.

At the same time, [11] implemented a hardware-based traceability mechanism for AI-generated images, making it possible to trace the origin and history of creations. Although limited to images, this project illustrates the benefit of integrating blockchain with technical protections. Finally, [12] studied the applied blockchain to media rights management, showing how licenses and payments can be automated through smart contracts, while reducing fraud (Table 1).

Table 1. Comparison of Some Recent Projects in Copyright Protection

Project	Content Type	Method	Limitations
REMARK LLM	Text	Watermarking	Sensitive to extreme transformations
RO SVD	Image	Hardware signature	Limited to images
Frontiers	Video	Blockchain + IPFS	Scalability issues

3 Impacts of AI on Copyright

3.1 Attribution and Originality

Generative artificial intelligence tends to blur the boundary between human creativity and automated production. In the textual domain, large language models (LLMs) are capable of producing writings close to existing texts without reproducing them identically [1]. In the field of images, systems such as Stable Diffusion generate works that are visually similar to protected creations, raising questions about originality and copyright [3,4]. Finally, in the domains of video and audio, synthesis models make it possible to reproduce voices or create audio-visual sequences that closely mimic pre-existing works, further complicating the distinction between copying and innovation [13].

3.2 Unauthorized Use of Works

The training of generative AI models is based on massive corpora collected online, which often include copyrighted works without authorization [2,5]. This practice raises questions of liability: is it the responsibility of model designers, operating companies, or end users? Several recent lawsuits illustrate this issue, such as that of writers and publishers against OpenAI for the exploitation of literary works [25]. These disputes reveal a gray area between the claimed notion of "fair use" and the direct violation of reproduction rights.

3.3 Transformation and Paraphrasing

AI systems are not limited to reproducing existing content: they apply various transformations, such as text paraphrasing, adapting the style of an author, or generating deepfakes for images and videos. These practices further blur the boundaries of copyright, as the resulting work may appear original while still being derived from protected content. This makes detecting unauthorized uses particularly complex. However, approaches from information retrieval (IR) and recent techniques in digital watermarking offer promising avenues to identify such transformations and restore the traceability of original works.

4 Technical Protection Methods

4.1 Watermarking

Watermarking is based on embedding invisible signatures or marks directly into digital content (texts, images, videos). These marks, imperceptible to the end user, make it possible to authenticate the origin of a work and ensure its traceability. In cases of dissemination or transformation, they provide a reliable way to prove the provenance of the content and to detect potential unauthorized uses (Fig. 1).

Fig. 1. Illustration of image transformations applied to assess the robustness of the watermarking mechanism in the SynthID system developed by Google DeepMind [7]. The embedded watermark remains detectable despite common image manipulations. The transformations were conducted by the authors on an original image sourced from freepik [26].

4.2 Fingerprinting

This technique consists of generating a distinctive digital fingerprint from the perceptual characteristics of the work (for example, its visual patterns or linguistic style). This fingerprint acts as a unique signature that remains recognizable even if the work has undergone modifications or transformations, thus facilitating the detection of derivative copies [14] (Fig. 2).

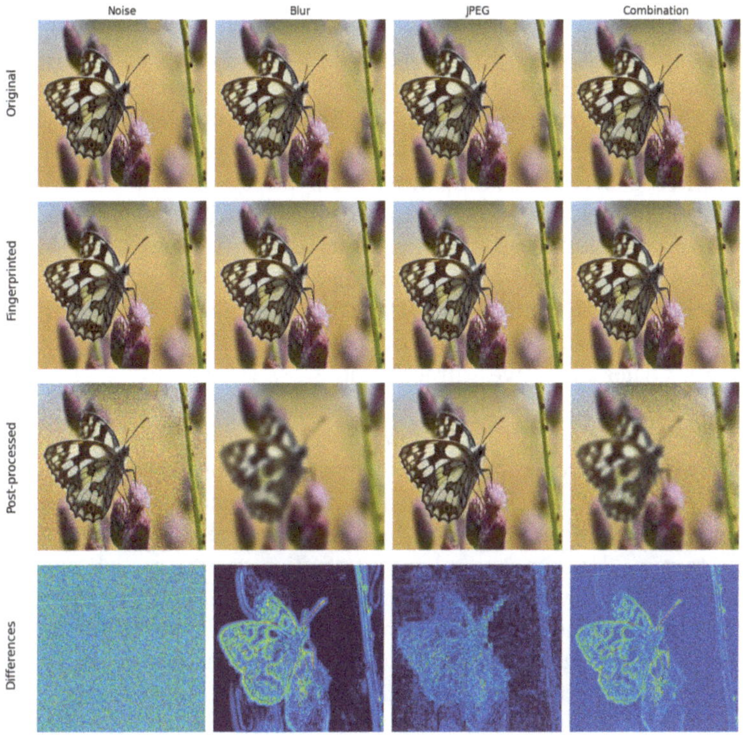

Fig. 2. Illustration of latent fingerprinting applied to evaluate both the generation quality and the robustness of the fingerprint against common post-processing operations. The embedded fingerprint remains detectable despite various transformations, including noise addition, blurring, JPEG compression, and combinations of these operations [15]. The transformations were conducted by the authors on an original image sourced from freepik.[26]

4.3 DRM and Metadata

Digital Rights Management (DRM) encompasses technical mechanisms designed to restrict the use of digital works according to predefined rules, such as access control, copy limitations, or distribution restrictions. In addition, metadata embedded in files (timestamping, author identity, creation history, generative model used, etc.) serve as evidentiary elements that establish authorship, ensure traceability, and strengthen the legal protection of content.

5 The Role of Information Retrieval

Information Retrieval (IR) is an essential tool for identifying potential copyright infringements. In the case of texts, advanced language models such as BERT and TinyBERT make it possible to measure semantic similarity between an original work and an AI-generated production, even after reformulations [10]. For images and videos, IR relies on the comparison of visual features extracted using descriptors (textures, shapes, colors) or deep neural networks, which makes it possible to identify copies or altered versions of protected works [16]. Several projects have highlighted the importance of IR in copyright protection in the era of artificial intelligence. For example, the CopyJudge project [16] uses advanced vision-language models to assess substantial similarity between AI-generated images and original works, providing an automated framework for infringement detection. In the video and audio domains, Google's Content ID system [13] has been enhanced with multimodal IR techniques capable of detecting infringements even when content has been transformed (such as by cropping or speed modification). Other initiatives directly integrate traceability and proof mechanisms.

The DataSafe project (2024) combines watermarking and information retrieval to track the reuse of protected images, even after modification [17]. Similarly, blockchain- and IPFS-based solutions ensure the authenticity and timestamping of the works, thereby facilitating verification in case of dispute [18]. In the textual domain, several recent studies exploit models such as BERT and its variants-already mentioned above-to detect plagiarism and identify AI-generated texts, going beyond simple copying to capture paraphrasing and automated rewritings [19]. In sum, these projects demonstrate that IR today constitutes a strategic lever for copyright protection. It not only ensures the indexing and comparison of works, but also provides authentication and the production of admissible evidence, in a context where AI-generated and AI-transformed content is multiplying at an unprecedented pace.

6 Blockchain and Copyright Management

Blockchain is a distributed ledger technology that enables transactions to be recorded in a decentralized, transparent, and immutable manner. Each block contains a set of validated data that is cryptographically linked to previous blocks, forming an unalterable chain. This technology was first introduced with Bitcoin in 2008 by Satoshi Nakamoto [20], and later evolved toward much broader uses, notably through smart contracts introduced by Ethereum [14].

Today, the blockchain goes far beyond cryptocurrencies and finds applications in healthcare, supply chains (logistics and procurement), and digital rights management, where it provides evidence of precedence and secure traceability of the life cycle of work [21].

Recent advances in blockchain applied to copyright highlight academic and industrial projects aimed at strengthening the protection, traceability, and management of digital works. Several scientific contributions have proposed protection architectures based on the combination of blockchain and cryptographic mechanisms. For example, [22] described a protection scheme based on CP-ABE encryption to ensure fine-grained access control to digital resources [1]. Meanwhile, [21] introduce BC-DERCP, a blockchain framework dedicated to protecting and tracking rights in multimedia content [2], while another project leverages Hyperledger Fabric [18] to design a reliable lifecycle management system for image copyrights, thereby reducing costs and improving efficiency compared to Ethereum [3]. Furthermore, specialized approaches for video have been developed, such as those that integrate smart contracts and perceptual hashing (pHash) to detect modifications and resolve disputes through decentralization [4,5].

Beyond research laboratories, industrial initiatives have also emerged. The most notable example is Fox Verify [23], launched in 2024 by Fox Corporation, which uses blockchain to authenticate the origin of digital content and facilitate license negotiations with platforms and artificial intelligence systems [6]. In the music industry, prototypes of blockchain-based decentralized rights management systems demonstrate the potential to ensure both transparency and fair remuneration for creators [7]. Finally, on the regulatory side, the USPTO & US Copyright Office [24] have examined the impact of NFTs and their connections to intellectual property, highlighting both their potential to establish proof of originality and persistent legal uncertainties [9].

These initiatives show that while blockchain is still in an experimental phase, it is becoming an essential tool for rethinking the governance and protection of copyright in the digital era (Fig. 3).

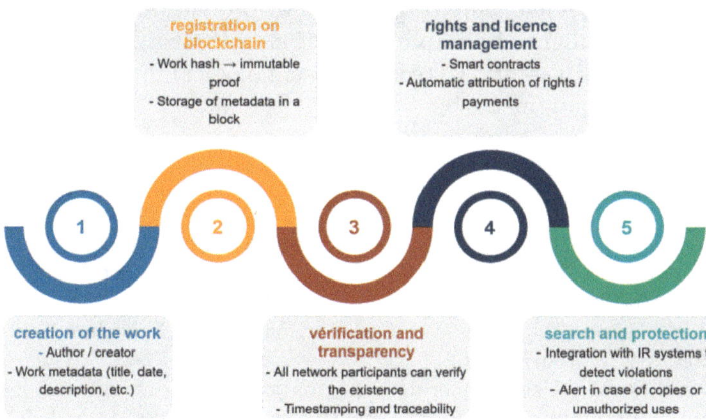

Fig. 3. Illustration of blockchain-based copyright protection process.

7 Future Perspectives

Future perspectives in copyright protection primarily rely on standardization and the establishment of a clear legal framework for works generated by artificial intelligence. At the same time, the integration of hybrid approaches (watermarking, fingerprinting, semantic information retrieval, blockchain) appears to be a promising path to strengthen the detection and traceability of protected content. The objective is to ensure increased robustness of the systems against practices such as paraphrasing, rewording, or visual and audio transformations. However, these developments must be accompanied by reflection on social acceptability and the protection of personal data, particularly through respect for the right to be forgotten.

Finally, energy optimization and the design of sustainable business models represent essential conditions to ensure large-scale adoption of these technologies while respecting environmental and ethical constraints.

8 Conclusion

The rise of generative artificial intelligence profoundly disrupts the foundations of copyright by making the creation, transformation, and dissemination of digital content more accessible but also more vulnerable to infringements. This article has shown that technical methods (watermarking, fingerprinting, DRM), information retrieval, and blockchain offer complementary levers to strengthen the traceability, authenticity, and legal protection of original works in the digital age.

However, none of these tools provides a single and definitive solution: only their integration into hybrid architectures, coupled with a clear and adapted regulatory framework, will ensure robust and fair protection for creators. Future perspectives also highlight the importance of standardization, energy sustainability, and social acceptability, to reconcile technological innovation, respect for rights, and ethical requirements.

References

1. Radford, A., Jong, L., Amodei, S., et al.: Improving Language Understanding by Generative Pre-Training. https://cdn.openai.com/research-covers/language-unsupervised/language_understanding_paper.pdf
2. Ramesh, A., Pavlov, M., Goh, G., Gray, S., Voss, C.: Zero-Shot Text-to-Image Generation. arXiv preprint (2021)
3. Tan, L., Luhrs, M.: Using generative AI midjourney to enhance divergent and convergent thinking in an architect's creative design process. Des. J. **27**, 677–699 (2024). https://doi.org/10.1080/14606925.2024.2353479
4. Esser, P.: Scaling Rectified Flow Transformers for High-Resolution Image Synthesis. arXiv preprint (2024)
5. Germanidis, A.: Scale, Speed and Stepping Stones: The Path to Gen-2 (2023). https://runwayml.com/research/scale-speed-and-stepping-stones-the-path-to-gen-2
6. Ren, J.: Copyright Protection in Generative AI: A Technical Perspective. arXiv preprint (2024)
7. Liu, A., et al.: A survey of text watermarking in the era of large language Models. ACM Comput. Surv. **57**(2), 47:1–36 (2024). https://doi.org/10.1145/3691626
8. Qu, W., et al.: Provably Robust Multi-bit Watermarking for AI-generated Text. arXiv preprint (2024)

9. Zhang, X.: REMARKLLM: Robust Watermarking for Large Language Models. arXiv preprint (2024)
10. Madapati, M.S.L., Pradhan, N.R.: Decentralizing video copyright protection: a novel blockchain-enabled framework with performance evaluation. Front. Artif. Intell. **8**, 1655709 (2025). https://doi.org/10.3389/frai.2025.1655709
11. Liu, X.: RO-SVD: A Reconfigurable Hardware Copyright Protection Framework for AIGC Applications. arXiv preprint (2024)
12. Frattolillo, F.: Blockchain and smart contracts for digital copyright protection. Future Internet **16**(5), 169 (2024). https://doi.org/10.3390/fi16050169
13. Centre d'aide Google: Content ID: Protecting Copyright on YouTube (2024). https://support.google.com/youtube/answer/2797370
14. Buterin, V.: Ethereum White Paper (2014). https://ethereum.org
15. Nie, H., et al.: Attributing image generative models using latent fingerprints. In: Proceedings of the 40th International Conference on Machine Learning (2023). https://proceedings.mlr.press/v202/nie23a/nie23a.pdf
16. Liu, S., et al.: CopyJudge: automated copyright infringement identification and mitigation in text-to-image diffusion models. In: ACM MM'25, Dublin, Ireland (2025). http://arxiv.org/pdf/2502.15278
17. Lian, S., Chen, X., Wang, J., et al.: Content distribution and copyright authentication based on combined indexing and watermarking. Multimedia Tools Appl. **57**(1), 49–66 (2012). https://doi.org/10.1007/s11042-010-0521-3
18. Madushanka, T., et al.: SecureRights: A Blockchain-Powered Trusted DRM Framework for Robust Protection and Asserting Digital Rights (2024). https://arxiv.org/html/2403.06094v1
19. Moravvej, S.V., et al.: A Novel Plagiarism Detection Approach Combining BERT-based Word Embedding, Attention-based LSTMs and an Improved Differential Evolution Algorithm (2023). arXiv preprint arXiv:2305.02374v1. https://doi.org/10.48550/arXiv.2305.02374
20. Nakamoto, S.: Bitcoin: A Peer-to-Peer Electronic Cash System (2008). https://bitcoin.org/bitcoin.pdf
21. Jiang, J., et al.: A blockchain copyright protection scheme based on CP-ABE scheme with policy update. Sensors **24**, 4493 (2024). https://doi.org/10.3390/s24144493
22. Zhao, L., Dong, G., Yuan, H.: A blockchain-based verifiable CP-ABE scheme for medical data privacy protection. Sci. Rep. **15**(1), 27325 (2025). https://doi.org/10.1038/s41598-025-13069
23. Koome, B.: Fox Corp. Launches Blockchain-Based Verify Tool for Content Authentication (2025). https://www.cryptopolitan.com/fox-corp-launches-verify-tool/
24. USPTO & US Copyright Office: NFTs and Intellectual Property: Report (2023). https://www.uspto.gov/sites/default/files/documents/NFT-Roundtable-TRADEMARK-Jan24-TRANSCRIPT.pdf
25. AP News: 'Game of Thrones' creator George R.R. Martin and other authors sue OpenAI for copyright infringement (2023). https://apnews.com/article/openai-lawsuit-authors-grisham-george-rr-martin-37f9073ab67ab25b7e6b2975b2a63bfe
26. https://www.freepik.com/

Bio-Inspired Feature Selection for Deep Learning-Based Intrusion Detection in IoMT

Karima Benhamza[1(✉)], Rania Benselim[2], and Hadil Merzougui[3]

[1] LabSTIC, University 8 May 1945, Guelma, Algeria
Benhamza.karima@univ-guelma.dz
[2] Telecommunication Institute, University 8 May 1945, Guelma, Algeria
[3] Computer Science Department, University 8 May 1945, Guelma, Algeria

Abstract. The rapid proliferation of the Internet of Medical Things (IoMT) has significantly transformed modern healthcare by enabling continuous patient monitoring, remote diagnostics, and intelligent medical services. Through interconnected medical sensors and smart devices, healthcare providers can collect and analyze real-time data, improving clinical decision-making and the quality of patient care. However, the increasing connectivity of IoMT devices also introduces critical cybersecurity challenges. These devices typically operate under limited computational and energy resources and are deployed within heterogeneous network environments, making them highly vulnerable to various cyberattacks that may compromise sensitive medical data and disrupt healthcare services. Consequently, the development of efficient and lightweight intrusion detection systems (IDS) is essential to ensure the security and reliability of IoMT infrastructures. In this study, a hybrid intrusion detection framework integrating the Clonal Selection Algorithm (CSA) and a Deep Neural Network (DNN) is proposed. The CSA is employed as a feature selection technique to identify the most relevant attributes from network traffic data, thereby reducing dimensionality and improving model performance. The optimized feature subset is then used to train a DNN classifier capable of learning complex patterns and accurately distinguishing between normal and malicious activities. Experimental evaluation using the WUSTL-2020 IoMT dataset demonstrates that the proposed CSADNN model achieves high detection accuracy, strong recall, and satisfactory computational efficiency. These findings highlight the effectiveness of combining bio-inspired optimization with deep learning methodologies for strengthening cybersecurity in IoMT-enabled healthcare systems.

Keywords: Internet of Medical Things (IoMT) · Intrusion Detection System (IDS) · Bio-inspired Algorithms · Clonal Selection Algorithm (CSA) · Deep Neural Network (DNN)

Z. Kouahla et al. (Eds.): IAM 2025, LNNS 1973, pp. 22–29, 2026.
https://doi.org/10.1007/978-3-032-26031-4_3

1 Introduction

The Internet of Medical Things (IoMT) refers to an integrated network of medical devices capable of collecting, transmitting, and analyzing health-related data through the Internet. This paradigm plays a pivotal role in modern healthcare by enabling applications such as real-time monitoring, remote diagnostics, and personalized treatment plans. By facilitating continuous data exchange between patients and healthcare providers, IoMT fosters a more patient- centric, data-driven healthcare model that enhances clinical outcomes and operational efficiency [1].

However, despite its transformative potential, IoMT also introduces critical security concerns. The high connectivity and limited computational resources of IoMT devices render them particularly vulnerable to a wide range of cyberattacks. These attacks can compromise diagnostic accuracy, disrupt communication between devices and practitioners, and expose sensitive patient information to unauthorized access or manipulation. Consequently, ensuring the security, reliability, and integrity of IoMT systems has become a priority for both the healthcare industry and the research community.

Indeed, Intrusion Detection Systems (IDSs) are crucial in safeguarding IoMT networks. By continuously monitoring system behavior and identifying deviations from normal patterns, IDSs help detect and mitigate cyber-threats in real time. Nevertheless, conventional IDSs often struggle in IoMT scenarios due to challenges such as high-dimensional data, class imbalance, and the rapidly evolving nature of cyber-threats. Although recent research efforts have explored various hybrid approaches to enhance IDS performance in IoMT environments, the application of bio-inspired algorithms remains relatively underexplored. Only a few studies have investigated their potential in optimizing detection processes and improving feature selection for anomaly identification.

For instance, in a study [1] employed a hybrid genetic algorithm-based random forest (GA-RF) for intrusion detection. By optimizing feature selection, the GA-RF approach demonstrated exceptional accuracy in detecting cyber-attacks within IoMT environments. Evaluations on the NSL-KDD and UNSW-2018-IoT-Botnet datasets reported an accuracy of 99%.

Another study [2], researchers combined the Lion Optimization Algorithm (LOA) and Grey Wolf Optimizer (GWO) to improve feature selection and boost IDS performance in detecting network anomalies. The model preprocesses data, selects relevant features, optimizes parameters, and utilizes a hybrid CNN-LSTM for intrusion detection. It was evaluated on the NSL-KDD and WUSTL- EOM 2020 simulation datasets, achieving an accuracy of 99.26%.

The research [3] introduced the Lion Salp Swarm Optimization Algorithm (LSSOA) for intrusion detection in IoMT networks. This approach integrates several metaheuristic techniques, including Lion, Whale, Spider Monkey, and Salp Swarm algorithms, to enhance online attack detection and defense. The model, tested using the IoT Flock software dataset, achieved an accuracy of 99.59%. In [4], a blockchain-assisted intrusion detection system was proposed for IoT healthcare networks. The system leverages the Ant Lion Optimizer (ALO) and Hybrid Deep Learning (HDL) to combine blockchain decentralized structure with IDS for secure medical data transmission. It was evaluated using the IoT-23 and NSL-KDD datasets, achieving an accuracy of 99.55%.

Additionally, in [5], the authors proposed a hybrid PSO- DNN method for intrusion detection in IoMT networks. The Particle Swarm Optimization (PSO) algorithm was used to identify the most relevant features, while a Deep Neural Network (DNN) was employed to classify and detect network intrusions. The model was validated using network traffic and patient sensing data, achieving an accuracy of 96%.

The last relevant work in [6], proposed a hybrid model combining Correlation-based Feature Selection and the Bat Optimization Algorithm (HCFS-BOA) with a Convolutional Neural Network (CNN) for intrusion detection in healthcare systems. This model improves the detection accuracy by selecting the most relevant features and classifying attacks using CNN. Tested on CIC-IDS2017 and NSL-KDD, the approach achieved an accuracy of 99.4%.

In this context, we propose in this work a hybrid IDS architecture that combines the Clonal Selection Algorithm (CSA) for intelligent feature selection with a Deep Neural Network (DNN) for accurate classification. Inspired by the adaptive mechanisms of the human immune system. CSA identifies the most informative features, thereby reducing the dimensionality of the data and enhancing learning efficiency. The selected features are then processed by a DNN capable of modeling complex patterns and detecting subtle anomalies in network behavior.

This paper is organized as follows: Section 2 presents the proposed CSA-DNN-IDS model. Section 3 details the experimental setup and results. Finally, Sect. 4 concludes the study and outlines future research directions.

2 Proposed Model

In this study, we propose a hybrid Intrusion Detection System (IDS) specifically designed for Internet of Medical Things (IoMT) environments. Our approach integrates CSA-based feature selection with DNN classification to efficiently detect intrusions. CSA is employed to select the most relevant features from the dataset, thereby reducing its dimensionality and enhancing the learning efficiency of the model.

To ensure consistency across features, we first apply data standardization to normalize the input data. Once the most important features are selected using CSA, they are input into the DNN, which handles the classification task and detects potential intrusions.

An overview of the framework is presented in Fig. 1 and further detailed in the following sections.

Fig. 1. Workflow of CSA-DNN intrusion detection model.

2.1 Pre-processing

To ensure consistency and improve learning efficiency, a pre-processing step was applied to the used datasets. In this work, the performance of the proposed CSA-DNN model was evaluated using the benchmark WUSTL 2020-IoMT dataset. This dataset, developed by Washington University in St. Louis, provides time-series data from smart medical devices. It contains both normal and malicious traffic, with detailed records such as timestamps, sensor readings, and communication logs, making it ideal for behavior-based intrusion detection. The WUSTL 2020-IoMT dataset is publicly available at https://www.cse.wustl.edu/~jain/ehms/index.html.

The following Table 1 summarizes the main specifications of this dataset.

Table 1. Characteristics of the used dataset

Dataset	Classes	Instances	Features
WUSTL 2020	2	16,318	44

This pre-processing phase included data standardization, a statistical technique that transforms input features to achieve a mean of zero and a standard deviation of one. This step is particularly important to ensure that all features contribute equally during the learning process, regardless of their original scales. By eliminating disparities in units and magnitudes, standardization facilitates faster model convergence and improves the overall stability and performance of the classification process.

2.2 Clonal Selection Algorithm for Features Selection

The Clonal Selection Algorithm (CSA) introduced by De Castro and Von Zuben [7] is a bio-inspired metaheuristic that emulates the adaptive behavior of B lymphocytes in the human immune system. In feature selection, CSA is employed to identify a sparse subset of input features that optimizes class separability while penalizing redundancy, thus reducing overfitting and enhancing generalization.

The CSA for feature selection consists of the following steps:

1. **Initialization:** The algorithm begins by generating an initial population of P individuals, each represented as a binary vector of length D, where D corresponds to the total number of features. A bit set to 1 indicates the selection of a feature, while a 0 indicates exclusion. If an individual does not select any features, it is corrected by ensuring that at least one feature is selected. This diversity in the initial population ensures a broad exploration of the search space.
2. **Affinity Evaluation:** Each individual is evaluated using an affinity function that measures the quality of selected features in terms of classification performance and sparsity. The affinity function used to evaluate the quality of each solution is defined as:

$$Affinity = \frac{\sum_{i<j} N_i N_j ||\mu_i - \mu_j||}{\sum_{i<j} N_i N_j} - \alpha \left(\frac{d}{D}\right)^{1.5} \tag{1}$$

   where:
   - d represents the number of selected features
   - D represents the total number of features
   - α is a regularization parameter
3. **Selection:** The top-performing individuals are selected based on their affinity scores to proceed to the cloning stage.
4. **Cloning and Mutation:** The clones undergo a binary mutation process, where the probability of mutation is inversely proportional to the original individual's affinity. Higher-affinity individuals mutate less, while those with lower affinity are perturbed more significantly, encouraging broader exploration of the search space. Mutation consists of flipping certain bits in the binary mask, thereby adding diversity to the population.
5. **Population Update:** The new population is formed by combining the mutated clones with the current population, and elitism is applied to retain the best solutions.
6. **Stopping Criterion:** The algorithm repeats the previous steps until a maximum number of iterations is reached.
7. **Final Solution:** The individual with the highest affinity is selected as the optimal feature subset for subsequent classification.

2.3 Classification Using a Deep Neural Network (DNN)

After selecting the optimal subset of features using the Clonal Selection Algorithm (CSA), the classification phase is performed using a Deep Neural Network (DNN). The DNN is designed to identify complex patterns in the selected feature space and distinguish between normal and malicious traffic, which is essential for efficient intrusion detection in IoMT environments, as demonstrated in [8]. The DNN consists of an input layer matching the selected features, two hidden layers with ReLU activation, and an output layer using sigmoid for binary classification. Dropout and batch normalization are applied to prevent overfitting and improve training stability. The model is trained using the Adam optimizer and cross-entropy loss, with stratified k-fold cross-validation ensuring robust generalization.

3 Results and Discussion

The performance of the proposed CSA-DNN model was assessed using four recognized classification metrics: Accuracy, Precision, Recall, and F1-score. These metrics are commonly employed in evaluating intrusion detection systems, particularly under conditions of class imbalance, which are characteristic of datasets in Internet of Medical Things (IoMT) environments.

3.1 Evaluation Metrics

Let TP (True Positives), TN (True Negatives), FP (False Positives), and FN (False Negatives) denote the classification outcomes. The evaluation metrics are mathematically defined and used to measure detection performance: Accuracy: Measures the proportion of correctly classified instances among the total number of samples.

$$Accuracy = \frac{TP + TN}{TP + TN + FP + FN} \tag{2}$$

Precision: Measures the ratio of correctly predicted positive observations to the total predicted positives.

$$Precision = \frac{TP}{TP + FP} \tag{3}$$

Recall (Sensitivity): Measures the ratio of correctly predicted positive observations to all actual positives.

$$Recall = \frac{TP}{TP + FN} \tag{4}$$

F1-score: Represents the harmonic mean of precision and recall.

$$F1 = \frac{2 \times Precision \times Recall}{Precision + Recall} \tag{5}$$

Table 2. Performance comparison on WUSTL dataset

Model	Accuracy	Precision	Recall	F1-score
CSA+DNN	0.9810	0.9880	0.9867	0.9874
Random Forest	0.9458	0.9546	0.7965	0.8534
PSO	0.9705	0.9530	0.9400	0.9464
DNN	0.9384	0.9909	0.5216	0.6835
CNN	0.9380	0.7605	0.8200	0.7893

3.2 Experimental Results

The proposed CSA-DNN hybrid model was evaluated both quantitatively and qualitatively based on the four previous key metrics. These metrics provide a comprehensive understanding of model performance, especially for identifying minority attack classes frequently encountered in IoMT environments.

The experimental results, summarized in Table 2 and visually represented in Fig. 2 compare the CSA-DNN model with four other approaches: Deep Neural Network (DNN), Convolutional Neural Network (CNN), Random Forest (RF), and PSO-based feature selection with DNN. The CSA-DNN consistently outperformed the other models. On the WUSTL-2020 dataset, the CSA-DNN model achieved an accuracy of 98.10%, precision of 98.80%, recall of 98.67%, and an F1-score of 98.74%, outperforming these baseline models.

The experimental results demonstrate that the hybrid CSA-DNN approach significantly improves intrusion detection performance compared to conventional machine learning and deep learning techniques.

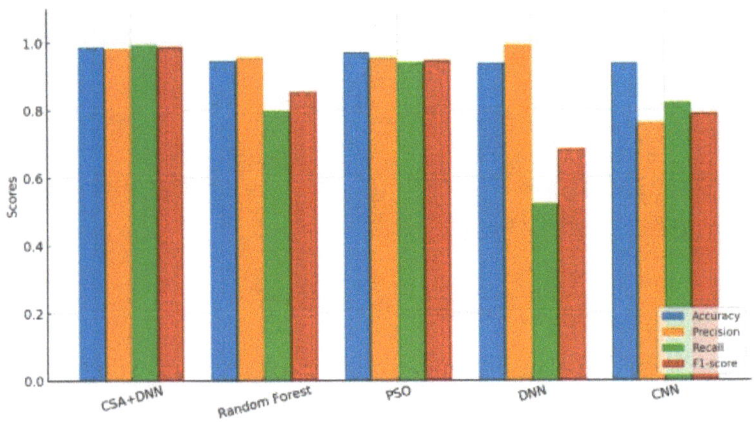

Fig. 2. Performance Comparison of IDS Models on WUSTL 2020 Dataset.

These results highlight the effectiveness and robustness of the CSA-DNN model in accurately detecting both normal and malicious activities in IoMT network environments.

4 Conclusion

This study proposed a hybrid intrusion detection framework for securing Internet of Medical Things (IoMT) environments by integrating the Clonal Selection Algorithm (CSA) with a Deep Neural Network (DNN). The CSA component was employed to perform efficient feature selection by identifying the most relevant attributes from high-dimensional network traffic data, thereby reducing redundancy and improving the quality of the input features. The optimized feature subset was then used to train a DNN classifier capable of learning complex traffic patterns and accurately distinguishing between normal and malicious activities within IoMT networks.

Experimental evaluations conducted on WUSTL-2020 benchmark dataset, demonstrated the effectiveness of the proposed approach. The hybrid CSADNN model achieved high detection performance, with accuracy values approaching 98%, along with strong precision, recall, and F1-score metrics. Comparative analysis with several conventional machine learning and deep learning techniques showed that the integration of bio-inspired optimization significantly improves detection accuracy, stability, and overall classification performance.

Although the results are promising, the evaluation was conducted on a single dataset. Future work will investigate the generalization capability of the model on multiple IoMT intrusion datasets and explore real-time deployment scenarios.

References

1. Norouzi, M., Gürkaş-Aydın, Z., Turna, Ö.C., Yağci, M.Y., Aydin, M.A., Souri, A.: A hybrid genetic algorithm-based random forest model for intrusion detection approach in internet of medical things. Appl. Sci. **13**(20), 11145 (2023)
2. Fadhil, H.M., Dawood, Z.O., Al Mhdawi, A.: Enhancing intrusion detection systems using metaheuristic algorithms. Diyala J. Eng. Sci. 15–31 (2024)
3. Goswami, N., et al.: Intrusion detection system for IoT-based healthcare intrusions with lion-SALP-swarm-optimization algorithm: metaheuristic-enabled hybrid intelligent approach. Eng. Sci. **25**(933), 83 (2023)
4. Alamro, H., et al.: Modeling of blockchain assisted intrusion detection on IoT healthcare system using ant lion optimizer with hybrid deep learning. IEEE Access **11**, 82199–82207 (2023)
5. Chaganti, R., Mourade, A., Ravi, V., Vemprala, N., Dua, A., Bhushan, B.: A particle swarm optimization and deep learning approach for intrusion detection system in internet of medical things. Sustainability **14**(19), 12828 (2022)
6. Bella, H.K., Vasundra, S.: Healthcare intrusion detection using hybrid correlation-based feature selection-bat optimization algorithm with convolutional neural network: a hybrid correlation-based feature selection for intrusion detection systems. Int. J. Adv. Comput. Sci. Appl. **15**(1), 1–9 (2024)
7. De Castro, L.N., Von Zuben, F.J.: Learning and optimization using the clonal selection principle. IEEE Trans. Evol. Comput. **6**(3), 239–251 (2002)
8. Areia, J., Bispo, I.A., Santos, L., Costa, R.L.D.C.: Iomt-trafficdata: dataset and tools for benchmarking intrusion detection in internet of medical things. IEEE Access **12**, 115370–115385 (2024)

Binary Moth-FlameBased Task Mapping for Energy Minimization in Heterogeneous MPSoCs

Farid Boumaza[1,2(✉)], Atmane Hadji[3], Djaafar Zouache[1], and Abdelkader Aroui[4]

[1] Department of Computer Science,University of Mohamed El Bachir El Ibrahimi, Bordj Bou Arreridj, Algeria
{f.boumaaza,djaafar.zouache}@univ-bba.dz

[2] Laboratory of Parallel, Embedded architectures and Intensive Computing, University of Oran1, 31000 Oran, Algeria

[3] LISI Laboratory, A. Boussouf Mila University, Mila, Algeria
a.hadji@centre-univ-mila.dz

[4] Center for Space Techniques, Palestine Avenue, 31200 Arzew Oran, Algeria

Abstract. The increasing complexity of real-time embedded applications has elevated heterogeneous Multi-Processor Systems-on-Chip (MPSoCs) as a primary platform for meeting stringent performance, energy, and scalability requirements. A central challenge is static task mapping, where precedence-constrained tasks must be placed onto heterogeneous processing elements (PEs) under hard timing constraints while minimizing system energy. This paper presents an energy-aware mapping framework based on a Binary Moth-Flame Algorithm (BMFA), a bio-inspired metaheuristic tailored to discrete assignment spaces. We encode each candidate solution as a binary task-to-PE vector and design feasibility-preserving operators that respect task precedence, end-to-end deadlines, and PE capacity constraints, while explicitly modeling both computation and communication energy, including network-on-chip (NoC) costs. The BMFA integrates adaptive flame reduction, elitist selection, and a lightweight local search to accelerate convergence and avoid premature stagnation. Across synthetic DAG/SDF workloads and embedded benchmarks, the proposed approach consistently yields lower total energy (compute + communication) and improved schedulability compared with representative baselines (Gentec algorithm (GA) , Particul Swarm Optimization (PSO)), and scales to problem sizes typical of modern MPSoCs. These results demonstrate that BMFA is an effective, robust, and scalable tool for energy-aware static task mapping in heterogeneous MPSoCs.

Keywords: Heterogeneous MPSoC · Static task mapping · Moth Flam Algorithm · Energy efficiency

Z. Kouahla et al. (Eds.): IAM 2025, LNNS 1973, pp. 30–43, 2026.
https://doi.org/10.1007/978-3-032-26031-4_4

1 Introduction

Embedded systems are specialized computing platforms engineered to execute domain-specific functions under tight resource, performance, and energy constraints [3,4,8]. Modern designs increasingly integrate one or more Systems-on-Chip (SoCs) [5,15] that are heterogeneous in both compute and memory organization, combining general-purpose CPUs with domain accelerators such as DSPs, GPUs, and reconfigurable logic (e.g., FPGAs [9]). This heterogeneity, along with deep memory hierarchies and on-chip interconnects (e.g., NoCs [7]), enables high performance per watt but makes design-space exploration (DSE) and task mapping notably complex.

Many embedded applications exhibit hierarchical structure and mixed forms of parallelism [13]. Representative workloads—e.g., multimedia pipelines (MPEG) [1], video object plane decoding (VOPD) [6], picture-in-picture (PiP) [11], and other streaming/vision stacks—interleave (i) irregular, control- and event-driven computation that is amenable to task/graph-level parallelism, with (ii) regular, data-parallel kernels that resemble high-performance computing (HPC) primitives [17]. A single mapping policy rarely serves both regimes equally well: data-parallel kernels benefit from throughput-centric placement close to accelerator-rich resources and high-bandwidth memory, whereas task-parallel stages require latency-aware placement that respects precedence and real-time deadlines.

This paper targets static task mapping for heterogeneous MPSoCs, focusing on energy minimization subject to schedulability and platform constraints. We consider applications modeled as precedence-constrained task graphs, and we explicitly account the computation energy (task execution on heterogeneous processing elements, PEs) and communication energy (NoC transfers induced by the mapping), while enforcing constraints such as end-to-end deadlines, PE capacity, and interconnect bandwidth/latency limits. Static mapping remains attractive for real-time embedded deployments because it yields predictable schedules and low runtime overheads, yet the combinatorial search space grows superlinearly with task and PE counts, rendering exact methods impractical for realistic systems [14] .

We propose a multi-task mapping framework that adapts the Moth-Flame Algorithm [12] to discrete assignment spaces and treats regular, data-parallel kernels via locality- and bandwidth-aware placement on available accelerators. Our Binary Moth-Flame Algorithm (BMFA) encodes task-to-PE assignments as binary variables and uses feasibility-preserving operators to enforce precedence and capacity constraints. The objective jointly models computation and communication energy, enabling co-optimization of placement and interconnect traffic rather than optimizing them in isolation.

The rest of the paper is organized as follows: Sect. 2 discusses the definitions and the necessary mathematical formulations for problem mapping onto the MPSoCs architecture. In Sect. 3, the proposed mapping strategies are presented. Experimental results are provided in Sect. 4, and the paper concludes with Sect. 5.

2 System Models and Problem Statement

We model the application and the heterogeneous MPSoC as directed graphs and define a constraint-aware mapping from tasks to processing elements (PEs) with single-objective energy minimization.

2.1 Application Model

Definition 1 (Task Graph [4]). *The application is a directed graph* $G = (T, E)$, *where each vertex* $t_i \in T$ *is a task/module and each directed edge* $e_{ij} = (t_i, t_j) \in E$ *represents a data dependency from* t_i *to* t_j. *The edge weight* $Q_{ij} > 0$ *denotes the data volume (in bits/bytes) transferred from* t_i *to* t_j *(See Fig. 1.).*

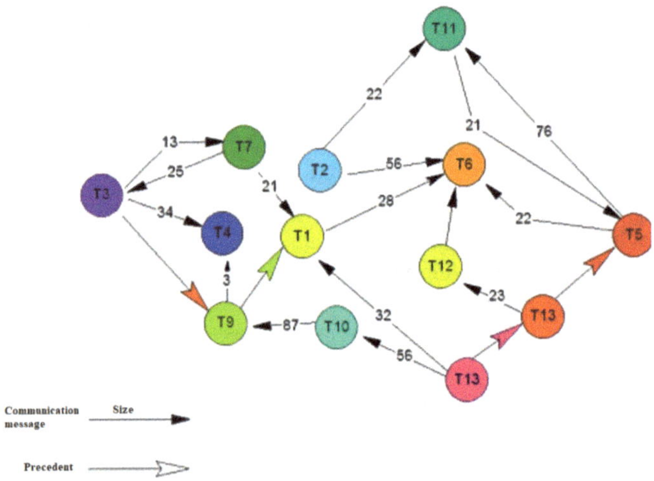

Fig. 1. Example task graph $G = (T, E)$.

2.2 Architecture Model

Definition 2 (Architecture Graph). *The MPSoC is a directed graph* $P = (S, F)$, *where each vertex* $s_k \in S$ *is a processing element (PE), and each directed edge* $f_{uv} = (s_u, s_v) \in F$ *is a physical NoC link. Each link* f_{uv} *is characterized by bandwidth* bw_{uv} *(bits/s), latency* ℓ_{uv} *(s), and per-bit energy* $\epsilon_{uv}^{\mathrm{comm}}$ *(J/bit).*

Each PE s_k supports a finite set of operating modes $\mathcal{M}_k$. A mode $m \in \mathcal{M}_k$ determines the frequency f_{mk}, energy-per-cycle e_{mk}, and (optionally) power P_{mk} (See Fig. 2).

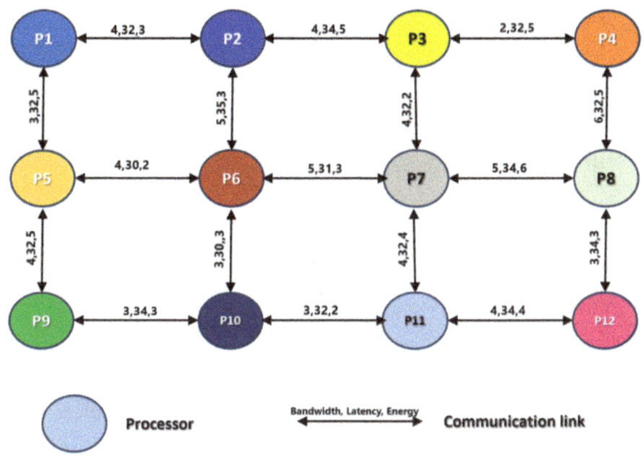

Fig. 2. Example architecture graph.

2.3 Mapping and Decision Variables

A mapping assigns each task to exactly one PE and selects one operating mode for each PE (See Fig. 3).

We use the binary indicator

$$X_m^{ip} = \begin{cases} 1, & \text{if task } t_i \text{ executes on PE } s_p \text{ in mode } m \in \mathcal{M}_p, \\ 0, & \text{otherwise,} \end{cases} \tag{1}$$

with the one-hot assignment constraint

$$\sum_p \sum_{m \in \mathcal{M}_p} X_m^{ip} = 1, \qquad \forall i. \tag{2}$$

The implied mapping function is $\text{map}(t_i) = s_p$ if $\exists m : X_m^{ip} = 1$. When two tasks t_i and t_j are mapped to PEs s_p and s_q, their NoC route is denoted $\pi(p, q) \subseteq F$ (an ordered set of links).

2.4 Execution and Communication Timing

Let Size_{ip} denote the number of cycles task t_i requires when executed on PE s_p (micro-architectural effects captured in Size_{ip}). If s_p operates in mode m with frequency f_{mk} (Hz), the execution time is

$$D_i = \sum_p \sum_{m \in \mathcal{M}_p} X_m^{ip} \frac{\text{Size}_{ip}}{f_{mk}}. \tag{3}$$

For a communication edge $e_{ij} \in E$ with volume Q_{ij}, and route $\pi(p, q)$ induced by the mapping, we model the end-to-end transfer time as serialization over the

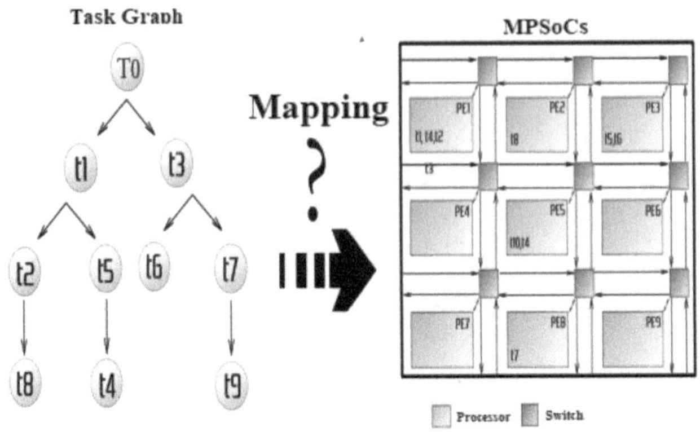

Fig. 3. Illustration of mapping from TG to AG.

bottleneck link plus accumulated per-hop latency:

$$D^{ij}_{\text{com}} = \sum_{p,q} \left(\sum_{m \in \mathcal{M}_p} X^{ip}_m \right) \left(\sum_{n \in \mathcal{M}_q} X^{jq}_n \right) \left[\frac{Q_{ij}}{\min\limits_{f_{uv} \in \pi(p,q)} \text{bw}_{uv}} + \sum_{f_{uv} \in \pi(p,q)} \ell_{uv} \right]. \tag{4}$$

If static schedulability is enforced, start/finish times $(\text{st}_i, \text{ft}_i)$ satisfy:

$$\text{ft}_i = \text{st}_i + D_i, \quad \forall i, \tag{5}$$

$$\text{st}_j \geq \text{ft}_i + D^{ij}_{\text{com}}, \quad \forall (i,j) \in E, \tag{6}$$

with optional releases/deadlines $r_i \leq \text{st}_i$ and $\text{ft}_i \leq d_i$.

2.5 Energy Consumption

Total energy is the sum of computation and communication contributions.

Execution Energy. Let e_{mk} be the energy per cycle for PE s_k in mode m. The execution energy of task t_i is

$$E^i_{\text{exec}} = \sum_p \sum_{m \in \mathcal{M}_p} X^{ip}_m \, \text{Size}_{ip} \, e_{mk}. \tag{7}$$

Communication Energy. Let $\epsilon^{\text{comm}}_{uv}$ be the per-bit energy on link f_{uv}. The energy to send Q_{ij} along $\pi(p,q)$ is

$$E^{ij}_{\text{com}} = \sum_{p,q} \left(\sum_{m \in \mathcal{M}_p} X^{ip}_m \right) \left(\sum_{n \in \mathcal{M}_q} X^{jq}_n \right) \left[\sum_{f_{uv} \in \pi(p,q)} Q_{ij} \, \epsilon^{\text{comm}}_{uv} \right]. \tag{8}$$

Total Energy.

$$E_{\text{total}} = \sum_i E^i_{\text{exec}} + \sum_{(i,j)\in E} E^{ij}_{\text{com}}. \tag{9}$$

2.6 The Standard Moth–Flame Optimization (MFO)

Moth–Flame Optimization (MFO) is a population-based, nature-inspired metaheuristic that models moths navigating by maintaining a fixed angular reference to moonlight and, consequently, following logarithmic spirals around "flames" (promising solutions) [12]. Let $\{\mathbf{M}_i\}_{i=1}^N$ be moth positions (candidate solutions) and $\{\mathbf{F}_j\}_{j=1}^N$ the flames (elite archive obtained by sorting moths by fitness) (Fig. 4).

Fig. 4. Behavior of the MFO Algorithm.

Update Rule (Logarithmic Spiral). Each moth is moved toward an assigned flame along a logarithmic spiral:

$$\mathbf{M}_i^{(t+1)} = \mathbf{S}\Big(\mathbf{M}_i^{(t)}, \mathbf{F}^{(t)}_{\sigma(i)}\Big) = \mathbf{D}_i\, e^{b\,\ell}\cos(2\pi\ell) + \mathbf{F}^{(t)}_{\sigma(i)}, \qquad \mathbf{D}_i = \big|\mathbf{F}^{(t)}_{\sigma(i)} - \mathbf{M}_i^{(t)}\big|, \tag{10}$$

where $b > 0$ controls spiral tightness and $\ell \sim \mathcal{U}(-1, 1)$ adds stochasticity. The mapping $\sigma(i)$ assigns the i-th moth to a flame (see below).

Flame Scheduling. At iteration t among T total iterations, the number of active flames decreases linearly to intensify exploitation:

$$F_{\text{active}}(t) = \left\lfloor N - t\cdot\frac{N-1}{T}\right\rfloor, \qquad \sigma(i) = \min\{i,\, F_{\text{active}}(t)\}. \tag{11}$$

Moths are evaluated, the population is merged with current flames, and the top N by fitness form the new flame set.

Algorithm 1. Standard MFO

Initialize moths $\{\mathbf{M}_i^{(0)}\}_{i=1}^{N}$ within bounds; evaluate fitness.
Set flames $\{\mathbf{F}_j^{(0)}\} \leftarrow$ top-N sorted by fitness.
for $t = 0$ **to** $T - 1$ **do**
  Compute $F_{\text{active}}(t)$ via (11).
  for $i = 1$ **to** N **do**
    Assign flame index $\sigma(i) = \min\{i, F_{\text{active}}(t)\}$.
    Update $\mathbf{M}_i^{(t+1)}$ using (10); repair bounds; evaluate fitness.
  end for
  Merge $\{\mathbf{M}_i^{(t+1)}\} \cup \{\mathbf{F}_j^{(t)}\}$; sort; keep top N as $\{\mathbf{F}_j^{(t+1)}\}$.
end for
return best flame.

Boundary Handling. Out-of-bounds dimensions are repaired by projection or reflection onto feasible limits. Algorithm 1 summarizes the main steps of MFO.

Standard MFO operates in continuous spaces [2,18]. In Sect. 3 we adopt a *binary* variant (BMFO) for discrete task-to-PE assignments by applying a transfer function to the continuous updates in (10) and enforcing one-hot/feasibility constraints during moves.

3 Proposed Solution: Binary Moth Flame Optimization

This section presents our Binary Moth Flame Optimization (BMFO) framework for energy-aware static task mapping on heterogeneous MPSoCs. The single objective is to *minimize total energy* (computation + communication), as defined in (9), subject to the feasibility constraints of Sect. 2 (one-hot assignment, precedence/schedulability, PE capacity, and link bandwidth).(See Fig. 5)

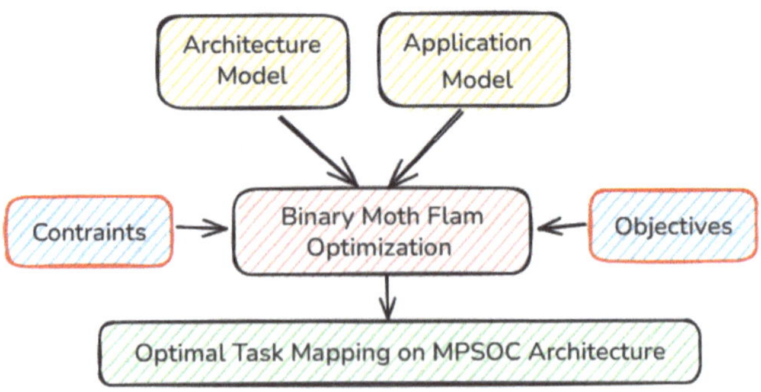

Fig. 5. Illustration of the proposed solution.

3.1 Objective Function

We search over discrete task-to-PE (and mode) assignments using a binary variant of MFO. Each candidate solution ("moth") represents a full mapping $\{X_m^{ip}\}$ (Eq. (1)), decoded from a latent continuous state. The fitness is

$$\mathcal{F}(\mathbf{X}) = E_{\text{total}}(\mathbf{X}) + \lambda_{\text{tim}}\,\Phi_{\text{tim}}(\mathbf{X}) + \lambda_{\text{bw}}\,\Phi_{\text{bw}}(\mathbf{X}) + \lambda_{\text{cap}}\,\Phi_{\text{cap}}(\mathbf{X}), \quad (12)$$

where E_{total} is from (9) and $\Phi_{\{\cdot\}}$ are penalty terms for timing (deadline/precedence), bandwidth, and PE-capacity violations; $\lambda_{\{\cdot\}} > 0$ are weights (annealed down during iterations). Feasible mappings have zero penalty, hence the optimizer drives energy minimization while maintaining feasibility.

3.2 Solution Encoding

Directly evolving a large binary tensor $\{X_m^{ip}\}$ is brittle under constraints. We adopt a *latent continuous* representation $\mathbf{Z} \in \mathbb{R}^d$ per moth and a deterministic *decoder* that yields a feasibility-preserving $\mathbf{X}$:

Block-Wise One-Hot Encoding. For each task t_i, we form a block $\mathbf{z}_i \in \mathbb{R}^{K_i}$ whose elements score the choices (p, m) available to t_i (i.e., $K_i = \sum_p |\mathcal{M}_p|$ options).(If mode selection is decided per-PE, K_i reduces to the number of PEs; a second block per PE can encode the mode $m \in \mathcal{M}_p$. Our implementation treats (p, m) jointly for simplicity.) The decoder enforces one-hot by

$$X_m^{ip} = \mathbf{1}\left[(p, m) = \arg\max_{(q,n)} z_{i,(q,n)}\right], \quad \forall i, \quad (13)$$

optionally breaking ties by energy-aware heuristics (e.g., smaller e_{mn}).

Feasibility-Aware Repair. After decoding, a light repair pass adjusts assignments that would violate hard constraints: (i) if link bandwidth is exceeded, re-route along an alternative admissible path or remap a minimal cut of communicating tasks; (ii) if PE capacity or schedulability is violated, greedily relocate offending tasks to the next-best (p, m) that reduces the violation while minimally increasing energy. Repairs are bounded in effort to preserve BMFO iteration speed.

3.3 Initialization

We hybridize diverse seeds to improve early coverage:

- **Random feasible**: sample $\mathbf{Z}$ i.i.d. (e.g., $\mathcal{U}[-1, 1]$), decode via (13), then repair.
- **Comm-aware clustering**: co-locate strongly communicating tasks (large Q_{ij}) on PEs with high intra-cluster bandwidth; initialize $\mathbf{z}_i$ accordingly.
- **Energy-greedy heuristic**: for each task, rank (p, m) by per-cycle energy e_{mk} and predicted comm energy to its predecessors/successors; set the top choice higher in $\mathbf{z}_i$.
- **DVFS diversity**: distribute modes across PEs to cover both low-power and high-performance points.

These seeds populate the initial moth set and the elite flame archive.

3.4 Adapting MFO to Discrete Mapping (BMFO)

Standard MFO operates in continuous space via logarithmic-spiral moves toward flames (Sect. 2.6). We retain the *continuous* spiral dynamics on $\mathbf{Z}$, then *binarize* through a transfer function and decode to $\mathbf{X}$.

Latent Spiral Update. Let $\mathbf{Z}_i^{(t)}$ be the latent vector of moth i and $\mathbf{F}_{\sigma(i)}^{(t)}$ the assigned flame. We update

$$\mathbf{Z}_i^{(t+1)} = \mathbf{F}_{\sigma(i)}^{(t)} + \mathbf{D}_i^{(t)} e^{b\,\ell} \cos(2\pi\ell), \quad \mathbf{D}_i^{(t)} = \left| \mathbf{F}_{\sigma(i)}^{(t)} - \mathbf{Z}_i^{(t)} \right|, \tag{14}$$

with $b > 0$ and $\ell \sim \mathcal{U}(-1, 1)$. Flame scheduling follows (11) (Sect. 2.6) to gradually reduce exploration.

Binary Transfer and Stochastic Flip. For each component z of $\mathbf{Z}_i^{(t+1)}$, compute a flip probability via an S- or V-shaped transfer (common in binary swarm/metaheuristics):

$$T_S(z) = \frac{1}{1 + e^{-\alpha z}} \quad \text{or} \quad T_V(z) = \left| \tanh(\alpha z) \right|, \tag{15}$$

with $\alpha > 0$ (we use $\alpha \in [1, 2]$). Given current discrete bit $b \in \{0, 1\}$, flip with probability $p = T(\cdot)$: $b \leftarrow 1 - b$ w.p. p. We apply this within each block $\mathbf{z}_i$ followed by the *block-wise argmax* constraint (13) to guarantee one-hot.

Constraint Handling and Local Search. Feasibility is reinforced by: (i) *penalties* in (12) with annealed $\lambda_{\{\cdot\}}$; (ii) *repair* after decoding; and (iii) a fast *local exchange* move: swap the (p, m) of two tasks on a congested link or overloaded PE if it reduces E_{total} or violation magnitude. We also use *elitist retention*: the best flame is copied forward each iteration.

The full BMFO procedure is summarized below.

Algorithm 2. BMFO-based Energy-Aware Task Mapping

Require: Task graph $G = (T, E)$ with volumes Q_{ij}; architecture $P = (S, F)$ with bw_{uv}, ℓ_{uv}, $\epsilon_{uv}^{\mathrm{comm}}$; per-task cycle counts Size_{ip}; per-mode $\{f_{mk}, e_{mk}\}$; population N, iterations T, spiral parameter b.
Ensure: Mapping $\mathbf{X}^{\star}$ minimizing $\mathcal{F}$ in (12).

```
Initialize moth latents {Z_i^(0)}_{i=1}^N using hybrid seeding; decode to {X_i^(0)} via (13); repair feasibility.
Evaluate fitness F(X_i^(0)); set flames {F_j^(0)} ← top-N latents sorted by fitness.
for t = 0 to T − 1 do
    Compute active flames F_active(t); set assignment σ(i) = min{i, F_active(t)}.
    for i = 1 to N do
        Spiral update latent Z_i^(t+1) using (14).
        Binary transfer using (15) within each task block; decode via block-wise argmax to get X_i^(t+1).
        Repair bandwidth/capacity/schedulability if needed; local exchange on hot spots.
        Evaluate F(X_i^(t+1)) by computing E_exec, E_com, and penalties.
    end for
    Merge {Z_i^(t+1)} ∪ {F_j^(t)}; sort by fitness; keep top N as new flames {F_j^(t+1)} (elitist retention).
    Anneal penalty weights λ_{·} ← η λ_{·} (e.g., η ∈ [0.9, 0.99]).
end for
return X* decoded from the best flame.
```

4 Results and Discussion

This section empirically assesses the proposed BMFO for single-objective, energy-aware static task mapping on heterogeneous MPSoCs. We quantify the total energy (computation + communication) under identical feasibility constraints for all methods and benchmark BMFO against two widely used meta-heuristics: Genetic Algorithm (GA) [10] and Particle Swarm Optimization (PSO) [16]. Our goals are to (i) establish whether BMFO consistently attains lower energy than GA/PSO across varying platform sizes and task counts, and (ii) interpret the observed trends in light of BMFO's flame scheduling.

4.1 Experimental Setup

All experiments were implemented in MATLAB R2019a and executed on a workstation equipped with an AMD Ryzen 5 1600 CPU running Microsoft Windows 10.

The proposed BMFO uses the configuration in Table 1;

4.2 Comparison of BMFO, GA, and PSO

We compare BMFO against GA and PSO. Each algorithm uses a population size 50 and 200 iterations. The primary metric is the total energy E_{total}; we also

Table 1. Baseline configuration and key parameters used by BMFO.

Category	Parameter	Value
Application/Platform	Architecture topology	2D mesh
	Link latency (per hop)	1 time unit
BMFO (search)	Population size N	50
	Maximum iterations T	200
	Spiral coefficient b	1.0
	Transfer function slope α	1.5
	Flame schedule	Linear decrease to 1 (Eq. (11))
	Elite retention	Best flame carried over
	Decoder	Block-wise argmax (Eq. (13))
	Binary flip	S/V transfer (Eq. (15))
Constraints/Penalties	Timing penalty $\lambda_{\text{tim}}^{(0)}$	10^6
	Bandwidth penalty $\lambda_{\text{bw}}^{(0)}$	10^6
	PE capacity penalty $\lambda_{\text{cap}}^{(0)}$	10^6
	Annealing factor η	0.95 per iteration

monitor feasibility (i.e., whether all constraints are satisfied) and the solver's wall-clock time. Unless otherwise stated, results in Table 2 are the best energies (in mJ) obtained per setting.

Table 2. Energy consumption (mJ) for BMFO, GA, and PSO on MPSoC mapping

Processors	Tasks	BMFO	GA	PSO
4	10	155.2	163.8	**154.9**
6	15	**148.1**	159.3	152.4
10	20	**146.3**	151.9	147.7
12	25	**149.0**	156.1	152.3
16	30	**145.9**	150.2	148.6
20	32	**139.4**	146.0	141.8

Across the six MPSoC configurations, BMFO attains the lowest energy in five cases, with PSO winning once at the smallest instance (4 processors / 10 tasks) by a negligible margin of 0.3 mJ ($\approx$ 0.19%). Averaged over all settings, BMFO achieves 147.32 mJ compared to 149.62 mJ for PSO and 154.55 mJ for GA, corresponding to mean improvements of $\approx$ 1.54% over PSO and $\approx$ 4.65% over GA. The bar chart in Fig. 6 visualizes these trends: GA consistently trails BMFO and PSO, while BMFO maintains a persistent advantage that slightly widens as the platform scales (e.g., at 20/32, BMFO 139.4 mJ vs. PSO 141.8 mJ vs. GA

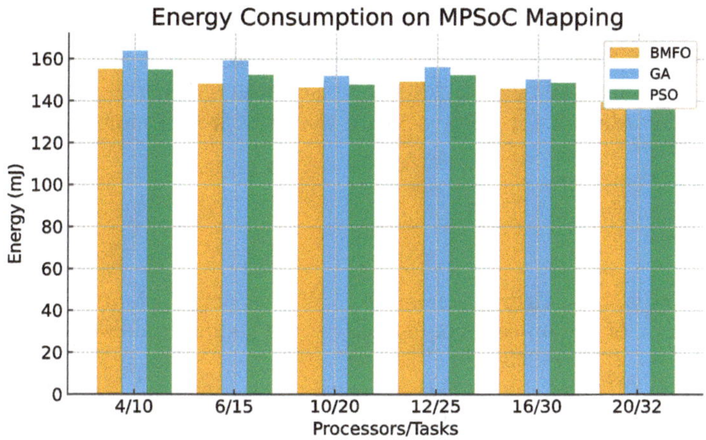

Fig. 6. Grouped bar chart of the energies in Table 2.

146.0 mJ). We attribute BMFO's gains to the synergy of continuous spiral guidance in the latent space, block-wise argmax decoding that enforces one-hot feasibility, and targeted local exchanges near congested links/overloaded PEs, which together sustain exploration while steering the search toward communication-aware, low-energy placements. In contrast, the standard crossover/mutation of GA appears less effective in this constrained discrete landscape, and PSO, while competitive, lacks the flame scheduling that enhances exploitation in later iterations.

5 Conclusion

This paper introduced a Binary Moth–Flame Optimization (BMFO) framework for single-objective, energy-aware *static* task mapping on heterogeneous MPSoCs. We formalized computation and communication energy models, captured timing and resource constraints (one-hot assignment, PE capacity, link bandwidth, and precedence/schedulability), and designed a feasibility-preserving encoding/decoding scheme that enables BMFO to operate effectively in a discrete, highly constrained search space. The method combines continuous spiral guidance in a latent space with block-wise argmax decoding, adaptive flame scheduling, repair mechanisms, and a lightweight local exchange. On a set of MPSoC configurations, BMFO consistently reduced total energy (computation + communication) relative to GA and PSO, achieving the best results in five of six cases and delivering average improvements of $\approx 4.65\%$ over GA and $\approx 1.54\%$ over PSO, while maintaining feasibility.

Future Directions. We will test BMFO on substantially larger task graphs and platforms (hundreds of tasks, tens of PEs), including chiplet and 3D-NoC fabrics, while preserving feasibility guarantees. In summary, BMFO provides a robust and scalable foundation for energy-aware static task mapping in heterogeneous MPSoCs.

References

1. Bhattacharyya, S.S., Eker, J., Janneck, J.W., Lucarz, C., Mattavelli, M., Raulet, M.: Overview of the mpeg reconfigurable video coding framework. J. Signal Process. Syst. **63**(2), 251–263 (2011)
2. Boumaza, F., Benyamina, A.E.H., Zouache, D., Abualigah, L., Alsayat, A.: An improved Harris hawks optimization algorithm based on bi-goal evolution and multi-leader selection strategy for multi-objective optimization. Ingénierie des Systèmes d'Information **28**(5) (2023)
3. Boumaza, F., Hadji, A., Aroui, A.: Efficient allocation of computationally intensive application tasks in noc-based heterogeneous mpsocs. Informatica **50**(6), 421–432 (2026). https://doi.org/10.31449/inf.v50i6.7895
4. Boumaza, F., Zouache, D., Belazzoug, M., Hadji, A., Aroui, A.: Binary grey wolf optimizer for mapping real time applications on mpsocs architecture (2024)
5. Boumaza, F., Hadji, A., Zouache, D.: Energy-efficient task allocation in heterogeneous noc-based mpsocs using binary chimp optimization. ITEGAM-JETIA **11**(56), 401–413 (2025)
6. Chen, Z., Li, Y., Zhang, Y.: Recent advances in omnidirectional video coding for virtual reality: projection and evaluation. Signal Process. **146**, 66–78 (2018)
7. Karthikeyan, V., Subbulakshmi, K.: 3d noc-enabled spiking neural networks: a high-performance computing paradigm. Evol. Syst. **16**(3), 1–18 (2025)
8. Ksira, Z., Mellit, A., Blasuttigh, N., Pavan, A.M.: A novel embedded system for real-time fault diagnosis of photovoltaic modules. IEEE J. Photovolt. **14**(2), 354–362 (2024)
9. Lahti, S., Hämäläinen, T.D.: High-level synthesis for fpgas-a hardware engineer's perspective. IEEE Access (2025)
10. Lambora, A., Gupta, K., Chopra, K.: Genetic algorithm-a literature review. In: 2019 International Conference on Machine Learning, Big Data, Cloud and Parallel Computing (COMITCon), pp. 380–384. IEEE (2019)
11. Li, Y.X., Luo, G., Xu, Y.K., He, Y., Zhang, F.L., Zhang, S.H.: Adapip: adaptive picture-in-picture guidance for 360° film watching. Comput. Visual Media **10**(3), 487–503 (2024)
12. Mirjalili, S.: Moth-flame optimization algorithm: a novel nature-inspired heuristic paradigm. Knowl.-Based Syst. **89**, 228–249 (2015)
13. Munir, A., Gordon-Ross, A., Ranka, S.: Modeling and Optimization of Parallel and Distributed Embedded Systems. John Wiley & Sons, Hoboken (2016)
14. Spieck, J., Wildermann, S., Teich, J.: A scenario-based dvfs-aware hybrid application mapping methodology for mpsocs. ACM Trans. Des. Autom. Electron. Syst. **29**(4), 1–43 (2024)
15. Valente, G., Muttillo, V., Pomante, L., Frigioni, D., Di Mascio, T.: A new hw/sw codesign approach for monitored systems-on-chip development. ACM Trans. Embed. Comput. Syst. (2025)

16. Wang, D., Tan, D., Liu, L.: Particle swarm optimization algorithm: an overview. Soft. Comput. **22**(2), 387–408 (2018)
17. Wang, H., Halak, B.: Tampml: tampering attack detection and malicious nodes localization in noc-based mpsoc. IEEE Trans. Emerg. Top. Comput. (2024)
18. Zouache, D., Abualigah, L., Boumaza, F.: A guided epsilon-dominance arithmetic optimization algorithm for effective multi-objective optimization in engineering design problems. Multimedia Tools Appl. **83**(11), 31673–31700 (2024)

Hybrid Deep Learning for Short-Term Load Forecasting: A Comparative Study on Algerian Buildings

Rania Farah(✉), Rahma Aissaoui, and Zineddine Kouahla

Department of Computer Science, LabStic Laboratory, University 8 May 1945, Guelma, Algeria
farah.rania@univ-guelma.dz

Abstract. The benefit of understanding future energy consumption is not just an advantage, but a requirement. When energy demand can be predicted in advance, it becomes possible to plan more efficiently, eliminate waste and make better use of available resources. This is particularly important in Algeria, where electricity is almost exclusively generated from fossil fuels, with natural gas accounting for over 98% of total production. As electricity cannot be stored, and demand peaks during extreme weather conditions often exceed the supply, the national grid managed by state enterprise Sonelgaz is under frequent tension, leading to power blackouts and operational efficiencies. In order to deal with this challenge, this study focuses on forecasting short-term energy consumption in Algerian buildings. Using Convolutional neural networks (CNNs) and long-term memory networks (LSTMs) are combined in a hybrid deep learning model, the task is approached as a time-series prediction challenge. The model was chosen for its ability to extract spatial patterns from structured data (via CNN) and to capture temporal dynamics (via LSTM). Trained on real consumption data from Algerian buildings, the model aims to provide accurate forecasts that enable more reliable and efficient energy management an essential step towards strengthening the national energy system and reducing dependence on non-renewable resources.

Keyword: Energy forecasting, convolutional neural networks, long-term memory networks, Deep Learning, Short-term Load Prediction, Time Series, Building Energy Consumption.

1 Introduction

As electricity consumption becomes increasingly volatile due to technological advancements, climate variability, and shifting societal behaviors, the need for accurate short-term load forecasting (STLF) has become critical for modern energy management. Forecasting demand not only enhances operational efficiency and cost control but also improves grid reliability by anticipating fluctuations that could otherwise result in overproduction losses or power outages [1]. Over the past decade, machine learning (ML) and deep learning

Z. Kouahla et al. (Eds.): IAM 2025, LNNS 1973, pp. 44–56, 2026.
https://doi.org/10.1007/978-3-032-26031-4_5

(DL) techniques have demonstrated substantial success across various building types and geographic settings. Non-parametric models like k-Nearest Neighbor (k-NN) have shown effectiveness in sensor-sensitive smart environments [1], while hybrid approaches combining Gradient Boosting Regressor (GBR) with Long Short-Term Memory (LSTM) networks have delivered high accuracy in educational buildings rich in IoT infrastructure [2]. Signal-processing-based models, such as those integrating Stationary Wavelet Transform (SWT) with GRU or Bi-LSTM, have enhanced noise reduction and prediction in residential energy datasets [3]. Comparative studies involving algorithms like Support Vector Machines (SVM), Multi-Layer Perceptrons (MLP), and XGBoost have underscored the importance of temporal features such as time of day and day of the week in structured commercial and institutional contexts [4]. In Algeria, deep learning models, especially Convolutional Neural Networks (CNNs), have achieved superior results on national utility datasets, with MAEs as low as 0.09 and accuracy exceeding 98%, outperforming traditional models like Random Forest and linear regression [5]. Recent research has emphasized hybrid models such as CNN-GRU, further enhanced with metaheuristics like Earth Worm Optimization (EWO), achieving accuracy over 96% [6], while explainable AI techniques like SHAP have added interpretability by identifying key consumption drivers in residential settings [7]. In several studies, CNN-GRU architectures have surpassed CNN-LSTM and Support Vector Regression (SVR), achieving MAEs as low as 0.24 [8]. Similarly, combining Variational Mode Decomposition (VMD) with LSTM has improved robustness in office forecasts, with R^2 reaching 0.97 and MAE as low as 21.36 kWh [9]. Ensemble approaches such as stacked LSTM with snapshot learning have achieved exceptional accuracy ($R^2 = 0.999$) on residential datasets [10], while hybrid frameworks combining first principles models with LightGBM have closed performance gaps by modeling both static building properties and dynamic load behavior [11]. Additionally, multi-model fusion techniques involving XGBoost, LightGBM, and Logistic Regression have demonstrated R^2 values above 0.99 in medium-term forecasting tasks, showcasing adaptability under seasonal variations [12]. Although no single model universally dominates all applications, hybrid architectures that merge CNNs for spatial feature extraction with LSTM or GRU for temporal modeling consistently deliver higher accuracy, flexibility, and interpretability. Based on these findings, this study proposes a CNN-LSTM hybrid model customized for the Algerian energy landscape, aiming to improve short-term electricity consumption forecasting and contribute to more responsive and intelligent energy grid operations.

2 Methodology

This section presents the dataset, the preprocessing steps, and the deep learning models used for accurate short-term electricity load forecasting. The dataset, provided by Sonelgaz, the Algerian national electricity supplier, spans from 2008 to 2020 and is structured in annual matrices where each row represents a single day and each column corresponds to an hourly electricity consumption value, giving 24 values per day [13]. Focused on smart buildings, the data enables detailed analysis of hourly load patterns and supports the development of forecasting models capable of capturing fine temporal variations in demand. As previous studies [5] have shown, electricity consumption in Algeria has steadily increased over time, with two main seasonal peaks observed, one in summer due to high temperatures and intensive use of air conditioning, and another in winter likely due to heating needs.

2.1 Data Visualization

Accurate time series forecasting begins with careful preparation of data and an in-depth exploration of the underlying trends. This section describes the approach used to construct meaningful features, preprocess electricity consumption data, and apply statistical and visual techniques to discover temporal patterns.

Data Preparation. The data preparation phase began with a comprehensive quality check to ensure the reliability of the building energy consumption dataset. All missing values and duplicate records were removed. The data set was then transformed from a wide format, where each row represented a day with 24 hourly columns, into a long time-stamped format. Each observation was restructured into a single row with one timestamp and one energy value, allowing temporal patterns to be more effectively captured. Mathematically, this process is represented as follows: The initial preprocessing steps include handling missing values, converting data types, and exploring basic statistics.

$$X_{d,h} \rightarrow (t_i, x_i) \tag{1}$$

Time Series Restructuring [14]

Where: $X_{d,h}$ is the data at day d and hour h; (t_i, x_i) are time-value pairs. The result of this transformation, as illustrated in Fig. 1, highlights how energy consumption varies over time. It can be seen that energy consumption is constantly on the rise.

Feature Extraction. To enhance the ability of the model to learn temporal patterns, several features were extracted. First, each observation was assigned a season based on the month: Winter (December–February), Spring (March–May), Summer (June–August), and Fall (September–November). These seasonal labels were then one hot encoded into binary variables to prevent the introduction of artificial order. Furthermore, cyclical encoding was applied to hours and months using sine and cosine transformations to reflect their periodic nature.

$$Hour_{\sin} = \sin\left(2\pi \cdot \frac{h}{24}\right), \qquad Hour_{\cos} = \cos\left(2\pi \cdot \frac{h}{24}\right) \tag{2}$$

$$Month_{\sin} = \sin\left(2\pi \cdot \frac{m}{12}\right), \qquad Month_{\cos} = \cos\left(2\pi \cdot \frac{m}{12}\right) \tag{3}$$

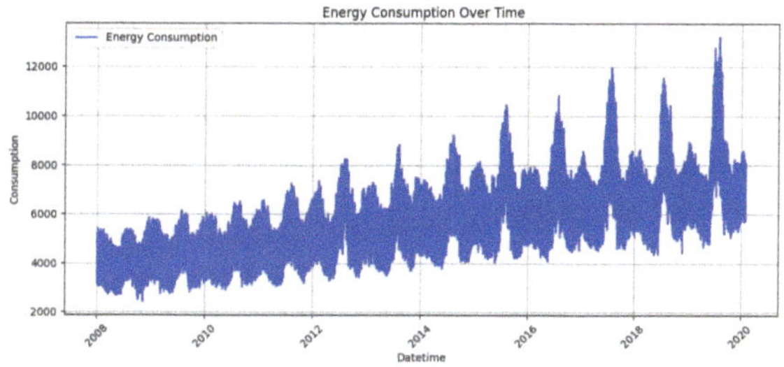

Fig. 1. Energy Consumption Over Time.

Cyclical Encoding [15]

Where: h is the hour (0–23),
m is the month (1–12);
sine and cosine encode cyclic time features.
This transformation helps the model understand that hour 0 and hour 23 are close in time due to the cyclical encoding applied to time-based features. Furthermore, seasonal aggregations such as the minimum and maximum values of variables like temperature and energy consumption were calculated to capture seasonal dynamics over time. Figure 2 illustrates the maximum energy consumption for each season across multiple years, clearly revealing consistent seasonal variations and highlighting seasonal peaks in summer and winter.

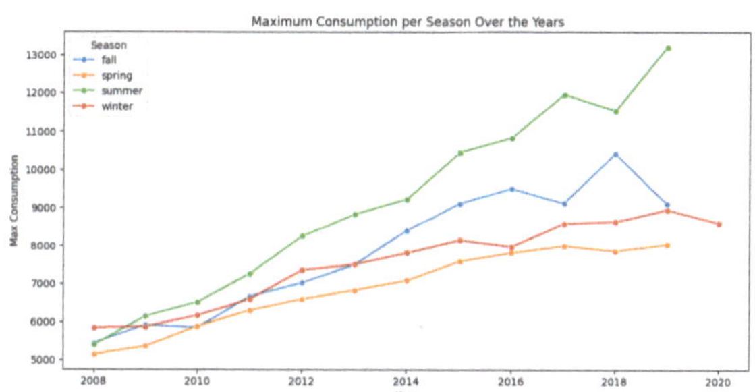

Fig. 2. Maximum Consumption per Season Over the Years

To evaluate the relevance of the different features to energy consumption, a correlation heat map was generated in Fig. 3. This visualization highlights the linear correlation coefficients between each input variable and the target variable. We can see that certain characteristics, such as `rolling_mean_6h`, `min_saison` and `max_saison`, show a significant positive correlation with energy consumption. In contrast, the majority of other variables, particularly cyclical time encodings such as `hour_cos` and `month_sin`, as well as one-hot encoded seasonal indicators, show weak or insignificant linear correlations. This suggests that these variables have no direct linear relationship with consumption, but could still have an influence that could be detected by models capable of capturing complex, non-linear relationships. These results therefore underline the relevance of using deep learning models, such as CNN-LSTM, which can effectively model these complex and non-linear patterns, in contrast to traditional linear approaches.

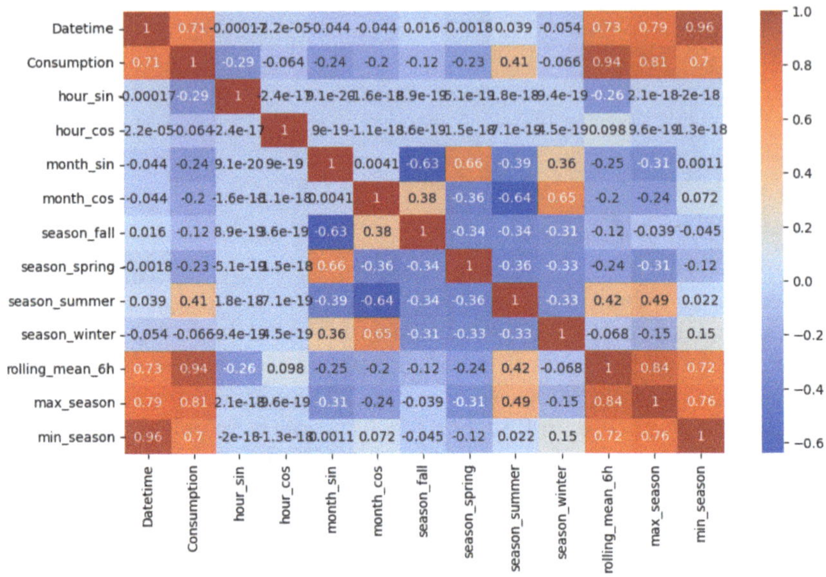

Fig. 3. Correlation Heatmap

Data Splitting and Scaling. The final preprocessing step involved dividing the dataset into training and test subsets using an 80/20 split, while preserving temporal order. To prepare the data for modeling, the input features and target variable were scaled using a `MinMaxScaler`. This ensured that all variables were scaled to a comparable scale between 0 and 1, preventing the predominance of features with larger numerical ranges.

Convolutional Neural Network (CNN). CNNs are traditionally used in image processing but have proven effective in time series forecasting due to their feature extraction capabilities. The core building block of CNNs is the convolutional layer, where a kernel (or filter) slides over the input data to extract local patterns. This process is defined as:

$$(f * h)(t) = \sum_{i=0}^{k} f(t-i) \cdot h(i) \tag{4}$$

Discrete Convolution [5]

Where: f is the input,
h is the kernel,
t is the time,
and$k$ is the kernel size.

CNNs can capture local dependencies in sequences and reduce dimensionality through pooling layers, making them efficient and suitable for extracting relevant features in temporal data.

Long Short-Term Memory (LSTM). LSTM networks are a specialized type of Recurrent Neural Network (RNN) designed to effectively capture long-term dependencies in sequential data. Unlike traditional RNNs, LSTMs incorporate gated mechanisms specifically, the input, forget, and output gates; which regulate the flow of information through the cell state. This architecture enables the model to selectively retain or discard information, making it particularly well-suited for modeling temporal patterns, such as those found in energy consumption data.

Mathematically, the internal operations of a single LSTM cell at time step t are defined as follows:

$$\begin{aligned}
f_t &= \sigma\left(W_f \cdot [h_{t-1}, x_t] + b_f\right) && \text{(forget gate)} \\
i_t &= \sigma\left(W_i \cdot [h_{t-1}, x_t] + b_i\right) && \text{(input gate)} \\
\tilde{C}_t &= \tanh\left(W_C \cdot [h_{t-1}, x_t] + b_C\right) && \text{(candidate cell state)} \\
C_t &= f_t \odot C_{t-1} + i_t \odot \tilde{C}_t && \text{(cell state update)} \\
o_t &= \sigma\left(W_o \cdot [h_{t-1}, x_t] + b_o\right) && \text{(output gate)} \\
h_t &= o_t \odot \tanh(C_t) && \text{(hidden state/output)}
\end{aligned} \tag{5}$$

LSTM Cell Computations

Where: x_t: input at time t
h_{t-1}: hidden state from previous time step
C_t: cell state at time t
σ: sigmoid activation function
tanh: hyperbolic tangent activation function
$\odot$: element-wise (Hadamard) product f_t, i_t, o_t: forget, input, and output gates.

Bidirectional LSTM (BiLSTM). BiLSTM extends the standard LSTM by processing the sequence in both forward and backward directions. This dual processing enables the model to have context from both past and future data points, improving accuracy in scenarios where understanding the full sequence context is important. This is particularly effective in building energy consumption where future conditions can influence current usage (e.g., thermostat preheating).

BiLSTM is essentially two LSTM layers, where the first reads the sequence as-is and the second reads it in reverse. Their outputs are then concatenated or combined to form a final representation used for prediction as shown in the Eqs. 6, 2.1 for the outputs and 8 for the concatenation:

$$\overrightarrow{h_t} = \text{LSTM}_{\text{forward}}(x_t), \tag{6}$$

$$\overleftarrow{h_t} = \text{LSTM}_{\text{backward}}(x_t) \tag{7}$$

Bidirectional LSTM Outputs [1]

Where: $\overrightarrow{h_t}$ and $\overleftarrow{h_t}$ are forward and backward LSTM outputs.

$$h_t = [\overrightarrow{h_t}; \overleftarrow{h_t}] \tag{8}$$

Concatenation of LSTM States[2]

Where: h_t is concatenation of forward and backward hidden states.

3 Results and Discussion

This section presents a comparative evaluation of four deep learning models-CNN-LSTM, CNN-BiLSTM, standalone CNN, and standalone LSTM for short-term electricity consumption forecasting in buildings. Each model was trained on the same dataset and assessed using consistent evaluation metrics (MAE, RMSE, R^2, MAPE) to ensure fair comparison. The aim is to determine which architecture offers the best performance in terms of accuracy, generalization, and robustness.

1) **Convolutional Neural Network - Long Short-Term Memory (CNN-LSTM):**
   Different configurations of the CNN-LSTM architecture were tested, combining convolutional layers for feature extraction with LSTM layers for capturing temporal dependencies. The most effective configuration included two `Conv1D` layers with 64 and 128 filters respectively, both using a kernel size of 5 and ReLU activation functions, followed by max pooling layers with a pool size of 2. These convolutional layers significantly contributed to the ability of the model to learn local patterns in the input sequences, and removing them led to a notable decrease in performance.
   The LSTM component consisted of two layers: the first with 128 units and `return_sequences=True`, and the second with 64 units and `return_sequences=False`. Each LSTM layer was followed by layer normalization and a dropout rate of 0.05 to enhance generalization and training stability. The output of the LSTM layers was passed through a dense layer with 64 neurons using ReLU activation, followed by a final output layer with a single neuron for regression. The model was trained using the Huber loss function with a delta value of 1.5, optimized with the Adam optimizer (learning rate = 0.001, `clipnorm` = 10.0). Training was conducted for a maximum

of 20 epochs with a batch size of 32, using early stopping with a patience of 3 epochs and validation loss monitoring to prevent overfitting.
As a result, training was halted early at epoch 9 due to the absence of further improvement in the validation loss, and the best-performing weights were restored. The entire training process took approximately 20 min. Although certain hyper-parameters such as the number of filters, the number of LSTM units, and the dropout rates were adjusted.

2) **Convolutional Neural Network - Bidirectional Long Short-Term Memory (CNN-BiLSTM):**
Different configurations of the CNN-BiLSTM architecture were tested, integrating convolutional layers for feature extraction with bidirectional LSTM layers to effectively capture both forward and backward temporal dependencies inherent in time series data.
The network architecture was carefully designed and systematically varied to assess the influence of key hyper-parameters, including the number of convolutional filters, kernel sizes, LSTM units, dropout rates, and learning rates, on predictive performance.
The most effective architecture consisted of two convolutional layers: the first with 64 filters and the second with 128 filters, both using a kernel size of 3 and ReLU activation functions. These layers were followed by a Bidirectional LSTM layer with 128 units and a standard LSTM layer with 64 units. A dropout rate of 0.3 was applied between layers to reduce overfitting, while layer normalization was employed to improve convergence and training stability. The model was compiled using the Huber loss function (with a delta value of 1.5) and optimized with the Adam optimizer using a learning rate of 0.001. Training was initially scheduled for 50 epochs; however, early stopping halted the process at the 18th epoch due to no further improvement in validation performance. The total training time was approximately 36 min. Sensitivity analysis demonstrated that increasing the dropout rate beyond 0.4 adversely impacted learning, while reducing it below 0.2 led to overfitting. Similarly, modifying the learning rate above 0.001 caused training instability, and values significantly lower than 0.0005 slowed down convergence without improving accuracy. The performance of the model was evaluated using standard regression metrics, including Mean Absolute Error (MAE), Root Mean Squared Error (RMSE), R^2 Score, and Mean Absolute Percentage Error (MAPE). Furthermore, removing the convolutional layers in ablation experiments caused a significant drop in performance, confirming their importance in capturing localized temporal features during early-stage processing.
3) **Convolutional Neural Network (CNN):**
An in-depth investigation of a one-dimensional convolutional neural network (CNN) model was conducted to evaluate its effectiveness in energy consumption forecasting based on time series data. The architecture consisted of two sequential Conv1D layers: the first with 64 filters and the second with 128 filters, both employing a kernel size of 3 and ReLU activation functions. These convolutional layers were followed by a MaxPooling1D layer (pool size = 2) to

reduce spatial dimensions, and a Dropout layer with a rate of 0.3 to enhance generalization and prevent overfitting. The feature maps were then flattened and passed through two dense layers with 64 and 32 neurons, respectively, before reaching a final output neuron for regression. The model was trained using the Huber loss function and the Adam optimizer with a learning rate of 0.001. Although training was scheduled for 50 epochs, early stopping halted the process at epoch 24 to avoid overfitting, with total training time averaging approximately 30 min per run on a standard GPU. A sensitivity analysis revealed that dropout rates below 0.2 led to overfitting, while rates above 0.4 reduced learning capacity; also, learning rates higher than 0.001 caused unstable convergence, whereas lower values prolonged training without clear performance improvement.

4) **Long Short-Term Memory (LSTM):**
An in-depth evaluation was conducted to examine the predictive performance of the LSTM neural network architecture for short-term load forecasting. The model was designed to capture temporal patterns in sequential data, and several architectural and training configurations were explored.
The final architecture consisted of a single LSTM layer with 128 hidden units, followed by a dropout layer with a rate of 0.2, and a fully connected dense output layer with one unit. The LSTM layer was responsible for learning temporal dependencies, while the dropout mechanism was employed to mitigate overfitting and enhance generalization.

The model was trained using the Adam optimizer with a learning rate of 0.001. The Mean Squared Error (MSE) loss function was used during training due to its sensitivity to large deviations, which is suitable for regression tasks. The training process was initially scheduled for 20 epochs with a batch size of 64; however, early stopping was activated at epoch 16, preventing overfitting once the validation loss ceased to improve. The total training duration was approximately 2 h and 36 min. To assess the robustness of the model and understand its sensitivity to key hyperparameters, a series of experiments were conducted. Variations in dropout rates (0.1, 0.2, and 0.3) were examined. A rate of 0.2 was found to deliver the best balance between underfitting and overfitting. Similarly, different LSTM layer sizes were tested, including 64, 128, and 256 units. The configuration with 128 units demonstrated superior performance, offering a favorable trade-off between computational complexity and accuracy. Furthermore, learning rates of 0.0005, 0.001, and 0.005 were analyzed, with 0.001 resulting in stable convergence and minimized validation error. Model evaluation was carried out using standard performance metrics: MAE, RMSE, coefficient R^2, and MAPE.

3.1 Discussion of the Results

The performance results presented in Table 1 clearly highlight the superiority of hybrid models that combine CNNs with LSTM architectures over individual models.

The CNN-LSTM model achieved the best overall performance, with the lowest MAE (232.74), RMSE (330.60), and MAPE (3.12%), along with the highest R^2 score (0.9549). These results demonstrate its effectiveness in capturing both spatial and temporal dependencies in the electricity consumption data. The convolutional layers help in extracting relevant features, while the LSTM layers learn the sequential dependencies, making this architecture particularly well-suited for time series forecasting tasks.

The CNN-BiLSTM model also performed well, ranking second across all metrics. Although slightly less accurate than CNN-LSTM, it still surpassed standalone models. The use of bidirectional LSTMs allows the model to learn from both past and future context, but the slight decrease in performance compared to CNN-LSTM could be attributed to the added complexity and potential overfitting.

The standalone LSTM model showed moderate performance, outperforming CNN but lagging behind the hybrid models. Its capacity to model temporal patterns is strong, but without convolutional feature extraction, it may be less effective in handling local patterns or noise in the input sequences.

Lastly, the CNN model exhibited the weakest performance among the four, with the highest MAE (393.69) and RMSE (531.05), and the lowest R^2 score (0.8836). This result suggests that while CNNs are good at capturing short-term spatial correlations, they are not sufficient on their own to model the long-term dependencies present in sequential data like energy consumption.

This results confirm that combining convolutional and recurrent layers yields better performance than using either architecture independently. This supports the growing body of literature advocating for hybrid deep learning approaches in time series forecasting tasks.

Table 1. Performance comparison of deep learning models

Model	MAE	RMSE	R^2 Score	MAPE
CNN-LSTM	232.7406	330.6046	0.9549	3.12%
CNN-BiLSTM	280.5946	381.7233	0.9399	3.78%
LSTM	320.6516	445.8256	0.9180	4.30%
CNN	393.6871	531.0480	0.8836	5.21%

Figures 4 and 5 complement the table by providing visual comparisons of these metrics. Among all tested architectures, the CNN-LSTM model consistently achieves the best performance, with the lowest MAE and RMSE, the highest R^2 score, and the lowest MAPE.

These results highlight the ability of the CNN-LSTM model to capture spatial and temporal dependencies in energy consumption data. Although the CNN-BiLSTM model also performs well, it is slightly weaker than the CNN-LSTM model, indicating that the bidirectional structure does not necessarily improve

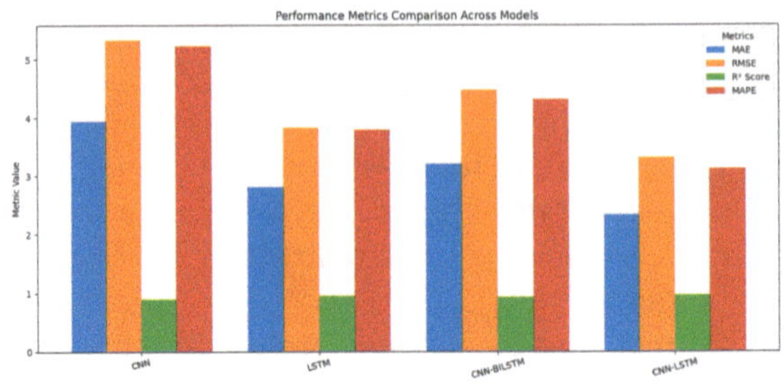

Fig. 4. Performance Metrics Comparison Across Models

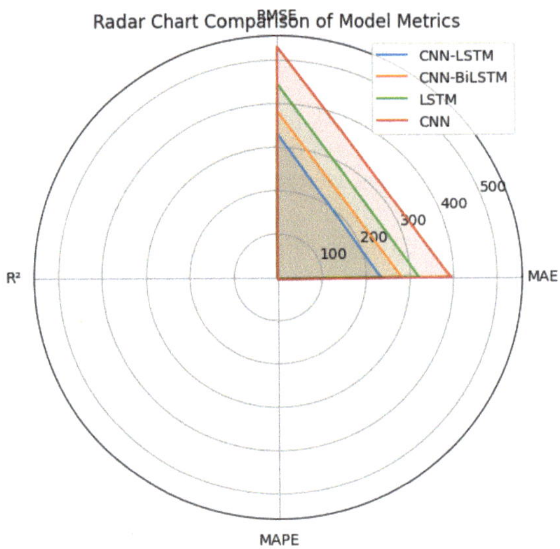

Fig. 5. Radar Chart Comparison of Model Metrics

forecast accuracy in this context. On the other hand, the standalone LSTM and CNN models perform significantly worse, confirming that hybrid architectures are better suited to complex, in time prediction tasks.

The radar chart in Fig. 5 visually illustrates this performance difference, clearly showing the advantage of hybrid models over individual components. This reinforces the conclusion that integrating convolutional layers, which are effective for local feature extraction, with LSTM layers designed for sequence modeling, delivers the most balanced and accurate results. The evaluation strongly supports the use of CNN-LSTM architectures in short-term load forecasting.

4 Conclusion

Understanding electricity consumption patterns is essential for enhancing energy efficiency and reducing operational costs, particularly in complex building environments. This study leveraged historical consumption data to identify daily and seasonal trends, as well as key relationships with external factors such as weather.

To address the growing demand for accurate short-term load forecasting, we evaluated a range of deep learning models, including CNNs, LSTMs, and hybrid architectures. These models were specifically chosen for their ability to capture both short-term variations and long-term dependencies in time series data.

Among the evaluated approaches, the CNN-LSTM hybrid model demonstrated superior performance, achieving an R^2 score exceeding 95%. This confirms the advantages of combining convolutional and recurrent layers, enabling the model to effectively extract local patterns while maintaining a strong memory of sequential dependencies.

Despite the inherent challenges of forecasting electricity demand such as sudden behavioral shifts and external variability-the results validate the robustness of deep learning methods. In conclusion, this work reinforces the potential of hybrid deep learning architectures as reliable tools for short-term electricity demand forecasting and energy management optimization.

References

1. Sari, M., et al.: Machine learning-based energy use prediction for the smart building energy management system. J. Inf. Technol. Constr. **28** (2023)
2. Elhabyb, K., et al.: Machine learning algorithms for predicting energy consumption in educational buildings. Int. J. Energy Res. **2024**(1), 6812425 (2024)
3. Frikha, M., et al.: Predicting power consumption using deep learning with stationary wavelet. Forecasting **6**(3) (2024)
4. Wu, J., et al.: A comparative analysis of machine learning-based Energy Baseline models across multiple building types. Energies **17**(6), 1285 (2024)
5. Rania, F., Brahim, F., Zineddine, K., et al.: A comparison of the finest electrical energy forecast models. In: 2023 International Conference on Decision Aid Sciences and Applications (DASA), pp. 570–575. IEEE (2023)
6. Ayub, N., et al.: Big data analytics for short and medium-term electricity load forecasting using an AI techniques ensembler. Energies **13**(19), 5193 (2020)
7. Moon, J., et al.: Advancing ensemble learning techniques for residential building electricity consumption forecasting: insight from explainable artificial intelligence. PLoS ONE **19**(11), e0307654 (2024)
8. Sajjad, M., et al.: A novel CNN-GRU-based hybrid approach for short-term residential load forecasting. IEEE Access **8**, 143759–143768 (2020)
9. Ruan, Y., et al.: A hybrid model for power consumption forecasting using VMD-LSTM. Appl. Energy **312** (2022)
10. Alghamdi, M., et al.: Predicting energy consumption using stacked LSTM snapshot ensemble (2024)
11. Chen, X., et al.: A hybrid-model approach for reducing the performance gap in building energy forecasting (2022)
12. Ji, Q., et al.: Short- and medium-term power demand forecasting with multi-model fusion (2022)
13. Farah, R., Farou, B., Kouahla, Z., Seridi, H.: Data-driven models on load forecasting: case study Algeria. Data Brief **52**, 109854 (2023)

14. Géron, A.: Hands-on Machine Learning with Scikit-Learn, Keras, and TensorFlow: Concepts. Google Books (2019)
15. Schuster, M., Paliwal, K.K.: Bidirectional recurrent neural networks. IEEE Trans. Signal Process. **45**(11), 2673–2681 (1997)
16. Sudre, G., Pomerleau, D., Palatucci, M., et al.: Tracking neural coding of perceptual and semantic features of concrete nouns. Neuroimage **62**(1), 451–463 (2012)

Evaluating Drinking Water Quality Through Machine Learning

Fatima Bouakkaz[1,2(✉)] and Wided Ali[1,2]

[1] Echahid Cheikh Larbi Tebessi University, Tebessa, Algeria
{fatima.bouakkez,wided.ali}@univ-tebessa.dz
[2] Laboratory of Vision and Artificial Intelligence (LAVIA), Tebessa, Algeria

Abstract. Access to clean drinking water remains a critical public health concern worldwide, as chemical and microbial contamination continues to affect millions. Machine learning (ML) provides a promising approach for water quality assessment, enabling rapid and data-driven predictions based on complex physicochemical measurements. This study evaluates the performance of multiple ML algorithms—including logistic regression, support vector machines, decision trees, random forests, gradient boosting, and neural networks—on a standard water quality dataset. Experimental results reveal that ensemble tree-based methods, such as Random Forest and Gradient Boosting, consistently outperform simpler models in accuracy and F1-score, while deep learning models offer limited benefits on small tabular datasets. The study highlights the importance of proper data preprocessing, handling of missing values, and techniques to address class imbalance. These findings underscore the potential of machine learning as a reliable tool for automated potable water detection and suggest directions for improving predictive accuracy and interpretability in real-world water management applications.

Keywords: Water Drinking · Machine Learning(ML) · Logistic Regression(LR) · Support Vector Machines (SVM) · Decision Trees(DT) · Random Forests(RF) · Neural Networks

1 Introduction

Access to safe drinking water is a fundamental requirement for public health and well-being. Despite global efforts, millions of people continue to suffer from waterborne diseases caused by chemical pollutants and microbial contamination. Traditional water quality monitoring relies on laboratory testing, which is often time-consuming, resource-intensive, and limited in scale.

Recent advances in machine learning (ML) offer an efficient alternative for water quality assessment. By analyzing physicochemical parameters such as pH, hardness, solids, chloramines, sulfate, conductivity, organic carbon, trihalomethanes, and turbidity, ML algorithms can provide rapid, accurate predictions of water potability. These methods can detect complex patterns and

Z. Kouahla et al. (Eds.): IAM 2025, LNNS 1973, pp. 57–64, 2026.
https://doi.org/10.1007/978-3-032-26031-4_6

relationships within data that may be difficult to capture using conventional statistical approaches.

This study proposes a comprehensive ML framework to automatically classify water samples as potable or non-potable. The framework integrates data preprocessing, exploratory analysis, model development using multiple supervised learning algorithms, and rigorous performance evaluation. The aim is to identify the most effective models and preprocessing strategies for reliable water quality assessment, thereby supporting public health initiatives and efficient water management

2 Related Work

Numerous studies have employed machine learning techniques to predict water quality using the Kaggle water potability dataset. Despite using the same dataset, differences in preprocessing methods, model selection, and evaluation protocols have produced widely varying results. For example, Chafloque et al. [1] implemented an Artificial Neural Network (ANN) with seven fully connected layers, using 10-fold cross-validation for evaluation. They achieved an accuracy of approximately 70%, illustrating the difficulty of applying deep learning methods to relatively small, structured tabular datasets, where overfitting and limited feature information can hinder performance.

Fen et al. [2] also worked with the nine standard physicochemical features of the dataset and compared multiple algorithms, including Decision Trees, Naïve Bayes, Support Vector Machines, linear models, and K-Nearest Neighbors (KNN). Their findings indicated that Decision Trees outperformed the others, reaching an accuracy of 86.67%, which highlights the robustness and suitability of tree-based models for water quality prediction in small to medium-sized tabular datasets.

Kurra et al. [3] focused on KNN and Decision Trees. They observed that KNN performed poorly (61.7% accuracy), emphasizing the sensitivity of distance-based algorithms to data scaling and preprocessing. In contrast, Decision Trees provided more stable predictions, further demonstrating the importance of proper feature preparation and model selection.

Patel [4] addressed class imbalance using the Synthetic Minority Oversampling Technique (SMOTE) and evaluated several models, including SVM, Decision Tree, Random Forest, Gradient Boosting, and AdaBoost. Random Forest and Gradient Boosting achieved around 81% accuracy, showing that ensemble methods can significantly improve predictive performance. Patel also incorporated Explainable AI techniques to identify key features contributing to model predictions, providing insight into the underlying relationships in water quality data.

Finally, Poudel et al. [5] investigated the impact of handling missing values on model performance. They applied median imputation for missing entries in features such as pH and trihalomethanes and trained a Random Forest model with 700 trees and a maximum depth of 30. This approach achieved 70% accuracy, highlighting how data imputation and preprocessing can critically affect the outcomes of machine learning models in water quality prediction tasks (Table 1).

Table 1. Comparison of Machine Learning Models for Drinking Water Quality Prediction

Ref	Datasets	Models	Accuracy	Notes
[1] (2021)	Kaggle dataset, 3276 samples, 9 physicochemical features	ANN (7 dense layers)	69%	Cross-validation with 10 folds; MinMax normalization
[2] (2021)	Kaggle dataset, 3276 samples	DT, NB, SVM, Linear, KNN	86.67%	Cross-validation with 10 folds; Decision Tree best performer
[3] (2022)	Kaggle dataset, 3276 samples	KNN, DT	61.7%	KNN underperformed; preprocessing affected results
[4] (2022)	Kaggle dataset, 3276 samples; SMOTE balancing	SVM, DT, RF, Gradient Boost, AdaBoost	81%	SMOTE + Explainable AI used
[5] (2023)	Kaggle dataset, 3276 samples; median imputation for missing data	LR, KNN, RF, ANN	70%	RF best (700 trees, depth=30); focus on missing data handling

3 Discussion of Related Work

Machine learning has been widely applied to predict water potability, often using the Kaggle dataset. Deep learning approaches, such as the ANN by Chafloque et al. [1], show potential but struggle with small tabular datasets due to overfitting. Tree-based models, including Decision Trees and Random Forests, consistently outperform simpler models, with Fen et al. [2] and Patel [4] reporting accuracies above 80%, particularly when combined with ensemble methods and hyperparameter tuning.

Data preprocessing is crucial: Kurra et al. [3] highlighted KNN's sensitivity to scaling, while SMOTE and median imputation in Patel [4] and Poudel et al. [5] improved model performance, emphasizing the importance of handling class imbalance and missing values.

Despite these advances, many studies rely on single metrics like accuracy and often neglect interpretability, limiting practical applicability. Overall, ensemble tree-based models with careful preprocessing and tuning provide the most reliable predictions, while future work should integrate explainable AI and robust evaluation to enhance both accuracy and usability in real-world water quality management.

4 Proposed System

This study presents a machine learning-based framework with Matlab 2021 to automatically determine whether water samples are potable or non-potable. The system is structured into four main components: data preprocessing, exploratory analysis, model development, and performance evaluation.

4.1 Data Preprocessing

The water quality dataset includes parameters such as pH, hardness, solids, chloramines, sulfate, conductivity, organic carbon, trihalomethanes, and turbidity. Initially, missing values are filled using the median for each feature, and any features with no variance are removed. Subsequently, all features are standardized through Z-score normalization to ensure uniform contribution during model training. (see Figs. 1, 2, 3).

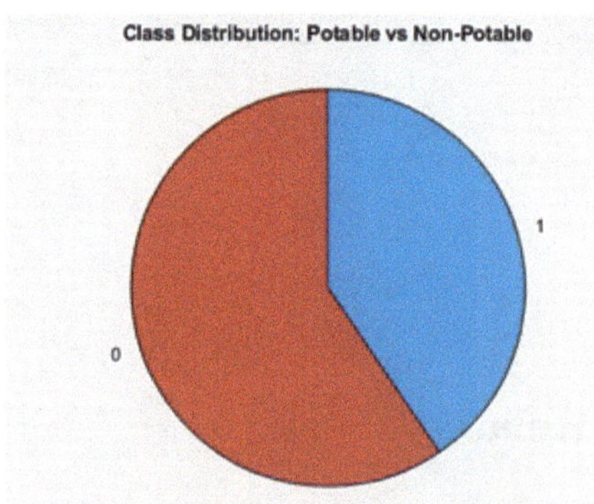

Fig. 1. Class Distribution Potable vs no Potable.

4.2 Model Development

Seven supervised machine learning models are employed and compared: Logistic Regression, Support Vector Machine (SVM), K-Nearest Neighbors (KNN), Naive Bayes, Random Forest, Gradient Boosting, and Artificial Neural Network (ANN). A 10-fold cross-validation procedure is applied to obtain reliable

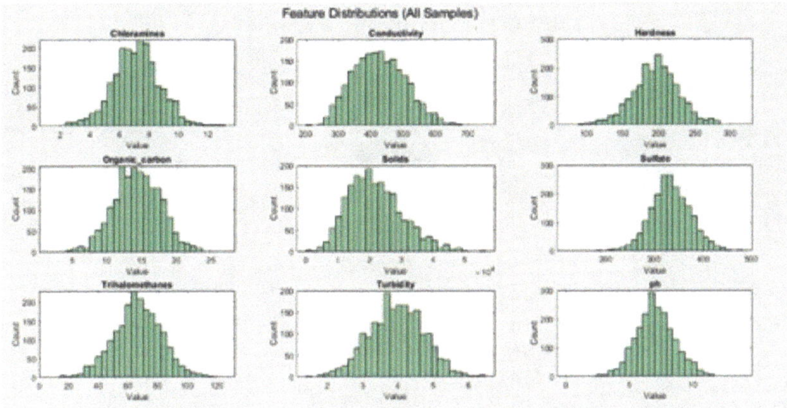

Fig. 2. Feature Distribution (all Sample).

performance estimates. The ANN consists of a two-layer feedforward network optimized with cross-entropy loss, whereas the ensemble models use multiple decision trees to capture complex relationships in the data.

4.3 Performance Evaluation

The models are assessed using Accuracy, Precision, Recall, F1-score, and Area Under the ROC Curve (AUC). Confusion matrices and ROC curves are generated for detailed insight into each model's classification capability. Results are summarized in tables and visualized with bar charts for easy comparison. Among the evaluated methods, ensemble approaches, particularly Random Forest and Gradient Boosting, consistently achieve superior performance by balancing high precision and recall, whereas simpler models like Logistic Regression underperform due to class imbalance challenges.

Overall, the proposed framework highlights the potential of ensemble learning techniques for effective potable water classification and provides a structured approach to evaluating multiple machine learning models on water quality datasets (Fig. 4).

4.4 Results Discussion

The goal of this study was to evaluate the effectiveness of various machine learning models in predicting water potability based on multiple water quality parameters, including pH, hardness, solids, chloramines, sulfate, conductivity, organic carbon, trihalomethanes, and turbidity.

The dataset was preprocessed with median imputation for missing values, standardization, and a 10-fold cross-validation framework to ensure reliable evaluation.

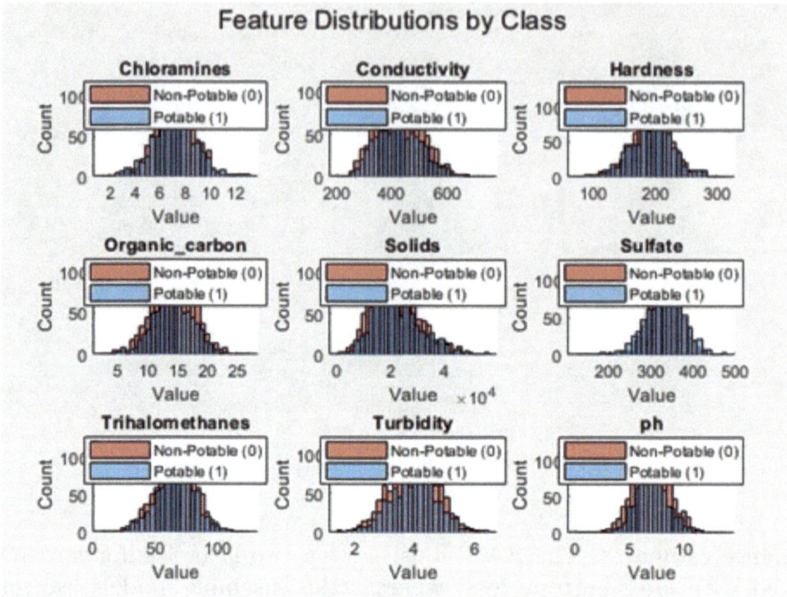

Fig. 3. Feature Distribution by class.

	1 Accuracy	2 Precision	3 Recall	4 F1	5 AUC
1 Logistic	0.5959	0.5000	0.0074	0.0146	0.5094
2 SVM	0.6064	0.5488	0.1455	0.2300	0.3654
3 KNN	0.6303	0.5594	0.4007	0.4670	0.3485
4 NaiveBayes	0.6228	0.5746	0.2565	0.3546	0.3946
5 RandomForest	0.6861	0.6737	0.4328	0.5270	NaN
6 GradBoost	0.6432	0.5689	0.4834	0.5227	0.3324
7 ANN	0.6512	0.6546	0.2898	0.4017	0.3496

Fig. 4. Seven supervised machine learning models.

The experimental results indicate that ensemble methods, particularly Random Forest and Gradient Boosting, outperform other models in terms of overall accuracy and F1-score.

Random Forest achieved the highest accuracy (0.6806) and precision (0.6660), indicating that it is relatively conservative and makes fewer false positive predictions for potable water.

Gradient Boosting, while slightly lower in accuracy, showed the highest recall (0.4809), meaning it is better at identifying actual potable water samples.

This suggests that ensemble methods effectively leverage multiple decision trees to capture complex patterns in water quality data.

Single models such as Logistic Regression and SVM showed relatively poor performance, especially in recall, indicating their limited ability to detect potable water in this dataset. Logistic Regression, for example, exhibited very low recall (0.0025), suggesting that it misclassified nearly all potable water samples as non-potable.

KNN and Naive Bayes performed moderately, with reasonable balance between precision and recall, but still lagged behind ensemble methods.

The low AUC values observed across most models (except for Random Forest, which produced a NaN due to output formatting) highlight the difficulty in discriminating potable from non-potable water based solely on the given features.

This is likely due to class imbalance, overlapping distributions between the two classes, and the presence of noisy or non-discriminative features.

Overall, the results indicate that ensemble learning approaches are more robust for water potability classification, especially when dealing with complex, imbalanced datasets. However, there is still significant room for improvement. Future work could focus on:

- Addressing class imbalance using techniques such as SMOTE or class-weighted training.
- Feature selection and engineering to identify the most informative water quality parameters.
- Hyperparameter optimization for ensemble and neural network models to further improve recall without sacrificing precision.

In conclusion, this study demonstrates the potential of machine learning for automated water quality assessment but also emphasizes the challenges posed by real-world water quality datasets. By leveraging advanced models and proper preprocessing strategies, potable water detection can be significantly improved, which is critical for public health and environmental monitoring.

5 Conclusion

This study demonstrates the potential of machine learning for automated assessment of drinking water quality. By analyzing multiple physicochemical parameters, several supervised learning models—including Logistic Regression, SVM, KNN, Naive Bayes, Random Forest, Gradient Boosting, and Artificial Neural Networks—were evaluated using a 10-fold cross-validation framework. The results indicate that ensemble tree-based methods, particularly Random Forest and Gradient Boosting, consistently outperform simpler models in terms of accuracy, precision, and F1-score, effectively handling the complexity and variability inherent in water quality data.

The study also highlights the critical role of data preprocessing, including handling missing values, standardization, and addressing class imbalance, in improving predictive performance. While simpler models struggled, ensemble methods provided robust predictions, suggesting their suitability for real-world applications.

Future research could focus on incorporating feature selection, hyperparameter optimization, and explainable AI techniques to enhance interpretability and predictive accuracy. Overall, the findings support the use of machine learning as a reliable tool for water quality monitoring, offering a scalable approach to safeguard public health and guide water management decisions.

References

1. Chafloque, R., Rodriguez, C., Pomachagua, Y., Hilario, M.: Predictive neural networks model for detection of water quality for human consumption. In: 2021 13th International Conference on Computational Intelligence and Communication Networks (CICN), pp. 172–176 (2021)
2. Fen, L., Lei, Z., Ting, C.: Study on potability water quality classification based on integrated learning. In: 16th International Conference on Intelligent Systems and Knowledge, vol. 2 (2021)
3. Kurra, S.S., Naidu, S.G., Chowdala, S., Yellanki, S.C., Sunanda, D.B.E.: Water quality prediction using machine learning. Int. Res. J. Modernizat. Eng. Technol. Sci. (2022)
4. Patel, J., et al.: A machine learning-based water potability prediction model by using synthetic minority oversampling technique and explainable AI. Comput. Intell. Neurosci. **2022**(1), 9283293 (2022)
5. Poudel, D., Shrestha, D., Bhattarai, S., Ghimire, A.: Comparison of machine learning algorithms in statistically imputed water potability dataset. J. Innov. Eng. Educ. **5**(1), 38–46 (2022)
6. Kadiwal, A.: Water Quality. Kaggle. https://www.kaggle.com/datasets/adityakadiwal/water-potability

Exploring Industry 5.0: Architecture, Threat Landscape, and Security Strategies

Djallel Hamouda(✉), Mohamed Amine Ferrag, and Zineddine Kouahla

Labstic Laboratory, Department of Computer Science, Guelma University, B.P. 401, 24000 Guelma, Algeria
{hamouda.djallel,ferrag.mohamedamine,kouahla.zineddine}@univ-guelma.dz

Abstract. Industry 5.0 marks a transformative shift in industrial systems, emphasizing human-centricity, sustainability, and resilience. By integrating advanced technologies such as artificial intelligence (AI), the Internet of Things (IoT), and robotics, Industry 5.0 seeks to enhance collaboration between humans and machines while addressing societal and environmental challenges. This paper provides a comprehensive review of the architecture underpinning Industry 5.0, analyzes its cyber threat landscape, and evaluates cybersecurity strategies essential for mitigating risks. Additionally, it highlights emergent technologies for their critical roles in safeguarding interconnected systems and explores future trends and research directions, offering a well-rounded perspective on this rapidly evolving paradigm.

Keywords: Industry 5.0 · Architecture · Threat Landscape · Security Strategies · Ethical Implications · Future Trends

1 Introduction

Industry 5.0 represents an evolution from Industry 4.0, focusing on human-centric design, sustainability, and enhanced collaboration between humans and machines. Unlike its predecessor, which emphasized automation, Industry 5.0 seeks to integrate human intelligence and creativity with advanced technologies such as AI, and robotics. This new paradigm addresses the growing need for resilient and adaptive systems capable of responding to dynamic global challenges, including sustainability and rapid technological disruption [17].

Recent studies highlight the transformative impact of Industry 5.0 on manufacturing, healthcare, and smart cities. For example, the integration of AI-driven systems with human oversight has demonstrated improved efficiency and safety in industrial processes [26]. Similarly, the transition from Industry 4.0 to Industry 5.0 has been framed as a shift toward addressing socio-economic challenges, leveraging AI and IoT for sustainability, and overcoming the limitations of fully automated systems [23]. However, as highlighted by Navale et al. (2023), this evolution also introduces critical security and privacy challenges. The increasing reliance on interconnected IoT devices, AI, and robotics in Industry 5.0 raises

Z. Kouahla et al. (Eds.): IAM 2025, LNNS 1973, pp. 65–76, 2026.
https://doi.org/10.1007/978-3-032-26031-4_7

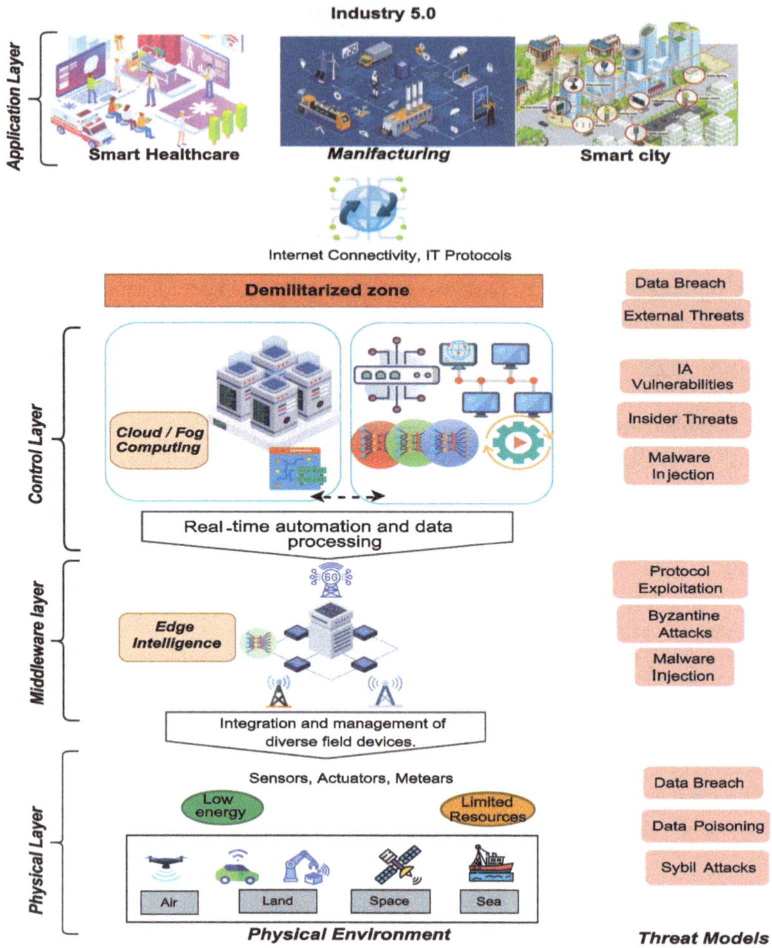

Fig. 1. Illustrative Industry 5.0 Architecture: Layers, Threats, and Defense Strategies.

concerns about data breaches, system vulnerabilities, and the trustworthiness of such systems. Addressing these challenges is essential to ensure dependable and secure systems, which are pivotal for the widespread adoption of Industry 5.0 technologies [21].

This paper examines the architecture, potential threats, and security strategies essential for realizing the vision of Industry 5.0. It also explores ethical implications and predicts future trends to provide a holistic understanding of this transformative era.

2 Industry 5.0 Architecture

The architecture of Industry 5.0 builds on foundational layers established in Industry 4.0, introducing enhanced human-machine interaction and sustainable design principles [20]. We synthesized the existing architectural frameworks by emphasizing a comprehensive interaction between its layers, threat models and real-world applications, as depicted in Fig. 1. The primary components include:

- **Physical Layer:** This foundational layer consists of IoT-enabled sensors, actuators, and metering devices deployed across diverse domains, including land, sea, air, and space. These devices face significant challenges such as energy constraints and security vulnerabilities, which necessitate cross-layer energy optimization to extend the lifetime of IoT nodes. Robust security measures, including protections against Sybil and data poisoning attacks, are essential to ensure the reliability of IoT systems. Balancing performance with complexity is critical to supporting resilient and efficient industrial operations [6].
- **Middleware Layer:** Acting as a bridge between the physical and control layers, this component facilitates edge and fog computing to enable real-time data processing and reduce latency. Enhanced edge intelligence manages distributed field devices efficiently, ensuring rapid data processing while minimizing delays. The integration of secure communication protocols and anomaly detection mechanisms is vital to counter threats such as protocol exploitation, Byzantine attacks, and malware injection. This layer further emphasizes data privacy and resilience, particularly in resource-constrained environments [4].
- **Control Layer**: Building on cloud and fog computing, this layer plays a crucial role in real-time automation, system coordination, and data processing. Cloud-fog automation (CFA) platforms provide flexibility and scalability, enabling real-time control and monitoring through edge devices. These platforms effectively coordinate production tasks, enhancing operational efficiency. Security measures in this layer include robust intrusion detection systems (IDS), such as decision fusion-based IDS for industrial control systems and a two-tiered anomaly detection architecture that leverages fog and cloud computing for real-time threat response. Addressing integration challenges with legacy infrastructure and continuously updating security protocols remain ongoing priorities [18,28].
- **Application Layer:** This layer prioritizes human-centric applications, driving advancements in domains such as smart healthcare, manufacturing, and smart cities. Personalized, data-driven solutions—ranging from autonomous systems to predictive maintenance and tailored healthcare services—align with Industry 5.0's goals for social and environmental impact. The integration of advanced AI and human-in-the-loop systems fosters sustainable and collaborative ecosystems [26].

We further integrate a ***Demilitarized Zone*** (DMZ) within the architecture to safeguard against both external and internal cyber threats, providing an additional security layer. This DMZ mitigates vulnerabilities by isolating critical systems from potentially compromised networks. The synthesis presented here underscores the holistic connectivity of the physical and cyber domains, ensuring robust security, low latency, and seamless data integration across all layers.

3 Threat Landscape

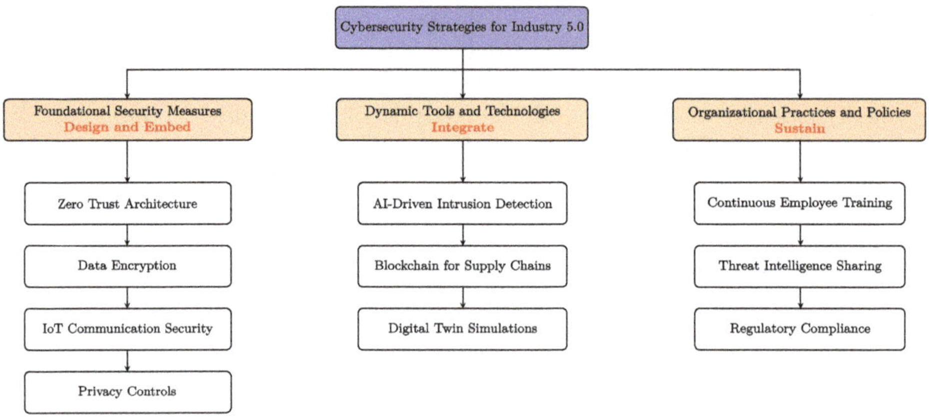

Fig. 2. Cybersecurity Strategies for Industry 5.0

The transition to Industry 5.0 necessitates a robust and multidimensional security framework that addresses functional safety, information security, and human-centric considerations to safeguard both individual and societal interests [27]. This paradigm shift not only inherits vulnerabilities from Industry 4.0 but also introduces novel and sophisticated threat models due to the integration of IoT, AI, and human-machine collaboration [11].

Table 1 outlines the major threat models targeting IoT network architecture in Industry 5.0, including Data Privacy Intrusion, Protocol Exploitation, Malware Injections, and AI-Associated Threats. These threats highlight vulnerabilities at various levels of the architecture, from data integrity and communication protocols to device security and AI systems. Notably, attackers exploit these weaknesses to manipulate system behavior, compromise data confidentiality, or disrupt network operations.

Addressing these threats involves not only recognizing their implications but also designing multi-layered countermeasures that mitigate risks while supporting the human-centric and sustainable goals of Industry 5.0.

Table 1. Threat Models Targeting IoT Network Architecture in Industry 5.0.

Threat Model	Description
Data Privacy Intrusion	IoT devices frequently collect and transmit sensitive data. Attackers may intercept or access this data, leading to privacy violations [8]
Protocol Exploitation	IoT devices rely on various communication protocols. Exploiting vulnerabilities in these protocols enables attackers to gain unauthorized access or manipulate device behavior [10]
Malware Injections	Injecting malware into IoT devices compromises their functionality and can transform them into tools for malicious purposes [12]
Data Poisoning	Attackers inject inaccurate or malicious data into IoT systems, causing incorrect decisions or actions by connected devices [8]
Sybil Attacks	An attacker generates multiple fake identities to overwhelm the system, disrupt operations, or gain unauthorized access to IoT devices [24]
Byzantine Attacks	Malicious or compromised nodes provide conflicting information within a network, resulting in system failures or flawed decision-making [24]
AI-Associated Threats	The integration of AI in IoT introduces new vulnerabilities, allowing attackers to manipulate AI algorithms to disrupt device behavior and decisions [7]
External Threats	Attackers exploit vulnerabilities in IoT device software, firmware, or communication channels to compromise device security

4 Cybersecurity Strategies for Industry 5.0

This section outlines a structured taxonomy of foundational measures, dynamic tools, and organizational practices that collectively strengthen the security posture of this new industrial paradigm, as illustrated in Fig. 2.

Foundational Security Measures

This category emphasizes embedding security at the core of Industry 5.0 systems to ensure resilience and protection from the outset. Examples include implementing Zero Trust Architectures, which verify all access requests regardless of their origin, and using advanced data encryption protocols to safeguard sensitive information during transmission and storage. Secure IoT communication standards such as TLS/SSL are crucial for protecting interconnected industrial systems from cyber threats [3].

Dynamic Tools and Technologies

Dynamic tools address the evolving threat landscape and provide adaptive defense mechanisms. AI-driven Intrusion Detection Systems (IDS) are pivotal in identifying and mitigating potential threats in real-time. Blockchain technology ensures data integrity and transparency across supply chain networks, while Digital Twin simulations enable proactive vulnerability testing of industrial systems [22]. These tools allow organizations to remain agile in their cybersecurity strategies.

Organizational Practices and Policies

Long-term cybersecurity relies on sustainable practices that maintain and enhance system security. Continuous employee training on cyber hygiene fosters a security-aware workforce. Real-time threat intelligence sharing platforms facilitate industry-wide collaboration to counter emerging threats. Regular security audits and compliance with global standards, such as IEC 62443, ensure adherence to best practices and regulatory requirements [2] .

5 Emergent Technologies for Securing Industry 5.0

Industry 5.0 introduces a range of interconnected systems that demand advanced technologies to ensure their security and privacy. This section evaluates these critical solutions for securing Industry 5.0, detailing their features, limitations, and practical applications (Table 2).

Federated Learning

Federated learning is a decentralized machine learning paradigm that processes data locally on edge devices, ensuring privacy by avoiding centralized data aggregation [8]. It enables privacy-preserving analytics in IoT networks, particularly for predictive maintenance. However, challenges include the reliance on robust edge devices and scalability issues in handling high-dimensional data or extensive industrial networks [14].

Differential Privacy

Differential privacy offers a framework for securing datasets by ensuring individual data points remain indistinguishable within aggregated data. This technology is crucial for collaborative industrial analytics, as it protects against re-identification attacks while maintaining data utility. Nevertheless, critics highlight the trade-off between privacy and accuracy, as improperly tuned parameters may compromise data utility [1].

Table 2. Evaluation of Emerging Technologies for Industry 5.0 Cybersecurity

Technology	Key Features and Advantages	Challenges and Limitations	Practical Use Cases
Federated Learning	Enhances privacy by processing data locally; reduces communication overhead; scalable for IoT networks	Requires reliable edge devices; may face challenges in handling high-dimensional data or large-scale networks.	Privacy-preserving analytics in IoT networks for predictive maintenance
Differential Privacy	Ensures individual data points remain secure; protects against re-identification attacks; maintains data utility	May reduce data accuracy; requires careful tuning of privacy parameters to balance utility and security.	Sharing industrial sensor data for collaborative analytics without compromising privacy
Quantum Cryptography	Provides unbreakable encryption; resilient against quantum computing threats; ensures secure key distribution	Highly resource-intensive; depends on the maturity of quantum technologies and infrastructure.	Secure communication between critical industrial control systems
Blockchain-Based Identity Management	Decentralized authentication; enhances trust and traceability; improves security in machine-to-machine interactions	Scalability and performance in large networks; potential regulatory and compliance challenges.	Authenticating devices and users in supply chain networks
Secure Multi-Party Computation (SMPC)	Enables collaborative computation without revealing inputs; ensures data confidentiality and integrity	Computationally intensive; requires high trust between participating parties and robust infrastructure.	Secure collaborative decision-making between industrial stakeholders
Zero-Trust Architecture	Eliminates implicit trust; provides real-time access control; reduces insider and external threats	Complex implementation; demands continuous monitoring and high-level trust models.	Dynamic access control in industrial environments to mitigate insider threats

Quantum Cryptography

Quantum cryptography leverages quantum mechanics to provide theoretically unbreakable encryption methods such as Quantum Key Distribution (QKD). It ensures secure communication for critical industrial systems against emerging quantum-computing threats. Despite its promise, the technology remains resource-intensive and dependent on the maturity of quantum infrastructures [15].

Table 3. Future Trends and Research Directions

Trend/Direction	Focus Area and Research Opportunities
Integration of AI for Predictive Security	AI-driven tools for real-time threat detection and mitigation using predictive analytics; algorithms to detect unknown attack vectors
Quantum-Resistant Cryptographic Algorithms	Development of post-quantum cryptographic protocols for securing systems against quantum computing threats
Cybersecurity for Digital Twins	Real-time security measures to detect and prevent tampering of virtual replicas of physical systems
Policy and Standardization	Creation of global cybersecurity standards and frameworks adaptable to diverse industrial contexts and regulations
Ethical AI and Autonomous Decision-Making	Development of governance models ensuring transparency, fairness, and accountability in AI-driven systems
IoT Security in Edge Computing	Lightweight encryption and intrusion detection for IoT devices operating at decentralized edge nodes
Resilience Against Cyber-Physical Attacks	Simulation environments to stress-test systems and enhance resilience to targeted disruptions
Green Cybersecurity	Exploration of low-power cryptographic techniques and energy-efficient blockchain solutions to support sustainability goals

Blockchain-Based Identity Management

Blockchain-based identity management utilizes decentralized systems for authenticating devices, machines, and humans in interconnected networks. It enhances trust, traceability, and security in supply chains and machine-to-machine communication [13]. However, scalability in large networks and regulatory challenges remain open issues.

Secure Multi-party Computation (SMPC)

SMPC enables multiple parties to collaboratively compute functions without revealing their individual inputs, preserving data confidentiality and integrity. It is particularly useful for secure collaboration between industrial stakeholders [13]. However, the computational intensity and requirement for trust among parties pose significant hurdles.

Zero-Trust Architecture

Zero-trust architecture eliminates implicit trust within industrial systems, implementing continuous verification of devices, users, and processes [3]. It is vital for real-time access control and mitigating both insider and external threats. However, its complexity and requirement for high-level trust models and monitoring make implementation challenging.

These technologies provide a foundation for addressing the cybersecurity needs of Industry 5.0 by ensuring privacy, data integrity, and operational resilience. However, issues such as scalability, resource intensity, and implementation complexity require further research to integrate these solutions into comprehensive security frameworks for Industry 5.0 systems.

6 Future Trends and Research Directions

To strengthen the security and privacy of Industry 5.0 systems, several future trends and research directions are emerging. These include leveraging **AI for predictive security**, developing **quantum-resistant cryptography**, and ensuring the integrity of **cybersecurity for digital twins**, as summarized in Table 3. Additionally, global efforts in **policy standardization** and the adoption of **ethical AI frameworks** will play pivotal roles in shaping secure autonomous decision-making processes. **IoT security in edge computing** focuses on mitigating vulnerabilities at decentralized processing nodes, while enhancing **resilience against cyber-physical attacks** is vital for safeguarding critical systems. Lastly, **green cybersecurity** aligns with Industry 5.0's sustainability goals, emphasizing energy-efficient security solutions.

7 Ethical Challenges in Industry 5.0

The integration of advanced technologies in Industry 5.0 introduces significant ethical challenges, including ensuring transparency and fairness in AI systems, reskilling the workforce, protecting privacy, promoting environmental sustainability, and establishing accountability for autonomous systems. Ethical frameworks are essential to mitigate risks such as algorithmic bias and privacy violations, with explainable AI and fairness metrics enhancing transparency [9] [19]. Workforce reskilling initiatives are crucial to address displacement caused by AI, requiring comprehensive training aligned with new technological demands [16] [25]. Privacy concerns arising from extensive data collection necessitate robust encryption and governance frameworks to ensure compliance and safeguard personal data [25]. Sustainability goals demand energy-efficient practices and resource monitoring through IoT technologies to mitigate environmental impact [5]. Finally, clear accountability frameworks are critical for managing autonomous systems, ensuring collaboration among stakeholders to foster responsibility [19]. Addressing these challenges not only mitigates risks but also fosters innovation, paving the way for equitable and responsible implementation of Industry 5.0.

8 Conclusion

Industry 5.0 offers immense potential to redefine industries by fostering a human-centric, intelligent, and sustainable approach to industrial processes. This paper has highlighted the transformative capabilities of Industry 5.0 while addressing critical challenges, including evolving cybersecurity threats, ethical considerations, and the need for robust architectural frameworks. Success in Industry 5.0 hinges on a balanced approach that integrates technological innovation with comprehensive security measures, ethical governance, and sustainable practices.

By leveraging cutting-edge technologies such as IoT, AI, and edge computing, and through collaboration between stakeholders, Industry 5.0 can overcome its inherent challenges. Continued research, proactive governance, and global cooperation will be instrumental in developing secure, equitable, and energy-efficient solutions that align with societal and environmental goals. Through this integrated approach, Industry 5.0 has the potential to pave the way for a more inclusive, resilient, and sustainable industrial future.

References

1. Abadi, M., et al.: Deep learning with differential privacy. In: Proceedings of the 2016 ACM SIGSAC Conference on Computer and Communications Security, pp. 308–318 (2016)
2. Akello, B.O.: Organizational information security threats: status and challenges. World J. Adv. Eng. Technol. Sci. (2024). https://doi.org/10.30574/wjaets.2024.11.1.0152
3. Alshra'a, A.S., Al Sardy, L., Dibaei, M., German, R.: Security principles in smart and agile cybersecurity for IoT and IIoT environments. In: Advances in Information Security, Privacy, and Ethics Book Series, pp. 1–26 (2024). https://doi.org/10.4018/979-8-3693-3451-5.ch001
4. Bravos, G., et al.: Cybersecurity for industrial internet of things: architecture, models and lessons learned. IEEE Access **10**, 124747–124765 (2022). https://doi.org/10.1109/ACCESS.2022.3225074
5. Chandre, P., Nalavade, J., Kanthe, A., Deshmukh, J.Y., Shendkar, B., Mahalle, P.N.: Industry 5.0 for society 5.0 a new transformation. Panamerican Math. J. (2024). https://doi.org/10.52783/pmj.v35.i1s.2188
6. Fang, Z., et al.: A model checking-based security analysis framework for IoT systems. High-Conf. Comput. **1**(1), 100004 (2021)
7. Ferrag, M.A., et al.: Edge learning for 6g-enabled internet of things: a comprehensive survey of vulnerabilities, datasets, and defenses. IEEE Commun. Surv. Tutor. (2023)
8. Ferrag, M.A., Friha, O., Maglaras, L., Janicke, H., Shu, L.: Federated deep learning for cyber security in the internet of things: concepts, applications, and experimental analysis. IEEE Access **9**, 138509–138542 (2021)
9. Firmansyah, G., Bansal, S., Walawalkar, A.M., Kumar, S., Chattopadhyay, S.: The future of ethical AI. In: Advances in Computational Intelligence and Robotics Book Series, pp. 145–177 (2024). https://doi.org/10.4018/979-8-3693-3860-5.ch005

10. Gerodimos, A., Maglaras, L., Ferrag, M.A., Ayres, N., Kantzavelou, I.: IoT: communication protocols and security threats. In: Internet of Things and Cyber-Physical Systems (2023)
11. Hamouda, D.: New Technologies for Security and Privacy Issues in the Era of Industry 5.0. Ph.D. thesis, University of Guelma, Algeria (2024). https://dspace.univ-guelma.dz/jspui/handle/123456789/15880
12. Hamouda, D., Ferrag, M.A., Benhamida, N., Kouahla, Z.E., Seridi, H.: Android Malware Detection Based on Network Analysis and Federated Learning, pp. 23–39. Springer, Cham (2024). https://doi.org/10.1007/978-3-031-34969-0_2
13. Hamouda, D., Ferrag, M.A., Benhamida, N., Seridi, H.: PPSS: a privacy-preserving secure framework using blockchain-enabled federated deep learning for industrial IoTs. Pervasive Mob. Comput. **88**, 101738 (2023)
14. Hamouda, D., Ferrag, M.A., Benhamida, N., Seridi, H., Ghanem, M.C.: Revolutionizing intrusion detection in industrial IoT with distributed learning and deep generative techniques. Internet Things 101149 (2024). https://doi.org/10.1016/j.iot.2024.101149
15. Hasan, M.: Quantum cryptography for secure communications and the enhancement of U.S. cybersecurity in the quantum age. Adv. Social Sci. Res. J. **11**(10), 313–320 (2024). https://doi.org/10.14738/assrj.1110.17764
16. Hasanah, U.: Data privacy and transparency: Ethical challenges in the implementation of artificial intelligence (AI) in the accounting field. Smart GOALS: Jurnal Bisnis Digital dan Manajemen **1**(1), 19–31 (2024). https://doi.org/10.36490/sg.v1i1.1330
17. Leng, J., et al.: Towards resilience in industry 5.0: a decentralized autonomous manufacturing paradigm. J. Manuf. Syst. **71**, 95–114 (2023)
18. Lyu, H., Yan, J., Zhang, J., Pang, Z., Geng, Y., Isaksson, A.: Cloud-fog automation: heterogenous applications over new generation infrastructure of virtualized computing and converged networks (2023). https://doi.org/10.36227/techrxiv.24062307
19. Meduri, K., Podicheti, S., Satish, S., Whig, P.: Accountability and transparency ensuring responsible AI development. In: Advances in Human and Social Aspects of Technology Book Series, pp. 83–102 (2024). https://doi.org/10.4018/979-8-3693-4147-6.ch004
20. Mourtzis, D., Angelopoulos, J., Panopoulos, N.: A literature review of the challenges and opportunities of the transition from industry 4.0 to society 5.0. Energies **15**(17), 6276 (2022)
21. Navale, G.S., Madala, R., Managuli, M., Jayalakshmi, N., Kadiravan, G., Rawat, R.: Research and innovation in next generation security and privacy in industry 5.0 Iot. In: 2023 6th International Conference on Contemporary Computing and Informatics (IC3I), vol. 6, pp. 1384–1390. IEEE (2023)
22. Ovabor, K., Sule-Odu, I.O., Atkison, T., Fabusoro, A.T., Benedict, J.O.: AI-driven threat intelligence for real-time cybersecurity: frameworks, tools, and future directions. Open Access Res. J. Sci. Technol. (2024). https://doi.org/10.53022/oarjst.2024.12.2.0135
23. Raja Santhi, A., Muthuswamy, P.: Industry 5.0 or industry 4.0 S? Introduction to industry 4.0 and a peek into the prospective industry 5.0 technologies. Int. J. Interact. Des. Manuf. (IJIDeM) **17**(2), 947–979 (2023)
24. Rajan, A., Jithish, J., Sankaran, S.: Sybil attack in IOT: modelling and defenses. In: 2017 International Conference on Advances in Computing, Communications and Informatics (ICACCI), pp. 2323–2327. IEEE (2017)

25. Reddy, K.S., Kethan, M., Mahabub Basha, S., Singh, A., Kumar, P., Ashalatha, D.: Ethical and legal implications of AI on business and employment: Privacy, bias, and accountability. In: 2024 International Conference on Knowledge Engineering and Communication Systems (ICKECS), vol. 1, pp. 1–6 (2024). https://doi.org/10.1109/ICKECS61492.2024.10616875
26. Rožanec, J.M., et al.: Human-centric artificial intelligence architecture for industry 5.0 applications. Int. J. Prod. Res. **61**(20), 6847–6872 (2023)
27. Wu, J., Hong, Y., Guan, X.: A survey of joint security-safety for function, information and human in industry 5.0. Security and safety (2024). https://doi.org/10.1051/sands/2024014
28. Xue, Y., Pan, J., Geng, Y., Yang, Z., Liu, M., Deng, R.: Real-time intrusion detection based on decision fusion in industrial control systems. IEEE Trans. Ind. Cyber-Phys. Syst. 1–11 (2024). https://doi.org/10.1109/ticps.2024.3406505

Towards Generative Context-Aware Communication in Smart Transport Environments

Roudeina Guerfi(✉), Ahmed-Chawki Chaouche, and Rania Zebiri

MISC Laboratory, University of Constantine 2 – Abdelhamid Mehri, Ali Mendjeli Campus, 25000 Constantine, Algeria
{roudeina.guerfi,ahmed.chaouche,rania.zebiri}@univ-constantine2.dz

Abstract. In modern transportation systems, contextual information is essential to enhance safety, efficiency, and passenger experience through adaptive communication. This paper introduces the smart announcement generative architecture, a multi-phase framework enabling real-time, context-aware communication in mobile smart environments. Integrating context harmonization, generative reasoning, and communication orchestration, this architecture transforms diverse contextual data into actionable announcements. It optimizes priority, timing, and delivery, adapting from instant reactions to refined, model-informed messages. A real-world urban transport case validates its efficiency, scalability, and impact in improving mobility and reducing congestion under dynamic conditions.

Keywords: Intelligent Transportation Systems · Mobile Smart Environments · Generative AI · Real-Time Communication · Urban Mobility · Announcement Management

1 Introduction

In recent years, the rapid evolution of Mobile Smart Environments (MSE) has reshaped the way information is generated, shared, and acted upon in intelligent urban systems. These environments integrate interconnected agents, sensors, and infrastructures that continuously exchange contextual data under dynamic conditions [6]. In such settings, real-time and context-aware communication has become essential to ensure that systems can adapt quickly to changing mobility patterns, user demands, and environmental fluctuations [16]. However, managing this continuous flow of heterogeneous and time-sensitive data remains a major challenge. Contextual information defined by its dependence on location, time, and user behavior must not only be captured but also interpreted and transformed into meaningful [10]. The key problem lies in how to process this information intelligently and deliver relevant, personalized, and timely responses within complex, fast changing mobility ecosystems.

Z. Kouahla et al. (Eds.): IAM 2025, LNNS 1973, pp. 77–94, 2026.
https://doi.org/10.1007/978-3-032-26031-4_8

Despite significant progress in IoT-driven smart infrastructure [14] and AI-based decision systems, current solutions remain limited in their ability to deliver timely and relevant information to end users [9]. Traditional approaches, such as [2–4,12], often based on static rule sets or reactive triggers struggle to handle the high dimensional, noisy, and redundant data streams typical of large scale MSE. Moreover, these methods rarely incorporate mechanisms for personalization, contextual learning, and adaptive communication, resulting in announcements that are either delayed, irrelevant, or overwhelming to the user.

To address these limitations, this paper introduces the Smart Announcement Generative (SAG) architecture, a multi-layered framework designed to produce real-time, adaptive, and user-oriented announcements in dynamic mobility environments. SAG architecture extends beyond conventional models by structuring its workflow into four synergistic components: Learning, context harmonization, generative reasoning, and communication orchestration. Each component contributes to transforming raw contextual data into actionable and meaningful announcements through a structured process of data learning, alignment, synthesis, and dissemination.

Unlike traditional single-step approaches, like [1,18], SAG architecture introduces modular adaptability across interconnected layers. The learning component builds a predictive model that captures correlations between contextual patterns and past outcomes, enabling proactive decision-making. The context harmonization layer refines real-time data through preparation, preference alignment, and clustering, ensuring that heterogeneous inputs are personalized and scalable. Then, the generative reasoning layer combines a prompt builder and generative synthesizer to create diverse, context-aware, and non-redundant announcements. Finally, the communication orchestration layer formats and dispatches these announcements through optimal delivery channels to ensure timeliness, clarity, and user relevance.

Through this modular design, SAG architecture strengthens four core dimensions of communication in Mobile Smart Environments: (1) relevance and context sensitivity through predictive learning and adaptive filtering, (2) scalability via independently deployable components, (3) user-centric interaction through preference alignment and adaptive delivery, and (4) robustness and adaptability by managing noise, redundancy, and dynamic conditions through clustering and diversification.

This study contributes to the growing body of research on AI-driven information dissemination in smart environments by bridging learning, reasoning, and generative modeling within a unified communication framework. While existing literature has explored context modeling, adaptive data processing, and generative AI independently, SAG architecture uniquely combines these dimensions to create a coherent, end-to-end pipeline for intelligent mobility communication. The real-world use case in Constantine's urban transport network illustrates the practical relevance of the SAG architecture. Faced with dynamic mobility challenges such as sudden passenger surges, congestion, and route disruptions the system demonstrated its capacity to interpret real time data, align user preferences, and generate adaptive announcements across multiple channels. These outcomes confirm the effectiveness of the SAG architecture in translating theoretical principles into tangible operational benefits, validating its role as a scalable and context-aware communication model for next generation transport ecosystems.

The remainder of this paper is organized as follows: Sect. 2 reviews related work on generative and context-aware communication systems in mobile smart environments. Section 3 presents the proposed SAG architecture and its operational workflow. Section 4 demonstrates the applicability of the proposed approach through a real-world dynamic mobility coordination scenario. Finally, Sect. 5 concludes with discussions and future research directions.

2 Related Work

The rapid development of intelligent transportation technologies has transformed how mobility systems are designed and managed [5]. The increasing fusion of physical and digital infrastructures has given rise to new paradigms that rethink traditional traffic management and decision-making [20]. According to Fei-Yue Wang [25], the founder of the concept of Parallel Transportation Systems (PTS), transportation in the era of intelligent mobility is no longer confined to the physical world but extends into its digital counterpart through cyber–physical integration.

Within the framework of Parallel Transportation Systems (PTS), Lv et al. [8] introduced the use of Generative Adversarial Networks (GAN) to simulate and predict traffic dynamics, demonstrating how generative models could reproduce realistic transport scenarios from limited sensor data. This integration marked a pivotal shift from traditional rule-based control toward data driven, predictive transport management. By generating synthetic yet contextually consistent traffic data, GAN enhanced network optimization, congestion forecasting, and system robustness laying the groundwork for subsequent generative approaches in intelligent transportation. However, this approach focuses mainly on traffic simulation and prediction, lacking real-time adaptability, contextual personalization, and communication mechanisms essential for dynamic urban mobility systems.

Building on the early generative transport models [13], Ruthotto and Haber [11] provided a broader theoretical foundation for deep generative modeling, introducing frameworks such as variational autoencoders and generative adversarial networks as key methods for learning complex, high-dimensional data distributions. Their work [11] formalized the mathematical principles that enable generative systems to infer structure, variability, and uncertainty from contextual data capabilities essential for adaptive mobility management. This theoretical grounding supports most architectures, where the transformation of contextual inputs into meaningful, real-time announcements relies on generative reasoning mechanisms capable of capturing dynamic, multimodal urban environments. However, their study remains largely conceptual, lacking direct implementation or validation in real-world mobility contexts, thereby limiting its applicability to operational intelligent transport systems.

Following this theoretical progression, Lin et al. [7] conducted a comprehensive survey exploring how generative adversarial networks have advanced the field of intelligent transportation systems. Their study demonstrated how generative adversarial networks contribute to data generation, trajectory prediction, anomaly detection, and spatio-temporal modeling, addressing challenges of incomplete or noisy transport data. By showing that generative models can replicate and enhance real-world mobility data, their work laid the groundwork for integrating generative reasoning into context-aware architectures, where real-time announcement accuracy depends on the quality and adaptability of generated contextual insights. However, despite its comprehensive scope, the study focuses primarily on data synthesis and prediction, without addressing how generative outputs can be integrated into real-time communication or decision-making processes within intelligent transport systems.

Connected vehicular networks represent a critical evolution in intelligent transportation, enabling continuous data exchange between vehicles, infrastructures, and users to enhance situational awareness and coordination. Zhang et al. [24] proposed a generative AI-enabled vehicular framework that integrates multi-modal data (text, image, and semantic layers) for real-time decision-making and communication in connected vehicle systems. Their architecture introduced a semantic-aware design that personalizes system outputs, optimizing guidance and usability across diverse agents and environments. This work highlights the potential of generative AI to deliver context-rich, user-specific announcements in vehicular networks—an important step toward more adaptive and scalable communication systems. However, while the framework improves communication richness, it remains limited in dynamic scalability and real-time adaptability when confronted with rapidly changing multi-agent contexts.

Multi-agent systems have long been recognized as a cornerstone of intelligent transportation research, enabling distributed coordination, cooperative perception, and autonomous decision-making among vehicles and infrastructures Building on this foundation, Xu et al. [22] extended the generative paradigm to multi-agent smart urban mobility systems, proposing the integration of Large Language Models (LLM) and Retrieval-Augmented Generation (RAG) into transport infrastructures. Their framework envisions conversational, context-sensitive, and personalized information delivery among agents, vehicles, and infrastructures, offering a scalable path toward real-time, user-centered mobility coordination. This conceptual advancement directly aligns with the vision of the proposed approach, where generative reasoning and adaptive communication orchestration collaboratively enhance decision-making and responsiveness in dynamic urban transport scenarios. However, while this approach enhances adaptability and semantic understanding, it remains primarily conceptual, lacking empirical validation in large-scale, real-world mobility environments.

The evolution of generative approaches in intelligent transportation has expanded across five major domains: real-time alert generation, edge intelligent vehicular systems, cooperative communication, large language model integration, and safety driven automation. Yan and Li [23] provided one of the most comprehensive overviews of Generative AI applications in road transportation, emphasizing its capacity to enhance perception, prediction, and adaptive alerting within dynamic traffic conditions. Building upon these foundations, Xie et al. [19] proposed the GAI-IoV framework, demonstrating how generative AI can operate efficiently at the network edge to support distributed and multi-modal announcement generation. In parallel, Tahir et al. [15] validated the practical deployment of cooperative Intelligent Transport Systems (C-ITS), showing that automated alerts can achieve high reliability and low latency in real world transport environments.

The role of cognitive intelligence was later reinforced by Xue et al. [21], who illustrated how Large Language Models (LLM) can be leveraged to produce semantically rich, context-aware mobility announcements that optimize urban flow and user communication. Complementing these perspectives, Vimal et al. [17] highlighted the integration of Internet of Things (IoT) and deep learning for automated injury detection and emergency alerting in public transport, underlining the potential of AI-driven responsiveness in passenger safety. Despite their progress, these approaches remain fragmented—often domain-specific, single-agent, or limited to isolated tasks such as perception or alerting. None provide a unified, adaptive framework that integrates contextual understanding, preference alignment, and generative reasoning for seamless, real-time communication in dynamic mobility environments.

Table 1 provides a comparative analysis of key generative and context-aware approaches in intelligent transportation systems, highlighting their methodological focus, adaptability, and domain of application.

Overall, these contributions demonstrate a transition toward intelligent and adaptive communication paradigms. Nevertheless, the absence of a unified architectural framework for generative announcement management in dynamic Mobile Smart Environments remains evident; a limitation tackled in this work by introducing Smart Announcement Generative (SAG), a modular and multi-level architecture for adaptive announcement generation.

3 Smart Announcement Generative Architecture

The proposed smart announcement generative architecture (SAG) in Mobile Smart Environments (MSE) aims to provide a structured methodology for managing data circulation and transforming it into meaningful, context-aware announcements. As illustrated in Fig. 1, the architecture unfolds in four complementary phases.

3.1 Learning

The Learning stage establishes the foundational knowledge of the SAG architecture by constructing the predictive model m from historical mobility data. It begins with the Data Preprocessing Unit (DPU), which applies the function $refine()$ to transform raw data D from the Knowledge Data Repository into a

Table 1. Comparative analysis of the key generative and context-aware approaches in intelligent transportation systems

Work	Generative Models	Context-awareness	Real-Time	Multi-agent	Application Domain
Lv et al. (2018) [8]	Generative Adversarial Networks (GAN) Theory	Basic	Simulation-driven	Single-system	Parallel Transport Systems
Ruthotto and Haber (2021) [11]	Variational Autoencoders (VAE)/ Generative Adversarial Networks (GAN)	None	Offline	None	Deep Generative Theory
Lin et al. (2023) [7]	Generative Adversarial Networks (GAN) Variants	Moderate	Analytical	Centralized	Intelligent Transport Survey
Zhang et al. (2024) [24]	Generative AI + RL Framework	Semantic	Adaptive	Cooperative	Vehicular Networks
Xie et al. (2024) [19]	Generative AI for Internet of Vehicles (IoV)	High	Edge-responsive	Distributed	Vehicular Edge Intelligence
Tahir et al. (2021) [15]	AI-based alerts	Contextual	Real-time	Cooperative	Cooperative ITS Deployment
Xue et al. (2025) [22]	Large Language Models (LLM) based Generation	Contextual	Dynamic	Multi-agent	Smart Urban Mobility
Yan and Li (2025) [23]	Generative AI Framework	Traffic-aware	Real-time	Limited	Road Transportation

refined dataset D^*. This step includes cleaning, normalization, feature extraction, and enrichment of data with metadata such as timestamps and reliability indicators.

The second component, the Training Unit, uses the refined dataset D^* together with ground-truth labels L and a set of hyper-parameters HP to train the predictive model m. The model captures correlations between contexts, events, and user responses, forming the system's knowledge base for all subsequent components.

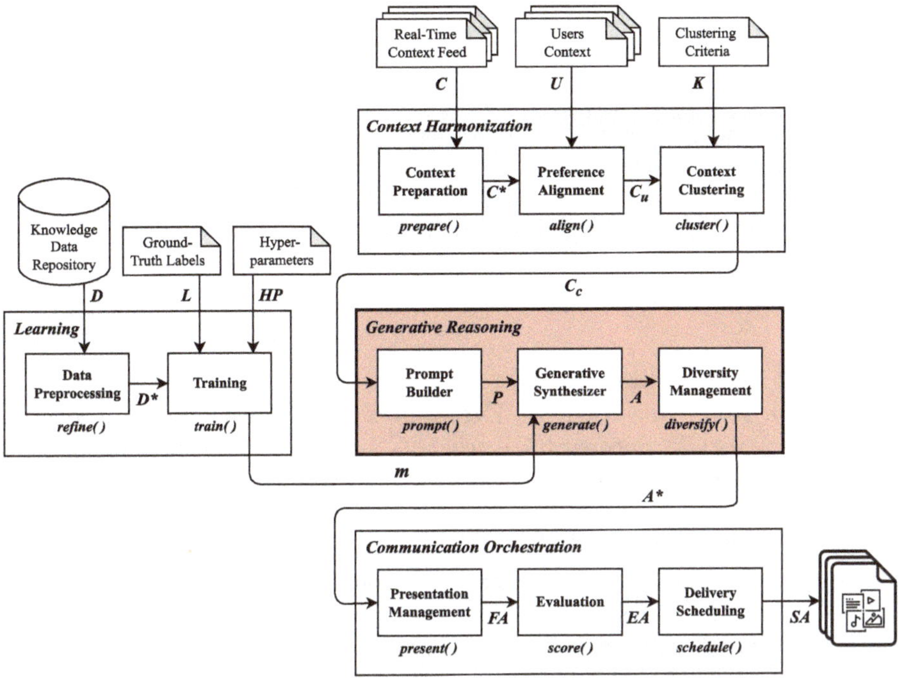

Fig. 1. Smart announcement generative (SAG) architecture.

3.2 Context Harmonization

The Context Harmonization stage acts as the bridge between real-time contextual inputs and the reasoning process. Its objective is to refine, personalize, and structure contextual data so that they can serve as coherent inputs for the generative mechanism. It includes three core functions: context preparation, preference alignment, and context clustering.

Context Preparation. The context preparation unit applies the function $prepare()$ to raw inputs C, which represent real-time contexts collected from heterogeneous sources (e.g., vehicles, sensors, mobile devices). Due to the variability of input quality, this component removes invalid or incomplete entries, normalizes measurement scales, and enriches contexts with metadata such as timestamps, source reliability, and freshness. The output of this step is a filtered and structured set of contexts C^*, which provides the foundation for subsequent personalization.

Algorithm 1. *learning*$(D, L, HP) : m$

1: **Require:**
2: D: set of historical data
3: L: set of ground-truth labels
4: HP: set of hyper-parameters

5: $D^* := refine(D)$ ▷ Data Preprocessing
6: $m := train(D^*, L, HP)$ ▷ Training Process

7: **return** m ▷ Trained predictive model

Definition 1. *(Prepared Context)*
Let C be the set of raw contexts. The context preparation process produces a refined context set $C^* = prepare(C)$, where C^* denotes the normalized, filtered, and semantically enriched contexts ready for subsequent processing.

Preference Alignment. The preference alignment unit applies the function *align*() to integrate user-specific requirements U into the prepared contextual data C^*. Positioned between raw context preparation and clustering, this component ensures that the system adapts its perception of the environment to the preferences and priorities of each user or operator.

This step allows the architecture to move from a purely objective context to one that is personalized and purpose-oriented. For example, while one commuter may prefer minimizing travel time, another may prioritize comfort, safety, or accessibility. These individualized constraints are injected directly into the contextual representation, making the subsequent clustering more meaningful and better suited for prompt construction.

By embedding such preference attributes, the resulting personalized context C_u represents a user-aware view of the mobility environment, where contextual data are no longer treated equally but weighted according to relevance for the intended user or service goal.

Definition 2. *(Personalized Context)*
Let C^* denote the set of prepared contexts and U the set of user preferences. The personalization process produces a refined set of user-aware contexts: $C_u = align(C, U)$, where C_u represents contexts enriched with preference-driven attributes, enabling adaptation to individual user needs.*

Context Clustering. The context clustering unit applies the function *cluster*() to group similar personalized contexts C_u into aggregated sets C_c. This step reduces redundancy (e.g., multiple agents reporting the same traffic jam), improves scalability by decreasing the number of contexts to process, and provides higher-level abstractions of the mobility environment.

This structured representation is crucial for the Prompt Constructor, as it provides a condensed and unified input, preventing redundant prompts and ensuring that the generative model receives only the most representative contextual data. Consequently, clustering improves both the computational efficiency and semantic quality of the prompts, leading to more relevant, diverse, and contextually consistent announcements.

Definition 3. *(Clustered Contexts)*
Let C_u be the set of personalized contexts and K the clustering criteria space. The context clustering process generates a structured grouping of related contexts: $C_c = cluster(C_u, K)$, where C_c denotes a compact and representative set of clustered contexts derived according to similarity measures or application-specific criteria.

3.3 Generative Reasoning

The generative reasoning constitutes the creative core of the architecture, where clustered contexts C_c are progressively transformed into meaningful and diverse candidate announcements. This phase ensures that announcements are not only context-aware but also adapted for relevance, variety, and downstream personalization. It unfolds through three main components: the Prompt Builder, the Generative Synthesizer, and the Diversity Manager.

Prompt Builder. The prompt builder applies the function *prompt*() to convert clustered contexts C_c into structured prompts P. Each prompt serves as an intermediary representation, encapsulating the key attributes of a context cluster (e.g., location, urgency, event type) into a form suitable for generative modeling. By formalizing raw contextual data into interpretable prompt templates, this unit provides the generative core with clear, structured instructions to guide announcement creation.

Definition 4. *(Prompt)*
Let C_c denote the set of clustered contexts. The prompt generation process produces a structured representation of relevant contextual information: $P = prompt(C_c)$, where each $p \in P$ encodes the contextual attributes required for subsequent announcement generation.

Generative Synthesizer. The generative synthesizer applies the function *generate*() to transform prompts P, in combination with the predictive model m, into a set of candidate announcements A. By leveraging the predictive capacity of m, this module ensures that generated announcements reflect not only the current contextual situation but also anticipated dynamics (e.g., evolution of traffic or impact on user behavior). The outcome is a pool of raw candidate announcements that are relevant, clear, and aligned with both context and prediction.

Definition 5. *(Candidate Announcement)*
Let P be the set of prompts and m the predictive model. The generative process produces a set of candidate announcements: $A = generate(P, m)$, where A represents the set of candidate announcements synthesized from prompts enriched with predictive knowledge.

Diversity Manager. The diversity manager applies the function $diversify()$ to refine the set of candidate announcements A into a more compact and balanced pool A^*. It eliminates duplicates or near-duplicates, mitigates redundancy, and ensures the preservation of multiple perspectives or communication strategies. This guarantees that the final pool of announcements remains representative, varied, and resistant to repetition, thus preventing information fatigue for end-users.

Definition 6. *(Refined Announcement Pool)*
Let A be the set of candidate announcements. The diversification process produces an enhanced set of varied and non-redundant announcements: $A^* = diversify(A)$, where A^* represents the refined announcement pool optimized for diversity and minimized redundancy.

3.4 Communication Orchestration

The communication orchestration is the final stage of the architecture, responsible for orchestrating the selection and dissemination of announcements to end-users. Its role is to adapt candidate announcements into user-ready formats, evaluate their quality and relevance, and finally schedule their delivery. This guarantees that the generated outputs are not only meaningful and diverse, but also actionable, timely, and aligned with system requirements. The phase unfolds through three sequential components: the presentation manager, the evaluation unit, and the delivery controller.

Presentation Manager. The presentation manager applies the function $present()$ to transform the refined pool of announcements A^* into user-adapted formats. This step ensures that each announcement is expressed in a form appropriate to its dissemination scope and medium, such as text notifications, audio alerts, or visual displays. It also sets preliminary attributes such as delivery priority and presentation order, which later guide the evaluation process.

Definition 7. *(Formatted Announcement)*
Let A^* *be the set of refined announcements and* F *the set of possible output formats. The presentation process produces delivery-ready announcements:* $FA = present(A^*, F)$, *where* FA *represents the set of formatted announcements adapted to specific delivery media and enriched with priority attributes.*

Evaluation Unit. The evaluation unit applies the function *evaluate*() to score and filter the formatted announcements FA. Unlike earlier stages that focus on content creation, this unit explicitly quantifies announcement quality according to predefined metrics, such as timeliness, clarity, redundancy, and relevance. Only the highest-scoring announcements are retained for dissemination, ensuring that system resources and user attention are dedicated to the most valuable information.

Definition 8. *(Evaluated Announcement)*
Let FA *be the set of formatted announcements and* mt *the evaluation metric space. The evaluation process determines which formatted announcements meet predefined quality criteria:* $EA = score(FA, mt)$, *where* EA *represents the subset of evaluated announcements selected for delivery based on the applied metrics.*

Delivery Controller. The delivery controller applies the function *schedule*() to manage the scheduling and dispatch of announcements EA. It determines the appropriate dissemination channel (e.g., personal, shared, or public), the timing of delivery (immediate or scheduled), and the delivery mode (notification, broadcast, or streaming). This ensures that the final set of announcements, denoted A_{del}, reaches end-users in a timely, context-aware, and effective manner.

Definition 9. *(Scheduled Announcement)*
Let EA *denote the set of evaluated announcements and* CH *the set of delivery channels. The scheduling process assigns announcements to appropriate channels and times:* $SA = schedule(EA, CH)$, *where* SA *represents the set of scheduled announcements ready for delivery according to channel-specific constraints.*

The overall functioning of the architecture, linking the different phases from Learning to the delivery of announcements is formalized in Algorithms 1 and 2.

4 Dynamic Urban Transport Management in Constantine City

A real-world use case in Constantine's urban transport system is presented to illustrate how SAG architecture dynamically adapts transport services in Constantine's primary corridor through coordinated sensing, decision-making, and communication during unexpected demand fluctuations.

Algorithm 2. $sag(C, U, K, m) : SA$

Require:
C: set of real-time context stream
U: set of user preferences
K: set of clustering criteria
m: pre-trained predictive model

$C^* := prepare(C)$ ▷ Context Preparation
$C_u := align(C^*, U)$ ▷ Preference Alignment
$C_c := cluster(C_u, K)$ ▷ Context Clustering

$P := prompt(C_c)$ ▷ Prompt Builder
$A := generate(P, m)$ ▷ Generative Synthesizer
$A^* := diversify(A)$ ▷ Diversity Manager

$FA := present(A^*)$ ▷ Presentation Manager
$EA := score(FA)$ ▷ Evaluation Unit
$SA := schedule(EA, CH)$ ▷ Delivery Controller

return SA ▷ Delivered announcements

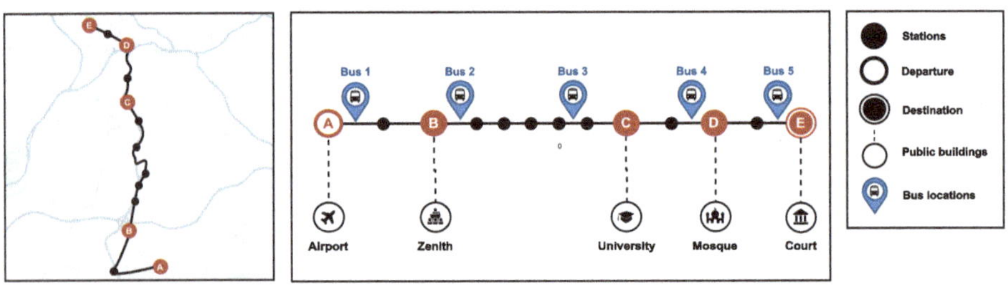

Fig. 2. Constantine urban use-case architecture.

4.1 Deployment Context

The SAG architecture is deployed along Constantine's transport corridor connecting Mohamed Boudiaf International Airport (A) to Downtown (E), as illustrated in Fig. 2. The route includes strategic contextual nodes:

- **Zenith Conference Room (B)** – commercial zone,
- **Mentouri University (C)** – educational zone,
- **Emir Abdelkader Mosque (D)** – cultural zone.

Each node continuously feeds real-time data (passenger flow, bus occupancy, traffic conditions) into the system for context-aware decision-making.

4.2 Dynamic Coordination and Real-Time Response

Initial Conditions

Flight AH101 arrives at the Airport (A) at 8:30 AM, creating an unexpected passenger surge. The SAG architecture must dynamically coordinate shuttle bus services to handle:

- 80 waiting passengers at the airport,
- Limited bus capacity (2 buses, 40 seats each, currently 70% and 80% full),
- Congested main route (20 km/h) vs. clear alternative route (40 km/h).

Operational Example: Approaching Zenith Conference Room (B)

To demonstrate the practical operation of the SAG system, consider the following scenario: At 9:15 AM, as Bus 2 approaches the Zenith Conference Room stop (See Fig. 2), SAG architecture identifies 15 waiting business travelers and recognizes the ongoing peak-hour demand in the commercial district. With 12 seats currently available and an alternative route that reduces travel time by 12 min, SAG broadcasts a station announcement: "Express service to Downtown via Avenue B departing shortly with an estimated arrival at 9:35 AM". Concurrently, targeted mobile notifications are dispatched to relevant users, highlighting real-time availability of meeting rooms. The system further refines passenger load forecasts for subsequent stops and dynamically adapts its communication protocols for the upcoming university zone, accounting for the bus's residual capacity.

System-Wide Response

When the airport surge hits at 8:30 AM, SAG quickly detects the capacity constraints and traffic conditions. It checks bus space and traffic, calls in a third bus from the depot, reroutes Bus 2 on the faster route, and shares clear updates at each stop. All morning, it keeps everyone informed by sending phone alerts, updating screens, and giving the right message in each area. Table 2 presents a detailed step-by-step breakdown of how SAG processes through its architecture.

4.3 System Dynamicity Demonstration

Beyond this specific scenario, SAG architecture continuously adapts its operations throughout the day to meet evolving urban demands. As passenger patterns change whether morning, midday, or evening passengers the system automatically adjusts its communication methods as described in Fig. 3. It delivers timely mobile alerts, public announcements, and messages tailored to each context. This ongoing, autonomous process ensures the system remains responsive and relevant, dynamically re-evaluating data and optimizing its strategies without the need for manual input.

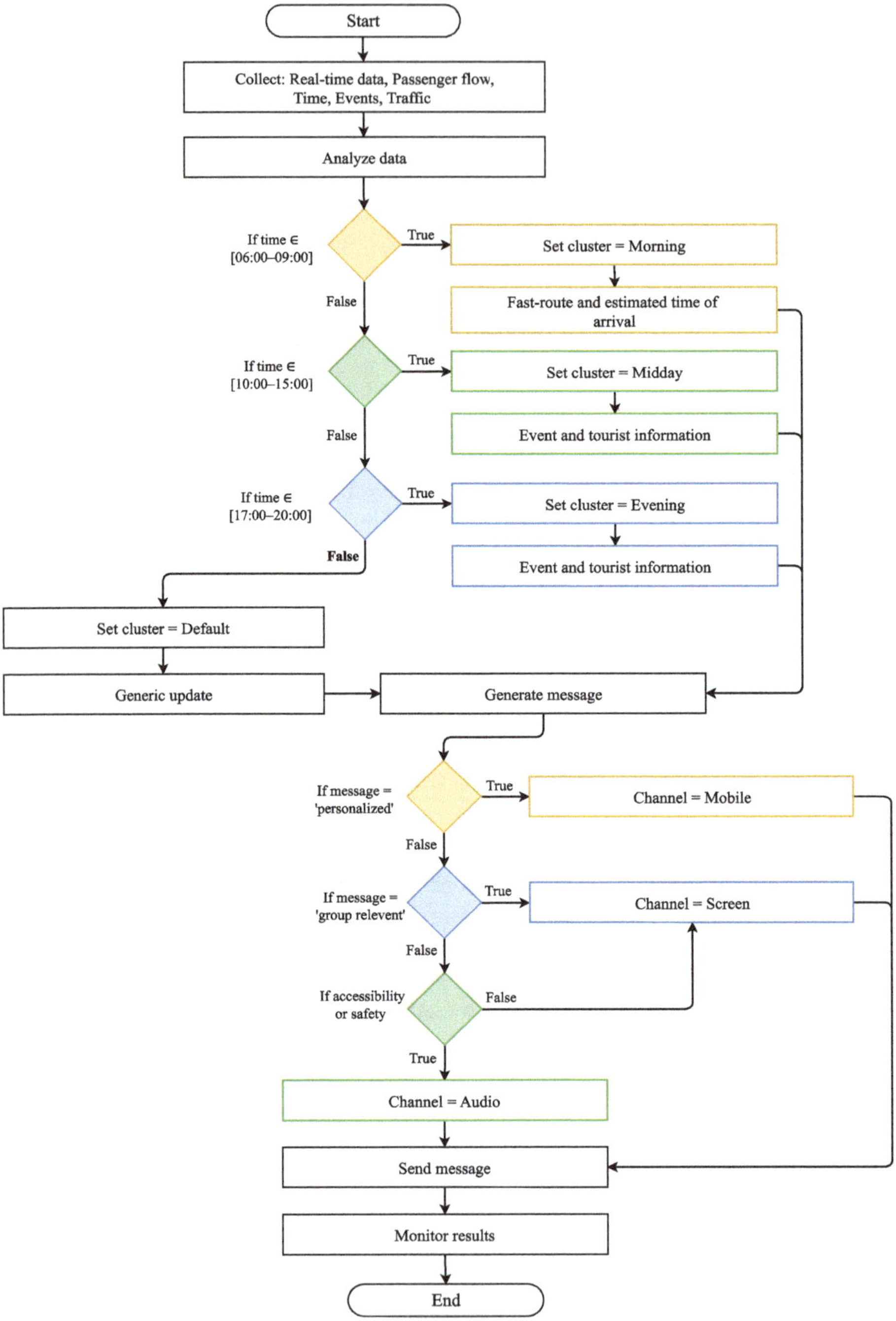

Fig. 3. Flowchart of dynamic context adaptation.

Table 2. Step-by-Step Processing Through the SAG Architecture

Step	Component/Function	Input	Process/Operation	Output
1. Learning	-	Historical data on traffic, demand, and announcements	Model training using correlations such as: - +40% passenger increase after morning flights - Delays >7 min when traffic <25 km/h - Extra bus reduces waiting by 30%	Trained model m predicting overcrowding and mitigation strategies
2a. Context Preparation	`prepare()`	$C =$ {passengers=80, bus1=70%, bus2=80%, main=20, alt=40, flight=AH101}	Cleaning, normalization, and metadata enrichment (freshness, source, reliability)	$C^* =$ {passengers:80, avg_bus_load:75%, main_speed:20, alt_speed:40, freshness:2 min, source:AirportSensor}
2b. Preference Alignment	`align()`	C^*, $U=$ {priority:min_wait _time, cost:low, comfort:medium}	Integrates user/operator preferences	$C_u =$ {passengers:80, avg_bus_load:75%, main_speed:20, alt_speed:40, goal:min_wait}
2c. Context Clustering	`cluster()`	C_u	Groups similar contexts into higher-level categories: - Cluster 1: Passenger Overload - Cluster 2: Route Congestion	$C_c =$ {cluster$_1$: Passenger Overload, cluster$_2$: Route Congestion}
3a. Prompt Builder	`prompt()`	C_c	Generates structured prompt describing key context variables and goals	$P =$ "High passenger load (80), limited capacity (2), congestion 20 km/h, alternate 40 km/h, objective: minimize waiting time."
3b. Generative Synthesizer	`generate()`	P, m	Produces candidate announcements	$A = \{A_1$: Extra bus in 3 min, A_2 : Delay notice, A_3 : Bike share suggestion$\}$
3c. Diversity Manager	`diversity()`	A	Removes redundant or overlapping messages	$A^* = \{A_1, A_3\}$
4a. Presentation Manager	`present()`	A^*	Formats announcements according to delivery channels	$A_1 \rightarrow$ Display screen; $A_3 \rightarrow$ Mobile app
4b. Evaluation	`score()`	A_{fmt}	Evaluates clarity, urgency, and contextual relevance	Score(A_1)=0.95; Score(A_3)=0.87
4c. Delivery Scheduling	`schedule()`	A_{eval}	Schedules and dispatches top announcements	$SA =$ {Delivered: A_1, A_3}

5 Discussion

Modern urban mobility systems have become increasingly complex, demanding real-time coordination among vehicles, infrastructures, and users. Yet, most existing frameworks remain reactive, limited to predefined rules and static AI models that struggle with data heterogeneity and fail to anticipate dynamic conditions such as congestion or passenger surges.

Unlike prior systems, SAG architecture performs deep contextual harmonization, aligning real-time environmental data with user preferences and situational priorities. Its strength lies in its ability to operationalize dynamicity as depicted in the flowchart through the coordinated action of context clustering, generative reasoning, and presentation management. As conditions evolve throughout the day, the system automatically reconfigures contextual clusters (e.g., morning commuters, midday tourists, evening residents) and adjusts both message content and delivery channels accordingly.

To the best of our knowledge, there is currently no unified architectural framework for generative announcement management in dynamic transport systems, making SAG architecture a novel and effective step forward in the evolution of intelligent urban mobility.

Despite its adaptability and contextual intelligence, SAG architecture faces several challenges. Its performance depends on the reliability of real-time data, making it sensitive to missing or inconsistent sensor inputs. The generative layer, although effective in controlled scenarios, still requires large-scale validation to assess scalability and linguistic diversity. Computational load during peak mobility and the need for privacy-preserving preference alignment also remain open issues. Future work will address these challenges to ensure robust, ethical, and scalable deployment in intelligent urban transport systems.

6 Conclusion

This study presented the Smart Announcement Generative (SAG) architecture as a generative communication framework for adaptive and context-aware information dissemination in urban mobility systems. Through its deployment in Constantine's transport network, SAG architecture demonstrated the ability to handle real-time variations in passenger flow and traffic conditions while maintaining clarity, relevance, and responsiveness in generated announcements. Experimental observations showed that SAG architecture improves communication efficiency through dynamic adjustment of announcements to evolving contexts and user preferences, effectively reducing waiting times and mitigating congestion. Its modular structure proved adaptable across diverse communication channels, supporting both personalized and collective message delivery.

Despite these promising results, several challenges persist. SAG's dependency on consistent sensor data and the computational demands of generative processing highlight the need for optimization and data-reliability mechanisms. Additionally, ensuring privacy in preference alignment remains an essential direction for ethical deployment. The current study includes a qualitative assessment of SAG's performance. Future work will incorporate simulated evaluations to quantitatively measure responsiveness, accuracy, scalability, and comparative efficiency.

Future work will focus on improving data reliability through redundancy-aware sensing and context interpolation to handle missing inputs. The generative layer will be expanded with multimodal training to enhance linguistic diversity and scalability. Lightweight model optimization and edge-deployment strategies will be explored to reduce computational overhead during peak mobility periods. Additionally, privacy-preserving preference alignment and federated-learning mechanisms will be developed to ensure ethical data management and secure personalization within large-scale smart transport systems.

References

1. Belli, L., et al.: IoT-enabled smart sustainable cities: challenges and approaches. Smart Cities **3**(3), 1039–1071 (2020)
2. Bibri, S.E., Huang, J.: Generative AI of things for sustainable smart cities: synergies in cognitive augmentation, resource efficiency, network traffic, and anomaly and threat detection for environmental optimization. Sustain. Cities Soc. 106826 (2025)
3. Chaouche, A.-C., El Fallah Seghrouchni, A., Ilié, J.-M., Saïdouni, D.E.: A formal approach for contextual planning management: application to smart campus environment. In: Bazzan, A.L.C., Pichara, K. (eds.) IBERAMIA 2014. LNCS (LNAI), vol. 8864, pp. 791–803. Springer, Cham (2014). https://doi.org/10.1007/978-3-319-12027-0_64
4. Chaouche, A.-C., Seghrouchni, A.E.F., Ilié, J.-M., Saïdouni, D.E.: Smart agent foundations: from planning to spatio-temporal guidance. Enablers Smart Cities 33–63 (2016)
5. Elassy, M., Al-Hattab, M., Takruri, M., Badawi, S.: Intelligent transportation systems for sustainable smart cities. Transp. Eng. **16**, 100252 (2024)
6. Kaluarachchi, Y.: Implementing data-driven smart city applications for future cities. Smart Cities **5**(2), 455–474 (2022)
7. Lin, H., Liu, Y., Li, S., Xiaobo, Q.: How generative adversarial networks promote the development of intelligent transportation systems: a survey. IEEE/CAA J. Automatica Sinica **10**(9), 1781–1796 (2023)
8. Lv, Y., Chen, Y., Li, L., Wang, F.-Y.: Generative adversarial networks for parallel transportation systems. IEEE Intell. Transp. Syst. Mag. **10**(3), 4–10 (2018)
9. Ning, M.A.: Artificial intelligence-driven decision support systems for sustainable energy management in smart cities. Int. J. Adv. Comput. Sci. Appl. **15**(9) (2024)
10. Rocha, N.P., et al.: Systematic literature review of context-awareness applications supported by smart cities' infrastructures. SN Appl. Sci. **4**(4), 90 (2022)
11. Ruthotto, L., Haber, E.: An introduction to deep generative modeling. GAMM-Mitteilungen **44**(2), e202100008 (2021)
12. Sarker, I.H.: Context-aware rule learning from smartphone data: survey, challenges and future directions. J. Big Data **6**(1), 1–25 (2019). https://doi.org/10.1186/s40537-019-0258-4
13. Shu, B., Zhang, Y., Shu, D.: Generative models and connected and automated vehicles: a survey in exploring the intersection of transportation and AI. arXiv preprint arXiv:2403.10559 (2024)
14. Syed, A.S., Sierra-Sosa, D., Kumar, A., Elmaghraby, A.: IoT in smart cities: a survey of technologies, practices and challenges. Smart Cities **4**(2), 429–475 (2021)
15. Tahir, M.N., Mäenpää, K., Sukuvaara, T., Leviäkangas, P.: Deployment and analysis of cooperative intelligent transport system pilot service alerts in real environment. IEEE Open J. Intell. Transp. Syst. **2**, 140–148 (2021)

16. Urbieta, A., González-Beltrán, A., Ben Mokhtar, S., Anwar Hossain, M., Capra, L.: Adaptive and context-aware service composition for IoT-based smart cities. Futur. Gener. Comput. Syst. **76**, 262–274 (2017)
17. Vimal, S.P., Peter, J.B.J., Kavitha, U., Bhuvaneswari, G., Manikandan, G., Thamizhamuthu, R.: Automated injury detection and alert systems in public transportation integrating IoT with convolutional neural networks. In: 2024 2nd International Conference on Self Sustainable Artificial Intelligence Systems (ICSSAS), pp. 1069–1074. IEEE (2024)
18. Whaiduzzaman, Md., et al.: A review of emerging technologies for IoT-based smart cities. Sensors **22**(23), 9271 (2022)
19. Xie, G., et al.: GAI-IoV: bridging generative ai and vehicular networks for ubiquitous edge intelligence. IEEE Trans. Wirel. Commun. **23**(10), 12799–12814 (2024)
20. Xu, H., et al.: Smart mobility in the cloud: enabling real-time situational awareness and cyber-physical control through a digital twin for traffic. IEEE Trans. Intell. Transp. Syst. **24**(3), 3145–3156 (2023)
21. Xu, H., et al.: GenAI-powered multi-agent paradigm for smart urban mobility: opportunities and challenges for integrating large language models (LLMs) and retrieval-augmented generation (rag) with intelligent transportation systems. arXiv preprint arXiv:2409.00494 (2024)
22. Xue, H., Jin, M., Pan, S., Salim, F.: Transforming urban dynamics: harnessing large language models for smarter mobility. IEEE Intell. Syst. **40**(2), 5–7 (2025)
23. Yan, H., Li, Y.: Generative AI for intelligent transportation systems: road transportation perspective. ACM Comput. Surv. (2025)
24. Zhang, R., et al.: Generative AI-enabled vehicular networks: fundamentals, framework, and case study. IEEE Netw. **38**(4), 259–267 (2024)
25. Zhu, F., Lv, Y., Chen, Y., Wang, X., Xiong, G., Wang, F.-Y.: Parallel transportation systems: toward IoT-enabled smart urban traffic control and management. IEEE Trans. Intell. Transp. Syst. **21**(10), 4063–4071 (2019)

A Q-Learning Weighted Cluster Based Routing Protocol for VANETs

Ahlam Boussadia[1,2](✉) and Manel Khelifi[1]

[1] LAOTI Laboratory, University of Mohamed Seddik Benyahia, Jijel, Algeria
manel.khelifi@univ-jijel.dz
[2] MISC Laboratory, University of AbdelHafid Boussouf, Mila, Algeria
ahlam.boussadia@centre-univ-mila.dz

Abstract. Vehicular Ad-Hoc Networks are an emerging paradigm within Intelligent Transportation Systems (ITS), enabling communication between vehicles (V2V) and/or vehicles and infrastructure (V2I). These networks aim to enhance road safety and improve the driving experience. However, due to the high mobility of vehicles and frequent changes in their geographical positions, ensuring reliable data delivery remains a significant challenge. Clustering has emerged as a promising technique to improve scalability, reduce overhead, and enhance routing stability in VANETs. This paper first introduces a new classification framework for clustering-based routing protocols according to their operational and decision parameters. It then proposes the Q-learning Weighted Cluster-Based Routing Protocol (Q-WeCBR), which combines clustering with reinforcement learning. Q-WeCBR improves cluster head (CH) selection through a weighted selection function and a maintenance phase that ensures cluster stability. In addition, the integration of Q-learning enables the protocol to adapt intelligently to topology changes by selecting the most reliable routes. Simulation results obtained with OMNeT++ and SUMO demonstrate that Q-WeCBR outperforms CBR, DSDV, and GPSR in terms of packet delivery ratio and throughput, confirming the effectiveness of clustering combined with learning-based routing for dynamic vehicular networks.

Keywords: VANETs · Cluster Based Routing · Q-learning

1 Introduction

In today's world, mobility has become an essential aspect of daily life. Each day, millions of private and public vehicles transport individuals and goods across road networks. However, this growing demand for mobility brings with it critical challenges, such as persistent traffic congestion and fatal road accidents. Consequently, road safety has become a central priority for governments worldwide [1].

In response to these issues, Vehicular Ad Hoc Networks (VANETs) have emerged as a key component of Intelligent Transportation Systems (ITS). These

Z. Kouahla et al. (Eds.): IAM 2025, LNNS 1973, pp. 95–113, 2026.
https://doi.org/10.1007/978-3-032-26031-4_9

networks aim to deliver a broad spectrum of services, including accident prevention, traffic optimization, and infotainment for drivers and passengers. This is made possible through the integration of On Board Units (OBUs) and fixed Road Side Units (RSUs), which rely primarily on wireless communication technologies, particularly those based on the IEEE 802.11 standard. These technologies enable real-time vehicle-to-vehicle (V2V) and vehicle-to-infrastructure (V2I) communication, allowing vehicles to react promptly to unforeseen events [2].

Despite progress, VANETs face several technical challenges, particularly in areas such as data routing, frequent topology changes, QoS, and security [3]. Routing remains especially difficult due to high vehicle mobility, prompting frequent network reconfiguration [4]. Common approaches include geographic routing protocols such as GPSR (Greedy Perimeter Stateless Routing), a geographic forwarding protocol, and proactive protocols such as DSDV (Destination-Sequenced Distance-Vector).

This paper focuses primarily on the routing process, which remains a major research concern due to the frequent topological changes caused by high vehicle mobility. Several studies have explored solutions to ensure efficient data transmission under these conditions. Among the various approaches, Clustering has proven to be a particularly promising technique. It aligns well with the inherent characteristics of VANETs, where traffic dynamics naturally support the formation of stable clusters. Furthermore, many transportation-related applications are localized in specific geographical zones, making geographic clustering a logical and effective organizational method. By dividing the network into clusters, network management is significantly improved, and control message exchange is confined to intra-cluster and inter-cluster communication, thereby reducing overhead. In this architecture [7], each cluster is typically managed by a Cluster Head (CH), which is responsible for coordinating intra-cluster communications and maintaining routing information. Regular nodes within the cluster function as Cluster Members (CMs), which communicate with the CH for data forwarding and cluster-related tasks. Communication between different clusters is facilitated by Gateway (GW) nodes, which act as bridges for inter-cluster data exchange. This hierarchical organization not only reduces routing complexity and control overhead but also enhances scalability in highly dynamic vehicular environments, making clustering a widely adopted strategy in VANETs [8,9].

The main contributions of this paper can be summarized in these points:

- First, we present a novel classification framework for clustering-based routing protocols specifically designed for VANETs, providing a systematic foundation for their analysis.
- Second, we introduce the Q-learning Weighted Cluster-Based Routing (Q-WeCBR) protocol, developed by extending the baseline Cluster-Based Routing (CBR) protocol [10] with three complementary enhancements: a mobility-aware weighted cluster head (CH) selection function that prioritizes nodes with low relative speed, thereby prolonging CH lifetime and reducing re-clustering; a dedicated maintenance phase that proactively stabilizes clusters by minimizing unnecessary CH re-elections and reducing control overhead;

and, as the main innovation, the integration of the Q-learning reinforcement learning model, which enables adaptive refinement of routing policies based on historical network states and performance feedback, thus supporting context-aware and dynamic decision-making in highly mobile VANET environments.
- Lastly, we conduct extensive simulations using OMNeT++ and SUMO, demonstrating that Q-WeCBR significantly outperforms the baseline CBR protocol, as well as established schemes such as DSDV and GPSR (Conventional protocols), in terms of packet delivery ratio and throughput.

The remainder of the paper is organized as follows. In Sect. 2, we review the related work. Section 3 presents a novel classification of clustering-based routing protocols in VANETs. Section 4 provides a detailed description of the implemented CBR protocol, while Sect. 5 introduces our proposed routing protocol; Q-WeCBR. Simulation results and discussions are reported in Sect. 6. Finally, Sect. 7 concludes the paper.

2 Related Work

Vehicular Ad Hoc Networks (VANETs) have attracted significant research attention due to their potential to enhance road safety, optimize traffic flow, and enable infotainment services. One of the main challenges in VANETs is the design of efficient and robust routing protocols capable of dealing with high mobility, frequent topology changes, and strict Quality of Service (QoS) requirements. Over the past two decades, three main families of routing approaches have been extensively investigated [14]: those based on the routing strategy (proactive, reactive, or hybrid), the type of routing (unicast, multicast/geocast, or broadcast), and the network structure (position-based or clustering-based protocols).

Unicast-based routing; classical approaches such as Destination-Sequenced Distance Vector (DSDV) [6] and Ad hoc On-Demand Distance Vector (AODV) rely on proactive or reactive route discovery and maintenance. While effective in static or moderately dynamic networks, these methods suffer from high control overhead and frequent route breakages in highly mobile VANET environments. Consequently, their scalability is limited in dense urban scenarios.

Position-based protocols, such as Greedy Perimeter Stateless Routing (GPSR) [5]; exploit node location information to forward packets. These schemes are more scalable than topology-based ones since they avoid maintaining global routes. However, their performance degrades in the presence of obstacles, sparse networks, or local minima, where greedy forwarding fails to find a next hop.

Clustering-based routing; to improve scalability and stability, many studies have introduced clustering as a hierarchical mechanism for routing [7]. In such schemes, nodes are organized into clusters managed by a Cluster Head (CH), which reduces control overhead and enhances robustness. Different clustering strategies have been proposed, including heuristic-based [11], weight-based, AI-based [7], and QoS-aware protocols [12]. Weight-based approaches are particularly promising as they combine several metrics (e.g., distance, mobility, connectivity) to ensure stable CH election. Nevertheless, cluster maintenance under high mobility remains a critical challenge, often leading to frequent re-clustering and increased overhead.

Intelligent routing with machine learning; recently, artificial intelligence (AI) techniques have been applied to VANET routing to improve adaptability. Fuzzy logic, bio-inspired algorithms, and reinforcement learning have been explored to support dynamic CH selection and routing decisions [8,9]. In particular, reinforcement learning (RL) has gained attention as it allows nodes to learn optimal routing strategies based on feedback from past transmissions [4]. RL-based protocols have shown improved packet delivery ratio and adaptability to mobility patterns, but many of them remain limited by high learning overhead, slow convergence, or lack of integration with clustering mechanisms.

While clustering improves scalability and RL enhances adaptability, few works have effectively combined both paradigms in a unified framework. Existing RL-based protocols generally focus on flat network architectures, neglecting the benefits of hierarchical clustering. Conversely, clustering-based schemes rarely leverage learning-based mechanisms for route optimization. This gap motivates our proposed Q-learning Weighted Cluster-Based Routing (Q-WeCBR), which integrates weighted CH selection, cluster stability maintenance, and Q-learning-based adaptive routing to achieve both stable cluster formation and intelligent path selection.

3 A Novel Classification of Clustering-Based Routing Protocols in VANETs

Clustering techniques play a pivotal role in ensuring efficient communication and scalable management in vehicular networks, particularly in highly dynamic VANET environments. By organizing nodes into hierarchical structures, clustering reduces routing overhead, enhances network stability, and improves resource utilization. Over the past decade, numerous studies have proposed various clustering approaches that rely on parameters such as the node's position relative to its Cluster Head (CH), network density, mobility patterns, or the node's destination. Building upon this body of work, we introduce a novel classification of clustering protocols specifically developed or adapted for VANETs (see Fig. 1), organized according to their operational paradigm and the decision-making parameters they employ.

3.1 According to Operational Paradigm

Hierarchical routing protocols can be classified according to their operational mode into three main categories: Quality of Service (QoS)-based protocols, position-based protocols, and lane-based protocols. It should be noted that some clustering protocols proposed in the literature may fall into one or more of these categories.

Quality of Service (QoS)-based protocols introduce the notion of QoS into the clustering process, aiming to satisfy specific service quality parameters during

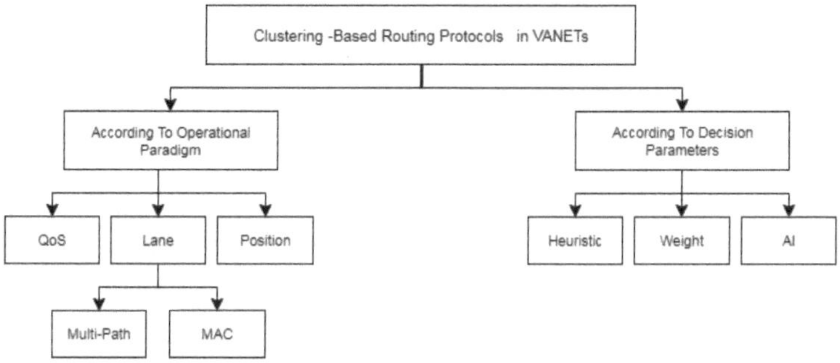

Fig. 1. Classification of Clustering Based Routing Protocols in VANETs.

cluster formation, such as transmission delay, bandwidth, data quality, and cluster stability. Minimizing relative mobility as well as the distance between each cluster head (CH) and its members improves cluster convergence and dynamics; however, vehicle density and speed influence mobility and must be considered. In position-based clustering [11], the cluster structure mainly relies on vehicles geographic location values, their spatial variations, and behaviors, where current positions, predicted future positions, and destinations contribute to cluster formation. Navigation systems may be used to determine destinations, and incorporating vehicle behavior improves both cluster stability and transmission efficiency, leading to longer cluster connection times since vehicles with similar destinations are less likely to leave the cluster; nonetheless, a large number of vehicles within a cluster may cause high transmission overhead due to message broadcasting. Lane-based protocols [12], on the other hand, exploit lane information availability for cluster formation and path selection, and can be divided into two subcategories: multipath protocols and MAC-based protocols. Multipath protocols aim to establish multiple routes between source and destination, possibly through one or more CHs, while also supporting intra-cluster communication between cluster members (CMs) and CHs. These protocols enhance fault tolerance by maintaining alternative active paths through periodic messaging, ensuring communication continuity in case of primary path failure. MAC-based clustering protocols [13], meanwhile, rely on medium access control mechanisms such as IEEE 802.11, TDMA, or SDMA; however, they often suffer from increased collision rates and higher average message delivery delays, resulting in reduced transmission efficiency and significant overhead due to channel contention, particularly in scenarios with high vehicle density or mobility.

3.2 According to Decision Parameters

Hierarchical routing protocols [7] can be divided according to their decision-making mode into three categories: heuristic-based, weight-based (weighted), and artificial intelligence (AI)-based protocols.

Heuristic-based protocols rely on mathematical optimization techniques to find solutions to a given problem, though without guaranteeing optimality. The obtained solutions are generally considered approximations, since these protocols are designed to reach results close to the best one within a short time. Moreover, they are not typically based on specific metrics. For this reason, many studies have investigated heuristic clustering solutions to achieve near-optimal performance within reasonable computational time. However, most existing heuristic clustering approaches do not ensure consistently high performance. Still, heuristic protocols remain effective for several graph theory problems, such as spanning tree construction and dominating set determination. In contrast, weight-based clustering protocols do not rely on a single metric to elect a cluster head (CH); instead, they use a function that combines multiple measures representing the weight of a node. This category aims to select a set of CHs better suited to environmental conditions and network requirements, thereby improving the overall performance of VANETs. More recently, artificial intelligence (AI)-based protocols have been introduced into hierarchical routing design, employing techniques such as fuzzy logic, bio-inspired algorithms, or learning methods (supervised, unsupervised, and reinforcement learning) to perform CH selection according to predefined criteria. For example, agents can interact with their environment to handle CH selection tasks, where node mobility direction is calculated dynamically, and agents are deployed at road intersections to monitor vehicle activity. Vehicle density and average speed are then used to divide the space into different zones, which in turn provide input to clustering agents. Additional parameters, such as inclination rates, are defined to enable adaptive decision-making. Each action taken by an agent is rewarded or penalized, with the learning rate being incremented or decremented until a maximum value is reached. Such AI-based protocols represent a promising alternative for future VANETs, particularly for applications requiring specific capabilities such as data security.

4 CBR Protocol

The CBR protocol [10] aims to eliminate the route discovery process by dividing the network into clusters. It operates in two main phases: the first phase focuses on cluster formation and the election of cluster heads (CHs), while the second phase is dedicated to data routing. The details of each phase are presented in the following section. In our work, the protocol has been implemented following our own approach and specific design choices.

4.1 Cluster Formation and CH Election

- Cluster Formation: in this sub-phase, the network is dynamically partitioned into square virtual cells (grids) of size $r \times r$ (see Fig. 2). These cells represent the clusters. Each node u can determine the cluster to which it belongs by using a grid identity (Grid ID) assigned to each cluster. The Grid ID calculation follows from Eq. 1:

$$Grid\_id = \sqrt{NC} \cdot \left\lfloor \frac{y}{r} \right\rfloor + \left\lfloor \frac{x}{r} \right\rfloor + 1 \tag{1}$$

where NC represents the number of clusters, (x, y) are the Cartesian coordinates of node u, and r is the side length of a grid.

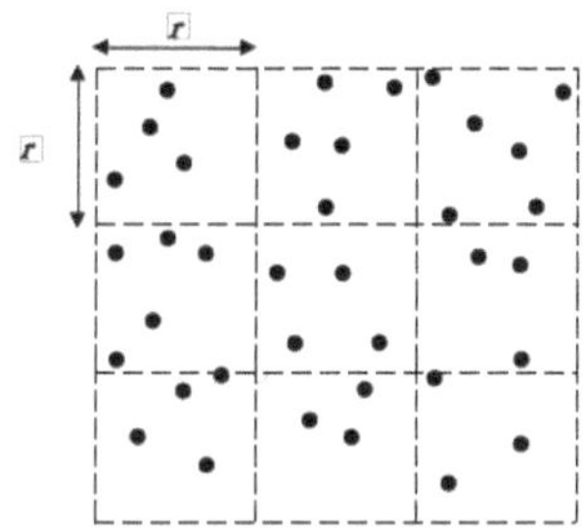

Fig. 2. Network Partitioning into Grids.

At the beginning, all vehicles exchange Hello messages containing their address, current position, velocity, and grid identity, as illustrated in Table 1.

Table 1. Hello Message Format.

Vehicle Address	Vehicle Address	Vehicle Velocity	Vehicle Grid
IP address	(x, y)	Vx, Vy	Grid_id

- Cluster Head (CH) Election: the election of a CH within each cluster is carried out in a competitive manner among the nodes belonging to the same cluster (i.e., having the same Grid ID). The node closest to the center of the cluster has the highest probability of being selected as CH. Each node computes the coordinates of its cluster's center point as well as its distance from that point. The node with the minimum distance is then identified as the CH. This CH subsequently broadcasts a *Lead* message (address, loc), where *address* and *loc* denote its IP address and geographic location, respectively. This message is sent to its cluster members (CMs) as well as to neighboring CHs, which reply with an acknowledgment (*Ack*) message.
  When a node enters a new grid, it broadcasts a *NewAccessGrid* message and executes the following steps:
  1. If the grid is empty (i.e., no node exists), the node designates itself as CH.
  2. If there is already a CH in the grid, the latter sends a *Lead* message simply to inform the newcomer of its existence.
  3. If there is no CH in the grid, the CMs initiate a new CH election by broadcasting a *Hello* message.

  When a CM leaves its cluster, it sends a *Leave CM* message to its CH in order to be removed from the membership list. Conversely, when a CH leaves, it broadcasts a *Leave CH* message to the members of its grid, thereby triggering the election of a new CH for the cluster (See pseudo-code in Algorithm 1).

Algorithm 1: Cluster Head Election Process

```
Input: Grid_j (cluster j), set of vehicles V_i in Grid_j
Output: Selected CH for Grid_j
Notations:
C_Grid(j) : center of Grid_j
V_i : vehicle i in Grid_j

Begin
foreach V_i ∈ Grid_j do
    distance = dist(V_i, C_Grid(j));
if dist(V_i) == minDistance then
    CH_j ← V_i;
Send_Packet(CH_j, Lead);
if V_i receives Lead from CH_j then
    if (Grid_id(V_i) == Grid_id(CH_j)) or (state(V_i) == CH) then
        Send_Packet(V_i, Ack);
if Grid_id(V_i) ≠ Update(Grid_id(V_i)) then
                                            // Vehicle leaves cluster
    if state(V_i) == CM then
        Send_Packet(V_i, Leave_CM);
    else if state(V_i) == CH then
        Send_Packet(V_i, Leave_CH);
        Select a new CH;
if Grid_id(V_i) ≠ Update(Grid_id(V_i)) then
                                        // Vehicle enters new cluster
    Send_Packet(V_i, NewAccessGrid);
    if exists(CH_j) then
        Send_Packet(CH_j, Lead);
    else
        if alone(V_i) then
            V_i becomes CH;
        else
            Select a new CH;
End
```

4.2 Data Routing

In this phase, clusters employ two routing methods [10]. The first is an intra-cluster routing method, which ensures communication among members of the same cluster. The second is an inter-cluster routing method, which enables communication between cluster heads (CHs) belonging to different clusters. The steps of this phase are detailed below (See pseudo-code in. Algorithm 2).

Intra-cluster communication: if a source node S wants to send data packets to a destination node D located within the same cluster, the transmission is performed directly. This is achieved using the intra-cluster communication channel, without requiring the CHs to provide the location of D, since the information of all cluster members is already stored.

Inter-cluster communication: if the destination node does not belong to the same cluster as the source node, the latter initiates the routing request procedure. It sends the data packet to the CH of its cluster, which is then responsible for selecting the optimal neighboring CH based on the minimum angle criterion, in order to efficiently forward packets from the source to the destination.

Algorithm 2: Selection of the Optimal CH and Data Routing

Input: V_{sender}: source node; $V_{destination}$: destination node;
Grid_id: grid identity; *listAngle*: list of computed angles; *listNeighbors*: list of neighbors.
Output: Successful delivery of Data packets.

```
Begin
while V_sender ≠ V_destination do
    Update position(V_destination);
    if Grid_id(V_sender) == Grid_id(V_destination) then
        Send_Data(V_sender, V_destination, Data);
    else
        if state(V_sender) ≠ CH then
            Send_Data(V_sender, CH(V_sender), Data);
            foreach CH ∈ listNeighbors do
                listAngle.add(computeAngle(CH(V_sender), V_destination,
                CH));
                min(CH_j) ← min(listAngle);
            next_CH ← CH_j;
            Send_Data(CH(V_sender), next_CH, Data);
            V_sender ← next_CH;
            CH(V_sender) ← next_CH;
        else
            foreach CH ∈ listNeighbors do
                listeAngle.add(computeAngle(V_sender, V_destination, CH));
                min(CH_j) ← min(listAngle);
            next_CH ← CH_j;
            Send_Data(V_sender, next_CH, Data);
            V_sender ← next_CH;
End
```

5 Q-WeCBR Protocol Design

The first improvement we proposed to the CBR protocol is called Weighted Cluster-Based Routing (WeCBR). The main objective of WeCBR is to ensure greater network stability by reducing the number of Cluster Head (CH) election processes and by providing a more efficient selection of CHs. In this regard, we introduced two major enhancements over the basic CBR protocol.

- We designed a weighted function that combines two key parameters to compute node weight $P(i)$: $D(i)$, representing distance from cluster center, and $V(i)$, representing velocity. To enhance cluster stability, our formulation intentionally prioritizes mobility over proximity. The weight function is defined as:

$$P(i) = \alpha \times D(i) + \beta \times V(i) \tag{2}$$

  where α and β are weighting coefficients with $\alpha + \beta = 1$. Empirical optimization led to selecting $\alpha = 1/4$ and $\beta = 3/4$, as this ratio maximizes cluster longevity in highly dynamic environments by favoring stable, low-mobility nodes for CH roles.
- We introduced a maintenance phase in WeCBR to ensure better utilization of the communication medium and to reduce collisions and interferences in the network. This phase allows for the designation of a relay node when the primary cluster-head (CH) leaves its cluster. To achieve this, we added a temporary cluster-head (backup CH) that will assume the role of the CH. The selection of the backup CH is performed during the CH election phase; specifically, it corresponds to the node with the second-best weight value after the primary CH. With this mechanism, a cluster can never remain without a CH, thereby eliminating the need to trigger the CH election process again. This improves both cluster and network management while also reducing the packet loss rate and alleviating network overhead by ensuring more reliable packet delivery. The mechanism for CH selection and maintenance is presented in the pseudo-code 3.

As a second improvement to the CBR protocol, we proposed Q-learning Weighted Cluster Based Routing (Q-WeCBR). Our proposal relies on a reinforcement learning approach, more specifically the Q-learning method. In fact, we integrated Q-learning into the routing process of our first improvement, WeCBR, in order to optimize data transmission by selecting the shortest path to the destination. In the remainder of this section, we define the Q-learning model that we employed, followed by the pseudo-code of the applied algorithms.

Algorithm 3: Cluster Head (CH) Selection and Maintenance

Input: $C_{Grid(j)}$: center of $Grid_j$; V_i: vehicle i; *velocity*: current velocity; *weight*: computed node weight

Output: Stable CH selection with backup maintenance.

Notations:

CH_j: Cluster Head of grid j;

$state(V)$: role of node V (CH or CM);

Lead, *Ack*, *Leave_CM*, *Leave_CH*, *CH_Tempo*, *NewAccessGrid*: control messages.

Begin

foreach $V_i \in Grid(j)$ **do**
- $distance \leftarrow distance(V_i, C_{Grid(j)})$;
- $velocity \leftarrow getCurrentVelocity(V_i)$;
- $weight \leftarrow \frac{1}{4} \times distance + \frac{3}{4} \times velocity$;

if $weight(V_i) == minWeight$ **then**
- $CH_j \leftarrow V_i$;
- sort(neighborList, ascending);

Send_Packet($CH_j(V_i)$, Lead);

if V_i *receives Lead from* CH_j **then**
- **if** $Grid\_id(V_i) == Grid\_id(CH_j)$ ***or*** $state(V_i) == CH$ **then**
  - Send_Packet(V_i, Ack);

if $Grid\_id(V_i) \neq Update(Grid\_id(V_i))$ **then**
- **if** $state(V_i) == CM$ **then**
  - Send_Packet(V_i, Leave_CM);
- **else if** $state(V_i) == CH$ **then**
  - Send_Packet(V_1, CH_Tempo) `// ` V_1 `is the 1st in neighbor list`
  - Send_Packet(V_i, Leave_CH);
  - selectNewCH();

if $Grid\_id(V_i) \neq Update(Grid\_id(V_i))$ **then**
- `// ` V_i `joins cluster`
- Send_Packet(V_i, NewAccessGrid);
- **if** *exists*(CH_j) **then**
  - Send_Packet(CH_j, Lead);
- **else**
  - **if** *alone*(V_i) **then**
    - becomeCH(V_i);
  - **else**
    - selectNewCH();

End

- To apply the Q-learning approach in our routing process, we modeled the problem of selecting the optimal next hop to reach the destination through the shortest path as follows in Table 2:

Table 2. Q-learning Model

Environment	The set of clusters in the network
Agent	Each Cluster Head (CH)
State space	The set of CHs in the network
Action space	Selecting a neighboring CH in one of four directions (Up, Down, Right, Left)
Reward	The value obtained by an agent for performing an action.

- We summarized the parameter values used in our Q-WeCBR model in Table 3. These values were selected based on both standard Q-learning practices and empirical validation in our VANET context.

Table 3. Q-learning Parameters

Epsilon	0.9	The exploration rate
Discount factor (γ)	0.9	The discount factor for future rewards
Learning rate (α)	0.9	The dynamic learning rate

- Agent Training
  The learning process of an AI agent within its environment is executed through the following steps:
  1. Select a random, non-terminal state for the agent to start the new episode.
  2. Choose an action (move up, right, down, or left) for the current state. The actions are selected using an epsilon-greedy algorithm; this method generally chooses the most promising action for all agents but occasionally selects a less promising option to encourage exploration of the environment.
  3. Perform the chosen action and transition to the next state (e.g., move to the next location).
  4. Receive the reward for reaching the new state and compute the temporal difference.
  5. Update the Q-value for the previous state–action pair.
  6. If the new (current) state is a terminal state, return to Step 1; otherwise, proceed to Step 2.

- The entire process will be repeated over 20,000 episodes, providing the agent with sufficient opportunities to learn the shortest paths between the destination cluster and all other grid locations.

Algorithm 4: Agent Training Algorithm

Input: Reward matrix R, learning rate α, discount factor γ, exploration rate ϵ
Output: Updated Q-Value table Q
Notations:
current_loc, action_index, previous_loc, reward;
old_q_value, new_q_value, temporal_difference;
current_loc: the current position of the agent;
action_index: the identifier of the next action;
previous_loc: the previous position.
Begin:

```
episode ← 0;
while episode < 20000 do
    current_loc ← find_start_loc();
    while isTerminal(current_sloc) == false do
        action_index ← find_next_action(current_loc, ε);
        previous_loc ← current_loc;
        current_loc ← find_next_action(current_loc, action_index);
        reward ← R[previous_loc][current_loc];
        old_q_value ← Q[previous_loc][action_index];
        temporal_difference ←
        reward + (γ · max Q(current_loc)) − old_q_value;
        new_q_value ← old_q_value + (α · temporal_difference);
        Q[previous_loc][action_index] ← new_q_value;
    episode++;
End
End
```

Algorithm 5: Shortest Path Selection Algorithm

Input: Start location *start_loc*, exploration rate ϵ
Output: List of nodes in the shortest path *shortest_path*
Notations:
shortest_path: list of visited nodes;
current_loc: current node;
action_index: identifier of the next action;
Begin
if *isTerminal(start_loc) == true* **then**
    shortest_path.add(start_loc);
else
    current_loc ← start_loc;
    shortest_path.add(current_loc);
    while *isTerminal(current_loc) == false* **do**
        action_index ← find_next_action(current_loc, ϵ);
        current_loc ← find_next_action(current_loc, action_index);
        shortest_path.add(current_loc);
End

6 Simulation Results and Analysis

In this section, we evaluate the performance of our proposed Q-WeCBR protocol by comparing it with traditional CBR and the conventional protocols DSDV and GPSR. The simulations were conducted under diverse scenarios, where each scenario varies a specific criterion. We used the OMNeT++ simulator along with INET, Veins, and SUMO to perform the simulations on the city map of Jijel, Algeria, as illustrated in Fig. 3. The main simulation parameters are summarized in Table 4.

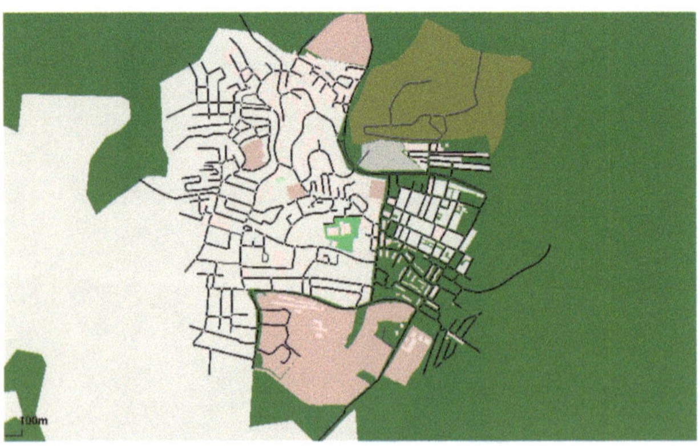

Fig. 3. Map used for simulation.

Table 4. Simulation parameters

Parameter	Value
OMNeT++	5.4.1
INET	4.1.0
Veins	5.0
SUMO	1.6.0
Simulation area dimensions	3000×3000 (for density) 900×900 (for speed)
Number of nodes	10, 20, 30, 40, 50, 60 (for density) 25 (for speed)
Number of clusters	9
Packet size	1000 bits
Simulation duration	100 s (for density) 60 s (for speed)
Propagation model	Free space model
Mobility model	VeinsInetMobility
Average speed	22 m/s (80 km/h) (for density) AVG(10, 20, 30, 40, 50) m/s (for speed)

Two scenarios are realized that to show the efficiency of the proposed protocol. We studied the impact of PDR (Packet Delivery Rate) and throughput on the Q-WeCBR performance. The protocol demonstrates consistent stability with standard deviations of (1.52–2.14%) for PDR and (0.02–0.04 $\times 10^4 bps$) for throughput across 5 independent runs at various speeds. Similarly, under variations in vehicle density, the protocol maintains comparable robustness, with PDR standard deviations of (1.67–2.31%) and throughput deviations of (0.03–0.067 $\times 10^4 bps$).

From the results obtained in Fig. 4 and Fig. 5, we observe that while the packet delivery ratio decreases as the number of vehicles increases, the throughput follows an increasing trend, with the implemented Q-WeCBR protocol achieving better performance than CBR, DSDV, and GPSR in both cases. This improvement is mainly due to the efficient cluster management strategy, where the weighted cluster head (CH) selection function ensures that stable nodes are elected as CHs and the maintenance phase preserves cluster stability under frequent topology changes. The superior performance of Q-WeCBR stems from Q-learning's ability to dynamically optimize routing decisions through continuous environmental feedback. Unlike traditional protocols that use static routing metrics, Q-learning enables each Cluster Head to:

- Learn optimal paths through experience: by continuously updating Q-values based on successful packet deliveries, CHs build knowledge about which neighboring clusters provide the most reliable routes to destinations.

- Adapt to network dynamics: the reward mechanism immediately penalizes route failures and rewards successful transmissions, allowing rapid adaptation to topology changes caused by vehicle mobility.
- Balance exploration and exploitation: The ϵ-greedy strategy ($\epsilon = 0.9$) ensures continuous discovery of new routes while predominantly using proven reliable paths.
- Optimize long-term performance: The discount factor ($\gamma = 0.9$) enables consideration of future route stability rather than just immediate next-hop quality.

This intelligent learning framework allows Q-WeCBR to proactively avoid unstable links and congestion points that persistently challenge conventional protocols. By leveraging historical transmission data to select more reliable routes, the protocol achieves enhanced stability, improved packet delivery, and superior throughput capacity under varying network densities.

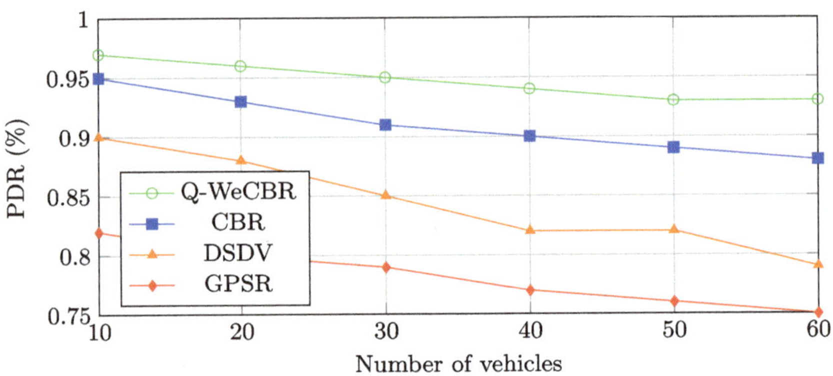

Fig. 4. PDR according the variation of number of vehicle.

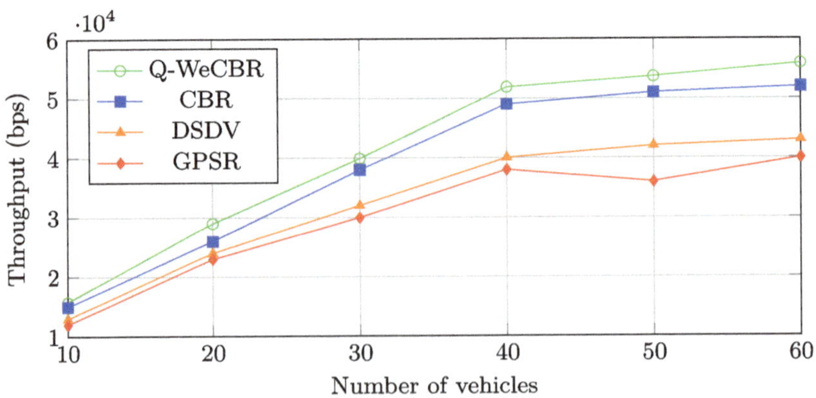

Fig. 5. Throughput according the variation of number of vehicle.

From the obtained results in Fig. 6 and Fig. 7, we observe that increasing vehicle speed leads to a decrease in both packet delivery ratio and throughput, with the proposed Q-WeCBR protocol achieving the best performance among the compared approaches. Its robustness stems from the integration of Q-learning, which enables intelligent adaptation to high mobility by selecting reliable paths, combined with clustering that ensures stability through efficient cluster head selection and maintenance. Consequently, Q-WeCBR clearly outperforms CBR, GPSR, and DSDV, which are more vulnerable to route breakages and higher overhead in dynamic environments.

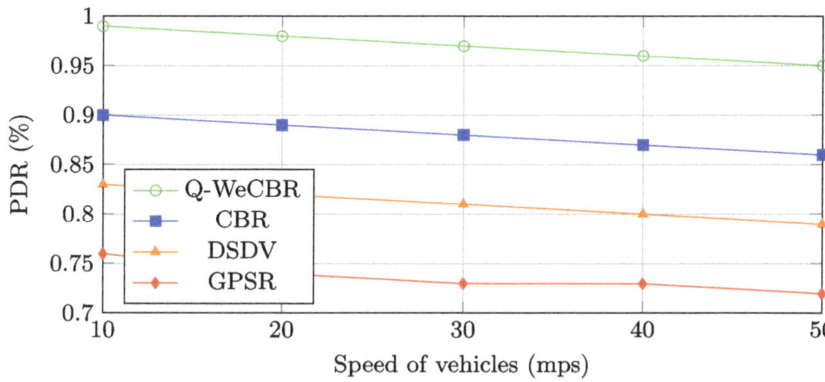

Fig. 6. Impact of speed on PDR.

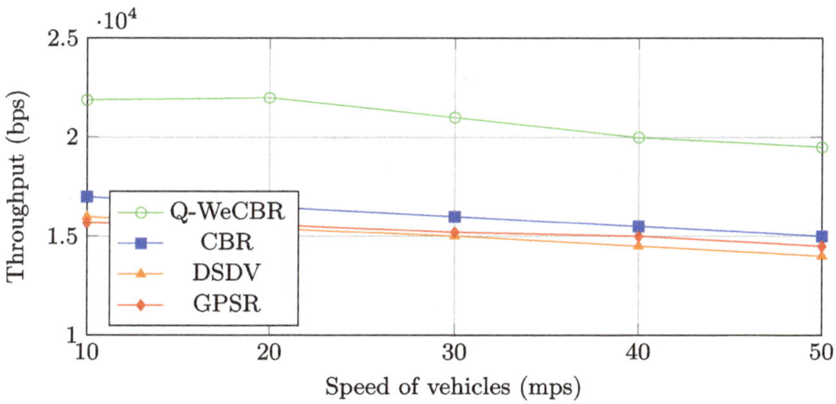

Fig. 7. Impact of speed on the throughput.

The simulation results demonstrate that Q-WeCBR consistently outperforms CBR, DSDV, and GPSR across varying network conditions. Beyond performance metrics, our protocol demonstrates practical feasibility for real-world deployment. The algorithm achieves $O(1)$ routing decisions after convergence at approximately 15,000 episodes, ensuring minimal computational overhead during operation. While the one-time training phase represents the main cost, this can be efficiently managed during network initialization. Furthermore, memory requirements scale linearly with cluster count, making Q-WeCBR suitable for implementation on modern On-Board Units (OBUs) without exceeding typical computational constraints.

This balance between performance gains and manageable complexity positions Q-WeCBR as a viable solution for intelligent routing in dynamic vehicular environments. The protocol's ability to maintain low overhead while adapting to network changes through Q-learning represents a significant advancement over static routing approaches.

7 Conclusion

In this paper, we addressed routing in Vehicular Ad Hoc Networks (VANETs) with a particular focus on clustering as a promising approach to improve stability and efficiency. We first proposed a new classification of clustering-based routing protocols according to their mode of operation and decision-making parameters. Then, we implemented the hierarchical CBR protocol and introduced its enhancement, named Q-WeCBR. To demonstrate the effectiveness of clustering compared to conventional approaches and to validate the improvement brought by Q-WeCBR, we evaluated its performance against CBR, GPSR, and DSDV in terms of packet delivery ratio and throughput in two scenarios. The obtained results clearly show that Q-WeCBR achieves superior performance, confirming both the advantages of clustering and the added value of integrating reinforcement learning for intelligent routing in highly dynamic vehicular environments.

References

1. Hartenstein, H., Laberteaux, L.P.: A tutorial survey on vehicular ad hoc networks. IEEE Commun. Mag. **46**(6), 164–171 (2008)
2. Arena, F., Pau, G., Severino, A.: A review on IEEE 802.11p for intelligent transportation systems. J. Sens. Actuator Netw. **9**(2), 22 (2020)
3. Vegni, A.M., Loscri, V.: A survey on vehicular social networks. IEEE Commun. Surv. Tutorials **17**(4), 2397–2419 (2015)
4. Nazib, R.A., Moh, S.: Reinforcement learning-based routing protocols for vehicular ad hoc networks: a comparative survey. IEEE Access **9**, 27552–27587 (2021)
5. Karp, B., Kung, H.T.: GPSR: greedy perimeter stateless routing for wireless networks. In: Proceedings of the 6th Annual International Conference on Mobile Computing and Networking (MobiCom), pp. 243–254 (2000)
6. Perkins, C.E., Bhagwat, P.: Highly dynamic destination-sequenced distance-vector routing (DSDV) for mobile computers. ACM SIGCOMM Comput. Commun. Rev. **24**(4), 234–244 (1994)
7. Senouci, O., Harous, S., Aliouat, Z.: Survey on vehicular ad hoc networks clustering algorithms: overview, taxonomy, challenges, and open research issues. Int. J. Commun. Syst. **33**(11), e4402 (2020)

8. Rahmani, A.M., et al.: A Q-learning and fuzzy logic-based hierarchical routing scheme in the intelligent transportation system for smart cities. Mathematics **10**(22), 4192 (2022)
9. HaghighiFard, M.S., Coleri, S.: Hierarchical federated learning in multi-hop cluster-based VANETs. IEEE Trans. Veh. Technol. (2025)
10. Luo, Y., Zhang, W., Hu, Y.: A new cluster based routing protocol for VANET. In: 2nd International Conference on Networks Security, Wireless Communications and Trusted Computing, vol. 1, pp. 176–180. IEEE (2010)
11. RaoDawande, J., Silakari, S., Deen, A.: A survey of all existing clustering protocols in VANETS: TCDGP emphasis. Int. J. Comput. Appl. **118**(6), 22–31 (2015). https://doi.org/10.5120/20751-3146
12. Purtoosi, R., et al.: A light-weight contention-based clustering algorithm for wireless ad hoc networks. In: 4th International Conference on Computing and Information Technology (CIT 2004), pp. 627–632. IEEE (2004). https://doi.org/10.1109/CIT.2004.1357265
13. Klir, G.J., Clair, U.S., Yuan, B.: Fuzzy Set Theory: Foundations and Applications. Prentice Hall, New Jersey (1997)
14. Venkatesh, A., Murali, R.: Routing protocols for vehicular adhoc networks (VANETs): a review. J. Emerg. Trends Comput. Inf. Sci. **5**(1), 25–43 (2014)

A Framework for Business Process Integration in Smart Grids Using Knowledge Graphs

Ibtissem Bouhaouche(✉) and Ali Khebizi

Department of Computer Science, LabSTIC Laboratory,
8 May 1945 University, P.O. Box 401, 24000 Guelma, Algeria
bouhaouche.ibtissem@univ-guelma.dz

Abstract. Large-scale deployment of Smart Grid (SG) technology requires advanced methods to ensure the integration and coordination of the underlying Business Processes (BPs). In SG ecosystems, multiple actors collaborate to achieve common goals, while each actor manages its own resources, operates under specific constraints, and follows distinct business logic. Consequently, designing a unified conceptual framework that fosters BP integration and cooperation to align with shared objectives inevitably raises diverse challenges of varying complexity.

This paper addresses the challenge of integrating BPs in SG environments by introducing a paradigm shift that merges traditional BPMN models with Knowledge Graph (KG) representations. Unlike traditional approaches, the proposed framework leverages the semantic richness and expressive power of KGs to enrich and integrate BPs specifications, enabling adaptive and context-aware business processes. Hence, the conceived framework promotes smarter BP integration and cooperation, while enhancing the entire BP management life-cycle in SG systems.

Keywords: Smart Grid (SG) · Business process (BP) · Business Processes integration · Knowledge representation · Knowledge graph (KG) · Mapping

1 Introduction

Recent developments in information and communication technologies (ICT) have significantly influenced nearly every aspect of human activity, reshaping both social interactions and economic processes. The widespread digitalization enabled by these advances has revolutionized the way organizations design, manage, and optimize their activities. In the energy sector, this digital transformation is particularly visible in the modern infrastructures, where the deployment of smart devices, sensors, and advanced communication infrastructures enables more efficient monitoring, control, and optimization of electricity generation, transmission, and distribution. Building on these technological developments, the concept of smart grids (SGs) has recently emerged as a novel paradigm in

Z. Kouahla et al. (Eds.): IAM 2025, LNNS 1973, pp. 114–131, 2026.
https://doi.org/10.1007/978-3-032-26031-4_10

energy systems engineering. Unlike traditional power grids, SGs are designed to intelligently integrate the behavior and actions of all parties (i.e.; producers, consumers, and prosumers) to ensure a reliable, sustainable, economical, and secure electricity supply.

Over the past two decades, SG infrastructures have substantially benefited from continuous advances in ICT, which have greatly enhanced their functionality and performance. This evolution has been further accelerated by the integration of emerging technologies such as Artificial Intelligence (**AI**), machine learning, and increasingly sophisticated hardware and software solutions. At the same time, the role of SG has become increasingly important with the growing penetration of renewable energy. By promoting the integration of distributed and variable generation, SG help balance supply and demand in real time. In addition, they enable the wider deployment of energy storage systems, which are essential for improving efficiency, enhancing operational flexibility, and ensuring the future-proofing of modern energy systems. These developments are increasing the complexity of operations across the energy value chain, reinforcing the need for advanced BP integration to coordinate the various actors, technologies, and activities within the SG ecosystem.

Alongside these technological advances, the concept of Business Processes (BPs) has emerged as a cornerstone of modern information systems. A BP specification captures the essential features and resources required to analyze, design, and implement appropriate software solutions that support organizational activities. A BP is generally defined as a set of structured activities carried out by one or more organizations to achieve specific objectives [1]. Integrating BPs within composition scenarios enhances interaction among activities, thereby improving efficiency, responsiveness, and task quality. Such integration also fosters information centralization and sharing among stakeholders, which significantly strengthens communication and collaboration both within and across organizations. In this context, stakeholders benefit from timely and consistent information, enabling more effective decision-making and process execution.

To the best of our knowledge, only a limited number of research efforts have addressed BP integration by explicitly incorporating their execution context [2,3]. In particular, this issue has not yet received the attention it deserves in the enhancement of SG systems. Thus, developing and deploying a comprehensive framework for BP integration and cooperation would provide significant added value to SG environments. In fact, tackling this challenge can substantially improve system management and monitoring by enabling seamless data exchange that reflects the coordinated execution of BPs across different stakeholders. The ultimate objective of such a coordination framework is to optimize overall system performance, promote more rational energy consumption, and ensure more efficient resource utilization. This challenge can be analyzed along two complementary dimensions: (i) the data dimension, which concerns the integration, consistency, and exchange of data; and (ii) the activities flow dimension, which involves coordinating tasks carried out by different operators to achieve common business objectives.

To illustrate the integration requirements, consider a scenario in which multiple utility companies within a given region integrate and exchange customer data in order to coordinate cooperative tasks aimed at optimizing energy distribution, particularly during shortages or peak demand periods. In such contexts, neighboring grids must be capable of supporting one another to ensure system stability and service continuity. Consequently, BP integration becomes a critical enabler for advanced functionalities such as dynamic pricing forecasting, real-time demand response, and the reduction of energy transportation costs. From a pragmatic perspective, addressing the above research issues requires perceiving the BPs involved in SG environments as basic building blocks that need to be formalized with adequate abstract models suitable for conducting effective reasoning capable of managing efficient mechanisms of integration between the conducted BPs.

In this article, we propose a comprehensive conceptual framework based on formal models, specifically abstract BPMN models and KG, to address BP integration and cooperation challenges. Within the proposed approach, KG representations are employed to enrich the description of activities contained in the considered BPs. This semantic enrichment provides valuable contextual information that strengthens both integration and cooperation objectives. In this perspective, the specifications expressed in BPMN diagrams are extended with domain-specific properties captured within the KG.

The remainder of this paper is organized as follows. Section 2 introduces the background concepts and definitions necessary for understanding the different aspects of the work. Section 3 reviews the related literature and discusses existing studies that have addressed integration and cooperation issues in SGs. Section 4 presents the various facets of the proposed framework for BP integration and cooperation. Section 5 explores implementation aspects and practical considerations. Finally, Sect. 6 summarizes the key findings and highlights promising avenues for future research.

2 Background

In this section, we present the key concepts, definitions, and background necessary to make the paper self-contained. We begin by describing the Smart Grid (SG) domain, then introduce the Business Process (BP) field and the related integration techniques. Finally, we focus on data management and knowledge representation, with particular emphasis on Knowledge Graphs (KGs).

2.1 Smart Grid Area

A smart grid (SG) is defined by the Department of Energy (**DOE**) as: "*the electricity delivery system, from point of generation to point of consumption, integrated with communications and information technology for enhanced grid operations, customer services, and environmental benefits*" [4].

According to this definition, different Business Processes (BPs) occur throughout the energy management life-cycle, each comprising a set of distinct but complementary activities. These processes cover the entire energy chain, from production to final consumption. As an illustration, these BPs may encompass electricity generation, transmission, distribution, and consumption, as well as demand response, trading, billing, customer service, regulatory compliance, and reporting.

Figure 1 bellow highlights the interaction between the different layers and components of the SG. Each layer includes different components used to achieve particular activities responding to the objectives of the considered level. The SG is structured into the following interactive layers: (i) the component layer, with physical assets such as generators, transmission lines, sensors, and smart meters; (ii) the communication layer, which enables real-time data exchange via wired and wireless networks; (iii) the information layer, which organizes and standardizes data; (iv) the functional layer, which supports services such as demand-side management and renewable energy integration; (v) and finally, the business layer, which governs market rules and policies. Together, these layers enable secure, efficient, and sustainable energy management through bidirectional flows of electricity and information. However, ensuring seamless interoperability between these heterogeneous layers and technologies remains a major challenge for SG deployment.

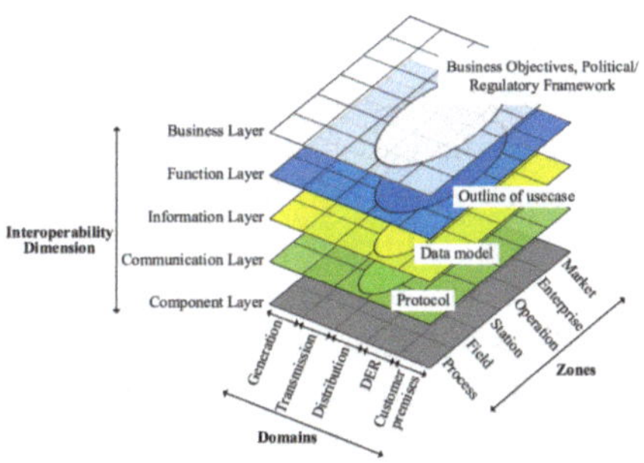

Fig. 1. SG infrastructures [5]

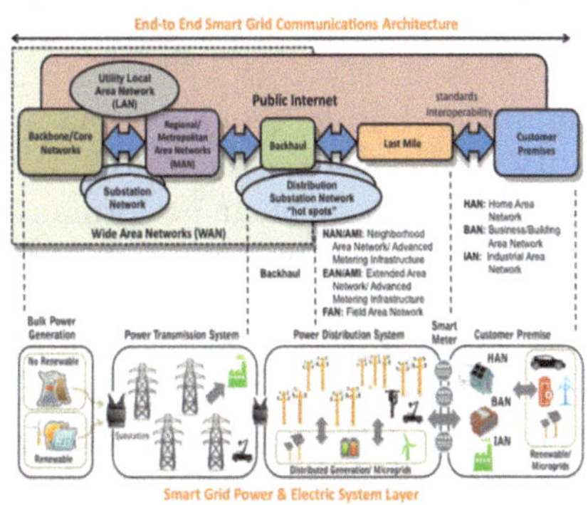

Fig. 2. SG communication layers [4]

Despite these advantages, the communication layer remains a significant challenge. Smart Grids rely on a wide variety of heterogeneous technologies and protocols, each characterized by different performance levels, security requirements, and interoperability capabilities. Ensuring reliable, low-latency, and secure communication within such a diverse and complex environment is therefore critical. Any shortcomings in this layer can negatively affect the efficiency and overall resilience of the SG.

As depicted in Fig. 2, communication systems realize message exchange and delivery between SG devices and actors. This layer is built on the base layer containing the physical equipment and is composed of different types of networks (*i.e.; Home Area Networks (HAN), Business Area Networks (BAN), Wide Area Network (WAN), and Neighborhood Area Networks (NAN))*.

On the other hand, information systems (***IS** for short*) play an important role in the SG. The main ISs in the grid are the following. Supervisory Control and Data Acquisition (SCADA) system, Customer IS (CIS), Geographic IS (GIS), Advanced Metering Infrastructure (AMI), Meter Data Management System (MDMS), Demand Response Management System (DRMS) and Outage Management System (OMS) [5]. The previous ISs manage useful information required by the function layer which in turn transmits them to the last layer relating to BPs.

Adopting new SG technology for energy management enables various parties to achieve the following objectives and to benefit from the advantages mentioned below [4].

- Enabling dynamic consumer participation and making system control less centralized due to the continuous interaction with the consumer.
- Improving energy system resiliency and flexibility during various energy management steps.
- Increasing operating efficiency.
- Anticipating and facing to system disturbances (*achieving self-healing and resilient system*).
- Providing power quality for the digital economy by ensuring continuous energy delivery.

Despite the considerable advantages offered by current SG technologies and the quality of the services they provide, several challenges remain to be addressed to further optimize their performance. The main challenges facing SGs include cybersecurity threats, data management and privacy concerns, interoperability issues, as well as integration and cooperation between different parties.

As illustrated in Fig. 1, the upper layer of the SG infrastructure is dedicated to supporting BPs and their supervision. This layer is mainly responsible for managing BP models and establishing appropriate cooperation mechanisms. In practice, this involves ensuring the effective integration and coordination of heterogeneous BP, executed by different interacting parties such as producers, distributors, operators, and consumers. This integration is necessary to align objectives, share resources, and synchronize activities throughout the energy value chain. However, an efficient integration and cooperation mechanisms raise several challenges, including the heterogeneity of the exchanged data and the semantic variety of BP models, interoperability between different information systems, and the need for dynamic adaptation to real-time operational conditions.

In this work, we focus on this important aspect by proposing a comprehensive framework for BP integration and cooperation in SG.
Below, we start by introducing the BP area and its underlying concepts.

2.2 BP Specification and Integration

This sub-section is dedicated to the notion of BP and the related concepts of BP integration.

"A BP refers to a set of activities or tasks performed by organizations to achieve a particular business goals" [6].

Example 1. Figure 3 below shows an example of Process model for monitoring temperature of a physical entity in the SG energy.

According to the previous definition, a BP is composed of a sequence of steps (*i.e.; Measure Temperature of Physical Entity, Reduce Price of Physical Entity, Update the Price, ...*) performed by one or a group of stakeholders (*i.e.; IoT (Sensor) Network, Cashier System, Electronic Shelf Label, ...*) to achieve a concrete business goal (*monitoring and managing physical entities and adjusting prices based on temperature and quality degradation.*).

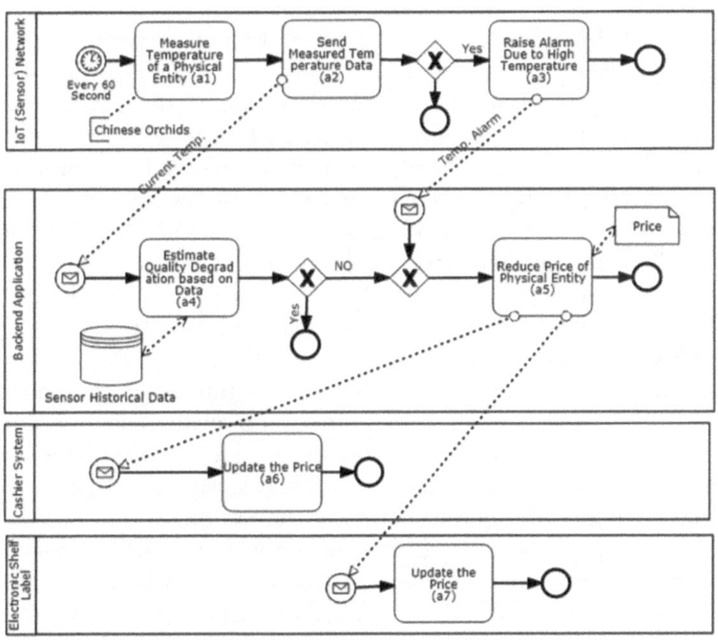

Fig. 3. An example of Process model for monitoring temperature of a physical entity in the SG energy system [7]

In the Fig. 3, a BPMN (Business Process Model and Notation) presenting:

Process: Measures temperature every 60 s using a sensor (IoT Sensor). "Measure Temperature" sends temperature data, if the temperature exceeds a threshold "Raise Alarm Due to High Temperature (a3)", If no alarm, the process loops back to measure temperature again.

Process: Estimates quality degradation based on sensor historical data: "Estimate Quality Degradation based on application (a4)" uses data from "Sensor Historical Data". If degradation is significant (not explicitly labeled as "NO"), it proceeds to "Reduce Price of Physical Entity (a5)". The process ends or loops back depending on the decision.

Process: Updates pricing and labels based on the estimated degradation: "Update the Price (a6)" and "Update the Price (a7)" are triggered, likely adjusting the price dynamically. The process involves a "Cashier" and "Electronic Label" (possibly for real-time price updates).

The process model demonstrates an IoT-based system for monitoring and managing physical entities (e.g., products) by adjusting prices based on temperature and quality degradation. The integration of real-time data (temperature) and historical data (sensor logs) suggests a data-driven approach. The diagram is part of a broader framework for energy-aware resource management, aligning with the paper's focus

From another point of view, the life cycle of a BP typically involves the following stages: development, deployment, execution, monitoring, optimization. Further, BP management and monitoring involves specifying inputs, actions, and outputs, as well as the set of constraints governing the BP execution. Such properties must be addressed in a rigorous manner through the use of appropriate representation models. Graphical models have emerged as powerful tools for specifying BPs and promoting their cooperation. In this context, the Business Process Model and Notation (**BPMN**) has emerged de facto as the most used standard. It offers a comprehensive set of graphical constructs enabling accurate and standardized representation of BPs, thus facilitating interoperability, analysis, as well as integration and cooperation of various BPs. In SG environments, BPMN is particularly suitable because it supports the modeling of complex and distributed processes, improves communication between the different parties involved, and enables the automation of critical energy management tasks. Moreover, by providing a unified representation, BPMN helps address key integration challenges such as semantic heterogeneity, system interoperability, and coordination between different actors in the SG ecosystem.

While significant advances have been achieved in the BP management field (*design, configuration, enactment and analysis*), the existing integration and cooperation techniques are still in a rudimentary state, especially in the field of SG. In fact, a huge size of data is generated during BPs execution and which are stored in adequate information systems to be efficiently used for various decision-making scenarios. However, BP interactions and coordination are often relegated to a secondary citizens class. In fact, the alignment and coordination of various organizational activities, require exploiting both execution data and activities description at the same time. In this perspective, both data and activities' flow reflecting instances execution must be considered uniformly and with the same degree of importance. Such a concern allows handling efficiently the integration and cooperation issues in SGs.

The integration concept refers to a mechanism which aims to incorporate different activities belonging to various BPs within and across organizations in a manner that the resulting BP acts as an unique holistic system. The ultimate goal of BP integration consists to streamline and orchestrate business activities. Three integration dimensions of BPs are distinguished: Data integration, interface integration and application integration.

After introducing BPs and their integration, we now focus on the knowledge representation aspect in SGs. In particular, we present KG models, which have proven to be a promising solution for addressing the challenges of managing heterogeneous, large-scale, as well as coordinating disparate activities in SG environments.

2.3 Deploying Knowledge Graph Models in SG

Since information and communication technologies (ICT) are widely used to manage Smart Grid (SG) systems, a large volume of data is continuously generated and stored within the underlying information systems of organizations. These data represent a valuable resource that requires careful management and effective exploitation. At the same time, various approaches and technologies have been developed to represent and reason about the complex data produced by SG systems more efficiently. The main objective of these approaches is to enhance operational efficiency, improve fault detection, and support the automated management of the entire grid, from power generation to the end consumer. In recent years, the field of knowledge representation has experienced significant progress, particularly with the rapid advancement of artificial intelligence. More recently, Knowledge Graphs (KGs) have transformed this domain by offering structured and scalable frameworks capable of integrating large volumes of heterogeneous data and their relationships. As a result, KGs have emerged as a highly promising technology for advanced data management.

Hereafter, the definition of a Kg is specified.

Definition 1. *A knowledge graph (**KG**) is defined as "a collection of interconnected specifications of entities (objects) and events (concepts)" [8].*

Thus, a KG incorporates the specification of data through links and semantic metadata. This model aims to construct an advanced network that integrates, unifies, and shares data in a contextualized way. According to [8], the rapidity of data storage and processing, conjugated with the ability to identify correlations between data, involves the construction of complicated queries on the manipulated data. Furthermore, KG have proven that they are very suitable data structures for addressing such concerns. In the SG field, KGs are widely used to model, store, and process data in order to improve the performance level of energy management by offering an advanced decision-support system (DSS). To better understand how KGs can be applied, we illustrate their usage with the following example.

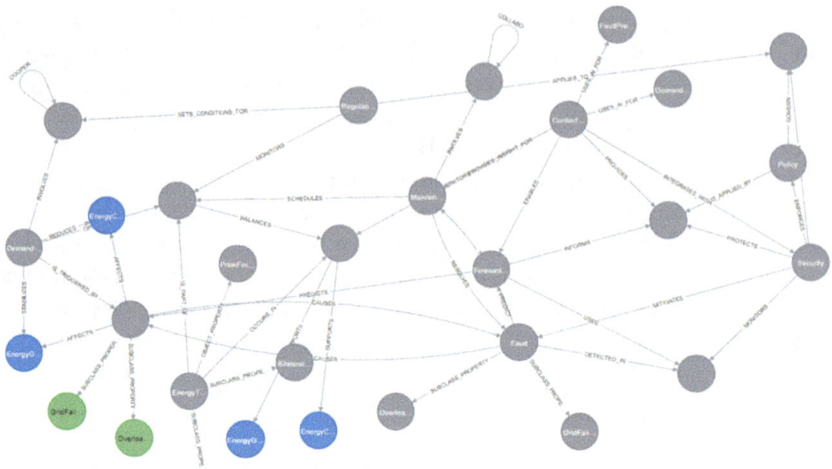

Fig. 4. Example of a KG representing the energy domain

Example 2. Figure 4 below depicts an overview of the key concepts and interactions within the SG energy domain. As it observed in the figure, nodes represent the entities of the energy domain (*i.e.; service, device, user, ...*) with their descriptive properties (*i.e.; forecast consumption, forecast weather, ...*), while arcs express the relationships between these entities.

Based on the aforementioned background, the following section provides a state-of-the-art overview of existing approaches, highlighting research efforts that address integration and cooperation in SGs.

3 State-of-the-Art of BP Integration Approaches

To address the integration issues, different techniques and approaches are proposed in the research literature which is very rich in works that deal in a consistent manner with different BP integration aspects. The most deployed techniques in this context range from Enterprise Resource Planing (ERP) [9], Enterprise Application Integration (EAI) [10], Enterprise serial Bus (ESB) [11] to the recent technology of Web services and its underlying Service Oriented Computing (SOC) paradigm [12].

The deep exploration of the research literature related to the deployment of KGs in the context of BPs, modeled as BPMN diagrams, especially in the context of SG depicts that only few research works have developed the concepts of the KG and BPMN at the same time. Here after, the relevant works in this context are analyzed and critically evaluated.

[13] developed a process-aware IIoT knowledge graph, focusing on the seamless integration of IoT data into process models, particularly within SG contexts.

Similarly, [14] and [15] emphasized the integration of dynamic execution representations into BPMN-based knowledge graphs, enabling real-time process connectivity. [16] advanced the integration of explainable AI with knowledge graphs for process analysis, enhancing transparency in BPM analytics through unified frameworks. [17] provided best practices for integrating enterprise knowledge graphs, critical for scalability in smart manufacturing and, by extension, SGs. [18] proposed higher-order transformations for integrating formal BPMN analysis with graph transformation systems, establishing a strong methodological basis for BPMN-to-KG integration and reasoning. [19] explored the integration of IoT-aware BPMN modeling, highlighting the necessity of incorporating and visualizing physical-world data within BPMN structures.

3.1 Using KGs in SGs

In what follows, we describe the works encountered in SG systems and having benefits from knowledge representation models, in general and KG ones, in particular.

- In [20] a transformation tool from BPMN2.0 process models into KGs is proposed. The developed system BPMN2KG is applied in smart manufacturing systems on quality management in plastic injection molding for the automotive sector.
- The authors in [21] suggested a project-based technical approach for building a hybrid knowledge base with reasoning rules, based on the interaction of agile modeling techniques and semantic technology.
- A modeled knowledge base for process analysis is described in [22] to automatically check the weaknesses of BPMN models.
- A framework for the interoperability of a knowledge graph-based is suggested in [16]. The proposed approach explains deeply AI approach deployed for business process analysis.
- The authors in [17] examine the selection of competitive, multi-source, intelligent business processes, and data collecting techniques for industrial enterprise knowledge-graph creation that enable smart manufacturing.
- In order to construct process-centric KGs that capture both BPM models and their actual implementations, a novel technique is described in [15]. Further, an in-depth study that focuses on use cases involving human-AI collaboration from a smart manufacturing project in Europe is developed.
- A KG that facilitates semantic interoperability in SGs and future electricity markets is proposed in [14]. It addresses a comprehensive approach for creating KGs.
- In [23] an energy KG, which provides the semantic integration of various energy services is proposed. In the same trend, in [24] an energy KG is proposed and the scenario specification of energy services is evaluated.

This first series of works addresses the general problem of using KGs in SGs. It's worth noting that, KG are not the only knowledge representation models used in SG. In fact, others categories of models are used in the research literature to address the knowledge representation issue in SG grid. Particularly, semantic networks [25,26] and ontologie [27,28] have been intensively used.

3.2 Existing Approaches to BP Integration and Cooperation

Hereafter, we examine the works having deal with BPs refinements and their associated diagrams to achieve integration and cooperation objectives.

- In [29] a knowledge-graph-driven method for intelligent decision-making is proposed on power communication equipment faults, enhancing fault retrieval and personalized recommendations. To promote the discovery of valuable data assets in big data managed by SGs, through data fusion inside and outside the grid, the authors in [30] proposed an intelligent classification that integrates KG and Internet of Things concepts in the same framework. The conceived framework is very suitable for integration and cooperation issues.
- The authors in [31] propose a process patterns which are used for achieving interoperability in BP integration projects and a framework for selecting and applying process patterns. It consists of a reusable templates that capture the best practices for integrating BPs. Furthermore, these templates provide a common language for describing and implementing integration, and they can help to reduce the cost, time, and risk of BP integration projects.
- A framework that emphasize the importance of communication and collaboration between stakeholders during the integration process is developed in [32]. The proposed framework highlights the need for flexible and adaptable description to achieve integration requirements.

As it can be observed in the previous analysis of the literature review, the challenge of integrating BPs belonging to the different stockholders of the SG have not benefits from all the interest it deserves. Furthermore, the cooperation issue was relegated to a second class level. To overcome this limitations, by improving BPs specifications of SGs, we propose in what follows a comprehensive framework that leverages KGs and BPMN diagrams into a hybrid model for ensuring integration and cooperation of BPs.

4 The Framework for BPs Integration and Cooperation

This section is dedicated to the description of the proposed approach for handling BPs integration and cooperation issues in SG systems.

4.1 Description of the Approach

BPs are conventionally modeled using the Business Process Model and Notation (BPMN) standard, and in our approach this formalism is employed to represent BP specifications. To address the limitations of existing approaches to BP integration and cooperation, we propose a paradigm allowing to enriching BP specifications by embedding execution context information directly into their corresponding Business Process Model and Notation (BPMN) models. In this perspective, our approach extends BPMN models of involved BPs of the SG with descriptive data extracted from the associated KG, thereby providing a context-aware specification that goes beyond traditional modeling of BPs. Specifically, the KG serves as a contextual layer for the BPMN model by dynamically extending tasks description with additional, context-specific properties.

Building on the two previous conceptual elements (*i.e.; the BPMN diagram and the KG structure*), the proposed approach introduces a mapping mechanism that enables BPMN descriptions (*i.e., BP models*) to directly interact with the associated KG to capture features required for integration and cooperation tasks. In this way, the execution context required by a specific BPMN activity is extracted from the KG features. Consequently, BPMN workflows can be linked to graph entities through annotations and descriptive attributes, thereby establishing a dynamic interaction between the workflow and the real-time data represented in the KG.

As an immediate consequence of the linking between static structures (*i.e., the KG*) with dynamic ones (*i.e., BPMN diagrams*), the overall functioning of SG systems is significantly enhanced. More formally, when a BPMN process requires resource allocation, the KG is able to automatically provide information on the available resources according to their current status.

To carry out the mapping mechanism, we establish a set of correspondences between BPMN elements and KG components, summarized below:

- **BPMN tasks** are mapped to relationships in the graph.
- **BPMN events** are mapped to the nodes of the KG.
- **BPMN actors**, such as operators, customers, autonomous agents correspond to nodes in the graph.

Figure 5 depicts the different steps of the proposed approach for BP integration and cooperation.

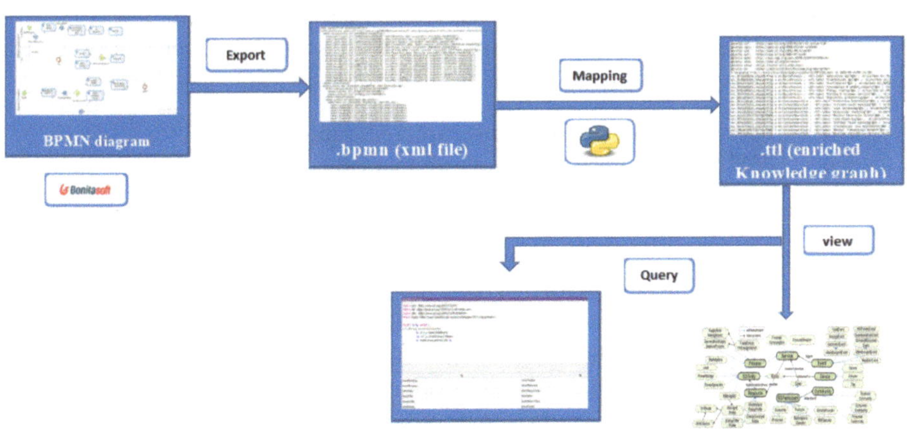

Fig. 5. Architecture and functionalities of the proposed approach

As it can be observed in the previous figure, the system architecture illustrates a comprehensive pipeline for transforming and enriching Business Process Model and Notation (BPMN) diagrams into a knowledge graph, facilitating advanced process analysis and querying. The procedure is starts with a BPMN diagram created using BonitaSoft [33], a platform for Business Process Management (BPM), which is exported into an XML file, encapsulating the process structure and metadata. This XML file undergoes a mapping phase, likely implemented using a Python-based tool, to convert the BPMN data into an enriched .ttl (Turtle) file representing a KG. The KG, visualized as a network of interconnected nodes (e.g., process, service, event, actor), enables the representation of complex relationships and entities within the considered BP. Subsequently, a query phase allows users to interact with this graph through a structured interface, supporting detailed analysis and retrieval of process-related information. This architecture underscores a robust integration of process modeling, semantic enrichment, and query-driven insights, offering a scalable framework for managing and optimizing BPs in dynamic environments.

The previous figure highlights the fact that the KG is a kernel of the conceived system and it plays a crucial role. Since it includes relevant properties of actors and devices of the SG, it facilitates automation of decision-making within BPMN models during their progression. Furthermore, it allows real-time monitoring of processes, tracking relationships between tasks, departments, and partners of the SG. On the other hand, it's forth noting that data synchronization and using shared entities ensure synchronization between data in both the BPMN engine and the KG by using shared unique identifiers and implementing a bidirectional update strategy. For example, all changes in BPMN (e.g., equipment status) should be reflected in the graph, and vice versa.

4.2 Advantages of the Approach

By enriching BPMN tasks and events with KG descriptive properties, a reactive and intelligent system is built in the SG. Hence, the adaptive and enriched BPs can satisfy real-time business data. Further, this descriptive extension offers the following advantages for the SG managers and decision-makers.

- **Context-aware process automation**: real-time data included in the KG enriches BPMN tasks with relevant information useful for the SG management.
- **Enhanced decision-making**: an automated decision-making process, based on contextual knowledge extracted from the graph, is realized.
- **Scalability**: the developed approach is flexible and can be applied to complex and interconnected corporate environments.
- **Customization and Flexibility**: the KG can be updated with new nodes and relationships expressing actors properties. Hence, BPMN processes can evolve and adapt to new context without needing a complete redesign from scratch.

5 Implementation

To show the feasibility and the applicability of the conceived approach, it has been implemented and experimented using the following adequate software environments and tools.

5.1 Used Technologies and Tools

Different tools and technologies are used to implement the proposed approach.

- The engine Bonita studio community edition 2023.2-u0 [34] is used as a BPM software that manages BPs life-cycle. It provides a Business Process Management (BPM) framework and a workflow automation platform.
- The Graph Database Management System (DBMS) Neo4j [35] is deployed for modeling entities and relationships in the SG system.
- The open-source message broker software RabbitMQ [36] is used to ensure the exchange of messages between different systems, applications, or services.
- Other associated technologies: various technologies like Node.js, Spring Boot, or Django are used for API development, and scripting languages like JavaScript, Python, and Groovy are employed for writing connectors in BPMN.

5.2 Case Study

The Fig. 6 bellow depicts a collaborative BPMN diagram that highlights the different interaction between the different business processes and actors, each playing a critical role.

Here's an overview of how these processes cooperate.

1. **Consumer Process:** The process starts with verifying the energy price from the energy provider and receiving price updates from the grid operator. It includes a decision point ("High price detected?") that triggers either reducing consumption or switching to backup energy, which is a practical response to price surges. The inclusion of an "energy shortage Alert" and applying modifications to devices shows a proactive approach to managing demand during crises. The loop for receiving new prices and adjusting demand ensures continuous optimization, which is a strong feature.
2. **Energy Provider Process:** Begins with analyzing energy demand and setting dynamic pricing based on demand exceeding supply, which is a standard demand-response strategy. The decision to notify the grid operator when demand exceeds supply, followed by optimizing grid balance, reflects a collaborative approach with grid management. The process ends with sending a new price, completing the feedback loop with consumers.

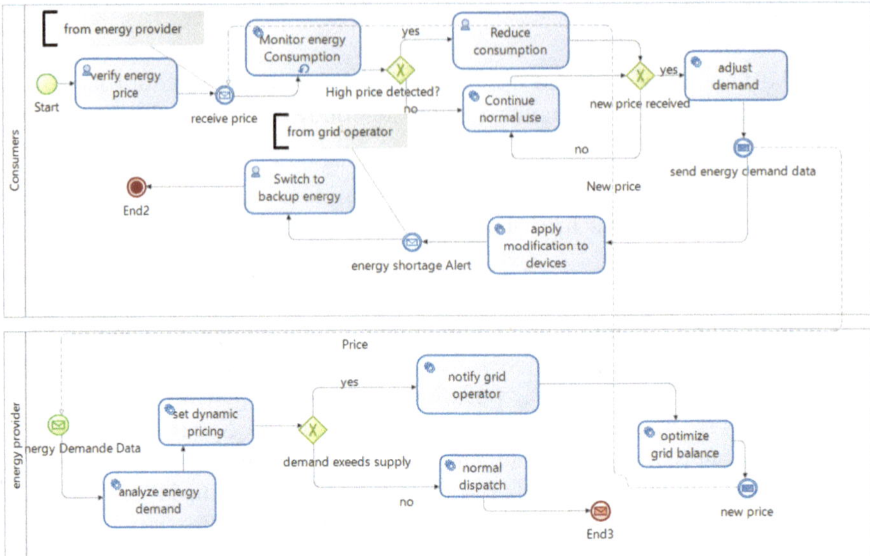

Fig. 6. Demand Response Management BP

The system effectively models a dynamic pricing and demand-response mechanism, which is critical for modern energy SG management. The dual-process approach (*Consumers and Energy Provider*) ensures a holistic view of the ecosystem. The integration of real-time price adjustments and grid balance optimization suggests a forward-thinking design, though implementation details (*e.g., algorithms, communication protocols...etc.*)

The Fig. 7 contains an excerpt of a BPMN model expressed in XML file, which defines a process model for energy price verification and demand response management. The XML elements correspond to the activities described in the workflow depicted in an earlier flowchart of Fig. 7. Each XML element is validated against the Bonita Connector schema, and outlines a collaborative process involving two primary participants: "Consumers" and "energy provider". The process initiates with a "verify energy price" user task, followed by monitoring energy consumption and decision points to either reduce consumption or continue normal use based on a "high price" condition. The inclusion of data objects such as "price" and input/output sets, along with service tasks and message flows, suggests a dynamic system that integrates real-time data and automated responses. This structured representation highlights the use of standardized loop characteristics and Java-based expressions, indicating a robust, programmable framework suitable for IoT-enabled energy management systems.

```xml
<?xml version="1.0" encoding="UTF-8"?>
<model:definitions xmlns:xsi="http://www.w3.org/2001/XMLSchema-instance"
    xmlns:bonitaConnector="http://www.bonitasoft.org/studio/connector/definition/6.0"
    xmlns:dc="http://www.omg.org/spec/DD/20100524/DC"
    xmlns:di="http://www.omg.org/spec/BPMN/20100524/DI"
    xmlns:di_1="http://www.omg.org/spec/DD/20100524/DI"
    xmlns:java="http://jcp.org/en/jsr/detail?id=270"
    xmlns:model="http://www.omg.org/spec/BPMN/20100524/MODEL"
    xsi:schemaLocation="schemaLocation http://www.omg.org/spec/BPMN/20100524/MODEL schemas/BPMN20.xsd"
    exporter="BonitaSoft"
    exporterVersion="10.2.0"
    expressionLanguage="http://groovy.apache.org/"
    targetNamespace="http://bonitasoft.com/_Aud2kPjZEe-904qtbcOItg">

  <model:collaboration id="_Aud2kPjZEe-904qtbcOItg">
    <model:participant id="_8NwisJzzEfCoG5jexwuEUQ" name="Consumers" processRef="_Aud2kfjZEe-904qtbcOItg"/>
    <model:participant id="_Aud2q_jZEe-904qtbcOItg" name="Employee actor"/>
    <model:participant id="_8OBBYZzzEfCoG5jexwuEUQ" name="energy provider" processRef="_3me14PzfEe-XCfMYiNrQMA"/>
    <model:messageFlow id="_tzueMPzmEe-XCfMYiNrQMA" sourceRef="_UCa4wPzlEe-XCfMYiNrQMA" targetRef="_hXLM4BlAEfCFN-jIwnjV3A"/>
    <model:messageFlow id="_wBuU4PzoEe-XCfMYiNrQMA" sourceRef="_fTfScPznEe-XCfMYiNrQMA" targetRef="_lvCZkPznEe-XCfMYiNrQMA"/>
  </model:collaboration>

  <!-- =================== Consumers =================== -->
  <model:process id="_Aud2kfjZEe-904qtbcOItg" name="Consumers">
    <model:startEvent id="_Aud2k_jZEe-904qtbcOItg" name="Start"/>
    <model:userTask id="_Aud2l_jZEe-904qtbcOItg" name="verify energy price"/>
    <model:serviceTask id="_1H9sYPzcEe-XCfMYiNrQMA" name="Monitor energy Consumption"/>
    <model:userTask id="_HgBo0PzdEe-XCfMYiNrQMA" name="Reduce consumption"/>
    <model:serviceTask id="_PaF8YPzdEe-XCfMYiNrQMA" name="Continue normal use"/>
    <model:exclusiveGateway id="_qrqQIPzdEe-XCfMYiNrQMA" name="High price detected?"/>
    <model:serviceTask id="_0OvI4PzdEe-XCfMYiNrQMA" name="adjust demand"/>
```


Fig. 7. An excerpt of The BPMN XML file

6 Conclusion and Future Works

In this paper a novel approach that integrates BPMN diagrams with KGs is proposed. The developed comprehensive framework enhances SG management by integrating KG data into BPMN models. This hybrid model enables agile decision-making, operational efficiency, and sustainable energy practices. The proposed approach was implemented using BPMN language, the Neo4j environment, and other adequate software tools and technologies. The first experimental results demonstrate that the integration and cooperation mechanisms are considerably facilitated and enhanced.

In our future works, we plan to exploit artificial intelligence and machine learning techniques for managing and monitoring the SG.

References

1. Afifi, C., Khebizi, A.: A road-map for mining business process models via artificial intelligence techniques. Int. J. IAM **5** (2022)
2. Jagannathan, V.: A context-aware approach for automation of end-user elements in the smart grid. Ph.D. thesis, University of California, Irvine (2016)
3. Souza e Silva, N., Castro, R., Ferrão, P.: Smart grids in the context of smart cities: a literature review and gap analysis. Energies **18**(5), 1186 (2025)
4. Smart Grid Architecture Overview, chapter 1, pp. 1–43. Wiley, Hoboken (2021)
5. Houda, D., et al.: Big data management in smart grid: concepts, requirements and implementation. J. Big Data **4** (2017)
6. Dumas, M., Rosa, M.L., Mendling, J., Reijers, H.A.: Fundamentals of Business Process Management. Springer, Heidelberg (2013)
7. Suri, K., Gaaloul, W., Cuccuru, A., Gerard, S.: Semantic framework for energy-aware resource management of IoT in business processes. Int. J. Syst. Serv.-Oriented Eng. **8**(1), 21–43 (2018)

8. Macedo, V., Thurler, L., Dias, E., Cavalcanti, M.: Knowledge graph: a strategy for knowledge management? (2023)
9. Chesang, A.: The effectiveness of enterprise resource planning systems in knowledge management. J. Inf. Knowl. Manage. (2024)
10. Formalisms for EAI: a survey of methodologies (2023)
11. Aziz, O., et al.: Research trends in enterprise service bus (ESB) applications: a systematic mapping study. IEEE Access (2020)
12. Kumara, I., van den Heuvel, W.-J., Tamburri, D.A.: QSOC: quantum service-oriented computing (2021)
13. Diamantini, C., Mircoli, A., Potena, D., Storti, E.: Process-aware IIoT knowledge graph: a semantic model for industrial IoT integration and analytics. Futur. Gener. Comput. Syst. **139**, 224–238 (2023)
14. Popadić, D., Iglesias, E., Sakor, A., Janev, V., Vidal, M.-E.: Toward a solution for an energy knowledge graph, pp. 3–12 (2023)
15. Franz, K., et al.: On the representation of dynamic BPMN process executions in knowledge graphs, pp. 97–105 (2023)
16. Füßl, A., et al.: Knowledge graph-based explainable artificial intelligence for business process analysis. Int. J. Semant. Comput. **17**, 173–197 (2023)
17. Jawad, M.S., et al.: Adoption of knowledge-graph best development practices for scalable and optimized manufacturing processes. MethodsX **10**, 102124 (2023)
18. Kräuter, T., Rutle, A., König, H., Lamo, Y.: A higher-order transformation approach to the formalization and analysis of BPMN using graph transformation systems. Logical Methods Comput. Sci. **20**(4) (2024)
19. Kirikkayis, Y., Winter, M., Reichert, M.: A user study on modeling IoT-aware processes with BPMN 2.0. Information **15**(4) (2024)
20. Stefan, B., et al.: Automated process knowledge graph construction from BPMN models, pp. 32–47 (2022)
21. Cinpoeru, M., et al.: Model-driven context configuration in BPM systems: an approach based on knowledge graphs. In: International Workshop on Bibliometric-enhanced Information Retrieval (2019)
22. Füßl, A., et al.: An inferential knowledge model for the digitalisation and automation of business process analysis. In: Wirtschaftsinformatik (2020)
23. Sejin, C., et al.: Knowledge graph modeling for semantic integration of energy services, pp. 732–735 (2018)
24. Chun, S., Jung, J., Jin, X., Seo, S., Lee, K.-H.: Designing an integrated knowledge graph for smart energy services. J. Supercomput. **76**(10), 8058–8085 (2018). https://doi.org/10.1007/s11227-018-2672-3
25. Sarica, S., Luo, J.: Design knowledge representation with technology semantic network. **1** (2021)
26. Chaccour, C., et al.: Less data, more knowledge: Building next generation semantic communication networks. arXiv, abs/2211.14343 (2022)
27. Artem, S., et al.: Towards ontology-based network model management and data integration for smart grids, pp. 1–6 (2017)
28. Joong, K.H., et al.: A comprehensive review of practical issues for interoperability using the cim model in smart grids. Energies **13**(6), 1435 (2020)
29. Wentao, L., et al.: Application of knowledge graph in smart grid fault diagnosis. Appl. Math. Nonlinear Sci. **8** (2022)
30. Ma, L., Dou, C., Hao, T.: Grid data asset relationship and intelligent classification integrating knowledge graph and internet of things. Appl. Math. Nonlinear Sci. **9** (2023)

31. Moufida, A., et al.: Business process integration: how to achieve interoperability through process pattern (2017)
32. Morrison, E., Menzies, A., Koliadis, G., Ghose, A.: Business process integration: method and analysis. Faculty of Informatics - Papers, 96 (2009)
33. Bonitasoft. Bonita BPM (2009). https://www.bonitasoft.com
34. Chabanoles, N., et al.: Bonita BPM: an innovative BPM-based application development platform to build engaging, user-oriented business applications (2015)
35. Performance of graph database management systems as route planning solutions for different data and usage characteristics (2023)
36. Ćatović, A., Buzađija, N., Lemeš, S.: Microservice development using rabbitmq message broker. Sci. Eng. Technol. (2022)

Sentiment-Enhanced Hybrid Models for Arabic Fake News Detection

Chatta Asmaa[1(✉)], Bouakkaz Mustapha[2], and Latreche Boubakeur[1]

[1] Computer Science and Applied Artificial Intelligence Laboratory, University Ziane Achour of Djelfa, Djelfa, Algeria
{Asma.chatta,b.latreche}@univ-djelfa.dz

[2] Laboratoire d'Informatique et de Mathématique, University of Laghouat, Laghouat, Algeria
m.bouakkaz@lagh-univ.dz

Abstract. The rapid spread of misleading Arabic news stories threatens societal harmony and facts-based choices, yet determining what is truthful remains complicated because of the language's distinctive linguistic qualities. Current techniques generally focus solely on surface textual patterns, neglecting the combined potential of merging advanced sentiment metrics with hybrid machine learning and deep learning frameworks. We introduce an innovative system for identifying deceptive Arabic news that combines sentiment analysis and several approaches, including both machine learning (RF, Naive Bayes, SVM, LR) as well as deep learning (LSTM, RNN). Experiments on a comprehensive corpus of over 6100 Arabic news show that our sentiment-analysis enhanced hybrid model achieves notably superior performance, with SVM attaining 96% accuracy, surpassing all baseline methods. These insights provide both a practical solution for detecting Arabic misinformation and a framework that can be adapted to other under-resourced languages.

Keywords: Fake news detection · Arabic NLP · sentiment analysis · lexicon-based features · hybrid models · LSTM

1 Introduction

The credibility of data in the Middle East and North Africa is now seriously threatened by the quick spread of false information in Arabic-language media. According to recent studies, false information spreads more quickly in Arabic social networks than on English-language ones [1]. This can have especially negative effects during public health emergencies [2] and political elections [3]. The complex nature of Arabic morphology a single root can produce dozens of derived words and the dearth of trustworthy natural language processing (NLP) tools for different Arabic dialects exacerbate this issue [4]. The three primary types of current Arabic fake news detection techniques every one come with major drawbacks. According to [5], machine learning techniques such as Support Vector Machines (SVM) struggle with semantic nuances but achieve moderate accuracy (78–84%). Although sentiment analysis methods are helpful for identifying

Z. Kouahla et al. (Eds.): IAM 2025, LNNS 1973, pp. 132–149, 2026.
https://doi.org/10.1007/978-3-032-26031-4_11

tones, they frequently depend on simplistic positive/negative categorizations [6]. For Arabic applications, deep learning models require unfeasible large datasets (>100,000 samples), despite their promising nature [7]. Importantly, there is a significant research gap because none of these methods successfully integrate multiple machine learning architectures with fine-grained sentiment analysis. This study makes four key contributions to address these limitations:

- We introduce a novel sentiment ratio metric derived from comprehensive Arabic lexicons, calculated as the normalized difference between positive and negative word counts in each article.
- We implement and compare six distinct models independently:
  - Traditional machine learning: Naïve Bayes (NB), Support Vector Machine (SVM), and Logistic Regression (LR)
  - Deep learning: Long Short-Term Memory (LSTM) and Recurrent Neural Network (RNN) architectures
- We demonstrate that incorporating sentiment ratios improves classification accuracy across all models by 3–6 percentage points, with SVM achieving the highest performance at 96.3
- We show that our approach achieves state-of-the-art results while using only 6,100 samples, making it practical for real-world applications where large labeled datasets are unavailable.

The AFND dataset [8], annotated news articles from various Arabic sources, is used in our methodology. Every article underwent several steps of processing: (1) preprocessing and normalization of the text for the particularities of Arabic, (2) sentiment analysis using acknowledged Arabic lexicons, and (3) independent training and evaluation of all six models. When sentiment features are added, the results show progressive improvement, with the SVM model performing especially well in identifying successive patterns of dishonesty. Both research and practice will be significantly impacted by these findings. Academically, they show how sentiment-aware methods are useful for Arabic natural language processing tasks, especially when resources are limited. In practice, Arabic fact-checking groups are already adopting our techniques, which may influence platform moderation guidelines. The rest of this paper describes related work (Sect. 2) in detection of fake news using sentiment analysis, our methodology (Sect. 3), experimental results (Sect. 4), and conclusions with future research directions (Sect. 5).

2 Related Works

This section explores recent research that integrates sentiment-based techniques with machine learning and deep learning models to enhance the accuracy of fake news detection across various datasets and domains.

[9] performed research on the detection of fake news in English based on a rule-based sentiment analysis (SA). To enable misinformation detection, they utilized a certain deep neural network. They evaluated their research on two splits of FakeNewsNet: PolitiFact is a political news fact-checking website that

verifies the truth of claims made by elected officials and other public figures. Both datasets were scraped from Twitter for collecting user comments. This subset contains 432 fake news posts and 624 real ones. GossipCop is a website solely dedicated to fact-checking that specifically addresses celebrity news facts. It verifies the truthfulness of entertainment stories published on magazines, newspapers, and social media platforms to classify them as true or false. The dataset consists of 5,323 samples of false information and 16,817 true news reports. Their model's performance was calculated through the F1-score, a score that is a combination of both recall and precision. Their F1-score in the PolitiFact dataset was 0.7724, and their F1-score in the GossipCop dataset was 0.8042. These are indicators of the effectiveness of their approach, which manifested good performance in fake news detection on both datasets. [10] conducted an experimental investigation aimed at detecting deceptive information in the English language through a lexicon-based approach combined with Support Vector Machines (SVM). Their study evaluated the model's performance on two different datasets: Silverman's Buzzfeed Political News dataset, where the SVM achieved an accuracy (Acc) of 0.77, and an ad-hoc collection of news articles, for which the accuracy was slightly lower (0.71). These results indicate that SVM performance can be extremely volatile depending on the dataset, reflecting the challenge of generalizing fake news detection across sources. [11] presented a study on rumor detection for the English language based on a lexicon-based sentiment analysis (SA) approach. In their attempt to identify false news, they employed two separate algorithms: the Support Vector Machine (SVM) and another hybrid neural network model known as LSTM-HAN (Long Short-Term Memory with Hierarchical Attention Network). The research by the authors was founded on the Rumors dataset, which comprised 5,800 tweets corresponding to 533 significant worldwide events that had been largely reported in electronic, print, and mainstream media. These events took place at different time periods between August 2014 and March 2015. Through the SVM algorithm, they recorded an accuracy (Acc), precision (P), recall (R), and F1-score, all of which were 0.86. The application of the LSTM-HAN model achieved an accuracy (Acc) of 0.86, precision (P) of 0.86, recall (R) of 0.82, and F1-score of 0.84. The outcomes show the success of both approaches in rumor detection, while also reflecting minimal differences in recall and F1-score specific to the LSTM-HAN model, suggesting divergence in the capacity of each model to deal with the complexity of text data. [12] conducted a study of Arabic fake news detection using different approaches. To conduct sentiment analysis, they utilized an Arabic lexicon and were therefore able to extract the emotional content of the texts. They then applied different classification techniques for fake news detection. With the Support Vector Machine (SVM) algorithm, they had an accuracy (Acc) of 0.64 in the "Hajj" dataset, which contained 700 real news articles: 244 Saudi Arabian, 220 Egyptian, and 236 Jordanian. With the Random Forest (RF) method, they had a lower accuracy of 0.61. And with the Naive Bayes (NB) algorithm, they had an accuracy of 0.496. [13] has conducted an investigation of fake news detection in English during the COVID-19 pandemic. They employed a lexicon-based sentiment analysis

(SA) and utilized some classification models for fake news detection: AdaBoost, Decision Tree, and K-Nearest Neighbors (KNN). Their experiment was conducted on a dataset completely devoted to fake news about COVID-19, The dataset consists of 586 instances of actual news and 578 instances of imitation news, both adding up to far more than 1,100 instances of news stories and social media postings about COVID-19. The actual news instances were gathered from sources deemed reliable, including Harvard Health Publishing, the World Health Organization (WHO), the Centers for Disease Control and Prevention (CDC), and The New York Times, to name but a few. On the other hand, instances of fabricated news were collected from social media platforms—namely Facebook status updates—and healthcare websites. The performance of each algorithm was calculated in terms of accuracy (Acc). The AdaBoost algorithm achieved an accuracy of 0.82, the Decision Tree achieved 0.77, and the KNN algorithm achieved 0.69. These results indicate that AdaBoost performed better than the other methods in terms of accuracy to detect fabricated news regarding COVID-19. [14] proposed an SA method based on an Arabic lexicon that effectively recognizes emotional nuances and polarities in the texts. For fake news detection itself, they utilized the Naive Bayes algorithm, which is well known for being simple and capable of managing enormous amounts of textual data. The corpus for this study was an ad hoc collection of tweets gathered from Twitter. They created a ground truth data set of 1,000 unique tweets from 700 unique accounts on a political topic (the Islamic State of Iraq and the Levant, ISIL) in the Arabic language. The credibility of the tweets was manually verified and analyzed. They took into account over 20 user and tweet status features. All the tweets having no pertinent news content were discarded for ensuring relevance to the study topic. The tweets were specially selected for relevance and diversity on the topic. The performance with this setup was remarkable: precision (P) of 0.8624, recall (R) of 0.988, F1-score of 0.926, and accuracy (Acc) of 0.903. The outcome demonstrated the model's ability to effectively detect fake news from real information in the Arabic language context as well as recognize the sentiments conveyed for analysis. [15] conducted an experiment on identification of fake news in English using a lexicon-based sentiment analysis (SA) method. For the identification of fake news, the K-Nearest Neighbors (KNN) algorithm was used. Their experiment utilized an ad hoc dataset of tweets collected from Twitter. They created a ground truth dataset of 1,000 unique tweets from 700 unique accounts on a political topic (the Islamic State of Iraq and the Levant, ISIL) in Arabic. The validity of the tweets was verified and hand-checked. More than 20 user- and tweet-status based features were investigated. In order to make the material applicable to their study, they removed all tweets that were not holding useful news. The precision obtained in this configuration was an accuracy (Acc) of 0.66. The results show that KNN algorithm with lexicon-based sentiment analysis can moderately detect fake news in the case of English tweets. [16] conducted an English fake news detection experiment with Naive Bayes-based sentiment analysis. They also employed the Random Forest algorithm to detect fake news. They relied on the LIAR dataset, which is a widely used benchmark dataset. The performance of their models was checked with the "Fake vs Real News" project. They achieved an accuracy (Acc) of 0.63 using the Naive Bayes algorithm, while the Random Forest algorithm achieved a higher accuracy (Acc) of 0.88. These results prove the better performance of the Random Forest algorithm in distinguishing between fake and real news in English compared to the Naive Bayes algorithm. [17] conducted a study on the Emotion Cognizance Improves Health Fake News Identification dataset in English using a lexicon-based approach. The authors used a variety of machine learning algorithms, including Naive Bayes, K-Nearest Neighbors (KNN), Support Vector Machines (SVM), Random Forests, Decision Trees, AdaBoost, Convolutional Neural Networks (CNN), and Long Short-Term Memory (LSTM) networks, to classify data from the HWB dataset. The models' performance differed greatly, with Decision Trees performing the best in terms of accuracy (Acc = 0.940), closely followed by AdaBoost (Acc = 0.965) and KNN (Acc = 0.925). SVM (Acc = 0.900), LSTM (Acc = 0.920), and CNN (Acc = 0.910) also showed good performance by the other models, whilst Naive Bayes performed moderately with an accuracy of 0.790. The results indicate that, among the models being tested, Decision Trees and AdaBoost are the best performing models for lexicon-based text classification in the HWB dataset. [18] experimented on English fact-checking with a lexicon-based method paired with machine learning algorithms. The experiment protocols utilized Maximum Entropy (MaxEnt) and Long Short-Term Memory (LSTM) networks for misinformation detection. The findings reported that the LSTM model gave a slightly higher F1 score (0.56) than MaxEnt (F1 = 0.55), representing a small improvement in the effectiveness of classification when applying deep learning techniques to the particular task. [19] presented a multilingual rumor detection study with a fusion of lexicon-based approaches and commercial AI tools. They employed the RumourEval-19c dataset for English data and utilized BiGRU, BERT, and NileTMRG models. The BiGRU model had an F1 score of 0.340; the BERT model, however, had a slight boost with an F1 score of 0.346, whereas the NileTMRG model had an F1 score of 0.342. For their Chinese section, the authors employed the Weibo-20 dataset along with an identical set of models, including BiGRU (F1 = 0.855), BERT (F1 = 0.867), and HSA-BLSTM, the latter of which performed better with an F1 score of 0.908. The findings indicate that although BERT provides a slight improvement over BiGRU in both English and Chinese datasets, the HSA-BLSTM model outshines all other methods when applied to Chinese text. This depicts the power of tailor-made architectures in addressing more complicated linguistic features. [20] conducted an experiment in fake news detection in Italian using a commercial tool (Dandelion API) and various machine learning models. The experiments were conducted on an ad-hoc dataset collected from Facebook, experimenting with models such as Linear Regression, Logistic Regression, K-Nearest Neighbors (KNN), and Decision Trees. The Logistic Regression model performed best with the highest precision (P = 0.91), recall (R = 0.91), and F1 score (0.91), while having a comparatively low false positive rate (FPR = 0.08). Linear Regression also performed exceptionally well with P = 0.90, R =

0.90, F1 = 0.90, and FPR = 0.11. Decision Trees performed a bit worse than Logistic Regression, achieving P = 0.89, R = 0.89, F1 = 0.89, and FPR = 0.12, whereas KNN performed the worst, achieving P = 0.87, R = 0.87, F1 = 0.87, and FPR = 0.13. These reflect how logistic regression performs for Italian fake news detection when combined with commercial text analysis software. [21] conducted an experiment on English misinformation detection with a Ranking SVM model. They employed an ad-hoc dataset collected from Twitter and focused on detecting the credibility of tweets. This approach leverages the ranking capability of SVM in giving precedence to more credible information over potentially misleading information and is, hence, highly suitable for fast-paced social media environments. [22] presented an English language fact-checking study based on a lexicon-driven method coupled with Conditional Random Fields (CRF) for the identification of falsity. In their experimental research, they employed the Snopes dataset and recorded an accuracy (Acc) rate of 84.02% and an area under the curve (AUC) score of 0.86. From these findings, it is apparent that CRF models are competent in grasping intricate linguistic features applicable to the fact verification field. [23] conducted research on the detection of English misinformation through the use of a commercial tool (Alchemy) coupled with a Support Vector Machine (SVM) model. The experiments were carried out on a specifically designed dataset derived from United States presidential debates transcripts using Alchemy's capability for the extraction process. The approach set out to outline the differences in linguistic form and rhetorical schemata generally encompassed in political communication and present the best possible means of detection of deceptive utterances in this particular setting. [24] performed a study on English misinformation identification through a rule-based system. The work was grounded on the BuzzFace dataset and encompassed various machine learning models such as K-Nearest Neighbors (KNN), Naive Bayes, Random Forest, Support Vector Machines (SVM), and XGBoost. The performance varied over these models, with XGBoost registering the highest AUC (0.86) and F1 score (0.81), followed by Random Forest with an AUC of 0.85 and an F1 score of 0.81. In comparison to this, SVM registered an AUC of 0.79 and an F1 score of 0.76, and KNN registered an AUC of 0.80 and an F1 score of 0.75. Naive Bayes, while less so, also did reasonably well with an AUC of 0.72 and an F1 score of 0.75. These results indicate that XGBoost and Random Forest are the better models among those tried for rule-based fake news detection for the BuzzFace dataset. [25] experimented to detect fake news in the English language based on a TI-CNN (Text-Image Convolutional Neural Network) model. Their experiment used the BS Detector dataset and reported excellent performance with a precision (P) of 0.9220, recall (R) of 0.9277, and F1 score of 0.9210. This shows that the TI-CNN model is extremely good at detecting fake news using both textual features and possibly image-related features to improve the correctness of classification.

3 Methodology

As seen in Fig. 1, the basic ideas of the suggested models have been covered in the parts that follow.

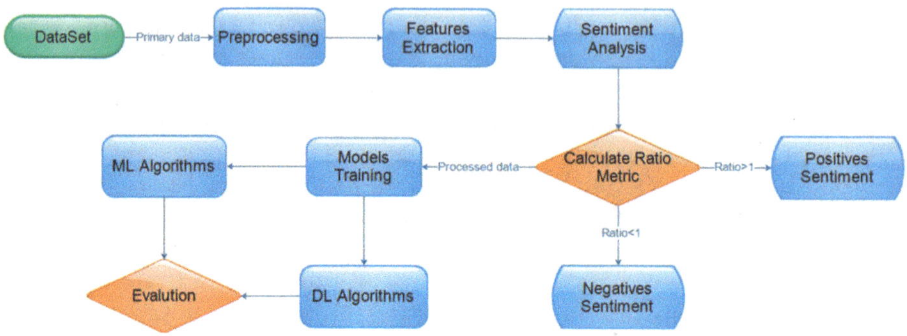

Fig. 1. Research methodology

3.1 Dataset

In this study, we applied utilisation of a subset of the Arabic Fake News Dataset (AFND) [8], which is an extensive collection of 606,912 Arabic news articles that were scraped over a six-month period from 134 public news websites in 19 different Arab countries. The dataset ensures wide geographic and thematic representation by drawing from a variety of sources. The chosen news sites cover the following nations: Jordan, Lebanon, Egypt, Iraq, United Arab Emirates, Saudi Arabia, Kuwait, Qatar, Bahrain, Syria, Oman, Tunisia, Morocco, Algeria, Libya, Palestine, Sudan, Yemen, and Mauritania, in order for a balanced distribution for reliable analysis, our curated subset of 6,175 news articles has 3,129 classified as fake ("1") and 3,046 as real ("0"). The label distribution is shown in Fig. 2, underscoring the dataset's applicability for identifying false information in Arabic media.

3.2 Data Preprocessing

Several cleaning techniques are used during the Arabic text data preprocessing phase to get the data ready for further analysis. First, since they can interfere with the analysis, special characters, punctuation, and numbers must be eliminated. Word standardisation follows to guarantee text consistency.

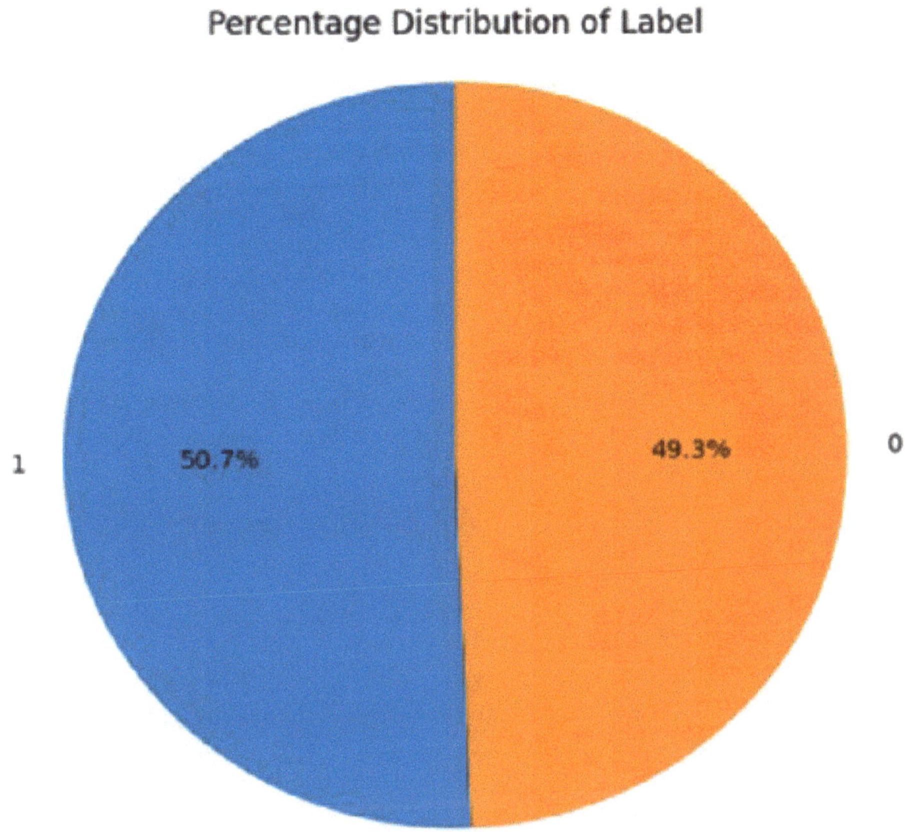

Fig. 2. Percentage distribution of label

- Tokenisation assists significantly, as it breaks the text into singular words or tokens.
- Stemming is applied to bring words to their root form so that it becomes convenient to perform the analytical process by gathering morphological variants. Along with this, removal of function words like prepositions and articles is also performed to minimise extraneous noise and enhance data integrity.
- Lemmatisation is utilised to bring back words to their canonical form, taking into consideration the linguistic context for better representation. [26,27]

3.3 Data Vectorization

In that phase, we analysed our Arabic datasets using the TF-IDF and embedding techniques [28,29].

- Term Frequency-Inverse Document Frequency By taking into account both the terms' rarity throughout the collection and their frequency within specific documents, this method enabled us to categorise terms based on their relative importance in the corpus.
  1. Term Frequency (TF): This metric measures how frequently a term appears in a document relative to the total number of terms in that document. It represents a relative frequency that can be calculated using different methods.
  2. Inverse Document Frequency (IDF): This component evaluates a term's importance based on its distribution throughout the entire corpus. Terms that appear frequently across all documents receive low IDF scores, while rare terms are assigned higher IDF values.
  3. TF-IDF: The final TF-IDF weight for each term is calculated as the product of its TF and IDF values, effectively balancing both local and global term importance.

$$TF - IDF(t, d) = TF(t, d) \times IDF(t) \tag{1}$$

3.4 Sentiment Analysis

We used a defined list of positive and negative words from an Arabic sentiment lexicon [30] to conduct sentiment analysis on the news articles that were gathered. We determined how many positive and negative words had been included in each news article. To measure the general polarity of each article, we next used a sentiment ratio metric, which is calculated as follows in Eq. (2):

$$RATIO = \frac{Numberofnegativewords}{Numberofpositivewords} \tag{2}$$

The values greater than one suggest a mainly negative sentiment, while values less than one provide a more positive sentiment. This ratio functions as a gauge of sentiment polarity. The sentiment ratio distribution for the datasets of fake and real news is shown in Fig. 3. Fake news shows an increased and right-skewed distribution, frequently reflecting exaggerated or emotionally charged content, whereas real news tends to show lower sentiment ratios.

3.5 Machine Learning Classification Models

In this study, for classifying fake news and real news, four machine learning models have been used. These models have been discussed in detail.

Support Vector Machine. We employed a Support Vector Machine (SVM) classifier as it has been demonstrated to work effectively for text classification problems, especially in cases of high-dimensional data like TF-IDF features. The model input was TF-IDF representations of Arabic news articles, optionally

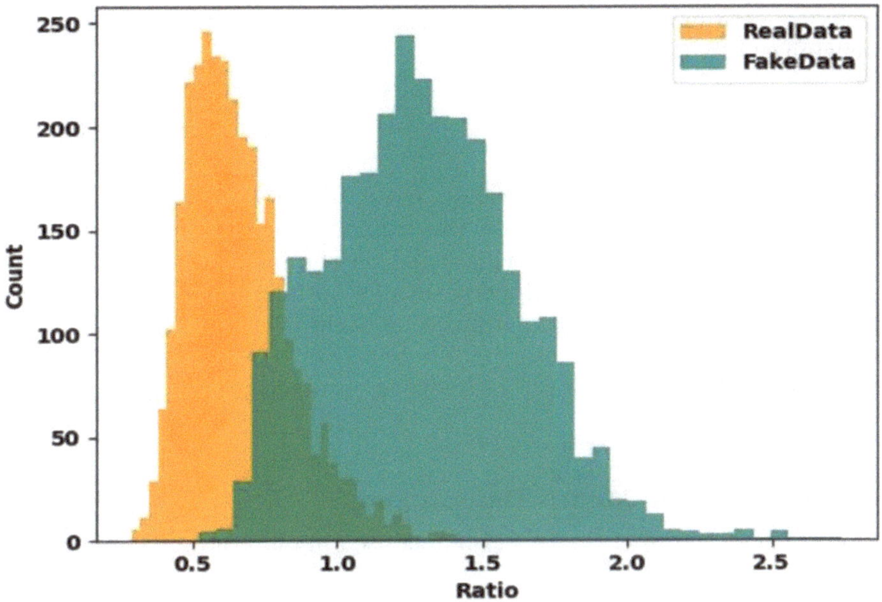

Fig. 3. Histogram showing the distribution of sentiment ratio values for real and fake news articles

enriched with sentiment polarity scores. Hyperparameter tuning was conducted using RandomizedSearchCV. The best configuration featured a regularization parameter of C = 1.57, employed a linear kernel, and utilized the gamma='scale' option, which is the default and adapts automatically to the variance of the input features. The linear kernel was suitable in this instance because of the data's sparsity and dimensionality.

Logistic Regression. Logistic Regression was chosen because of its simplicity and strong baseline performance on text classification tasks. We used TF-IDF vectors from Arabic news articles and sentiment scores as input features. Hyperparameter tuning was performed using RandomizedSearchCV. The best results were achieved with C at 7.85, a penalty of 'l2', and the 'lbfgs' solver. This setup allowed for good generalization while reducing the chances of overfitting, hence making it well-suited for sparse input datasets.

Random Forest The Random Forest classifier was tried as a non-linear ensemble approach that is adept at dealing with feature interactions and preventing overfitting. It was trained on the same TF-IDF and sentiment features. Hyperparameter optimization was conducted with RandomizedSearchCV and arrived at a configuration of 158 estimators, max depth of 48, min samples leaf of 9, min samples split of 2, and bootstrap of False. This configuration gave the model capacity to learn more complicated patterns in the data without compromising its generalizability.

Naïve Bais. As a probabilistic baseline, we employed a Multinomial Naive Bayes classifier, which is commonly utilized in the task of text classification because of its simplicity and efficiency, particularly when dealing with word count or TF-IDF features. The model was trained on TF-IDF representations of Arabic news articles, with the optional addition of sentiment polarity scores for detecting tone-related markers. Hyperparameter optimization was conducted via RandomizedSearchCV, with the optimization of the smoothing parameter alpha being the primary focus. The best performance was achieved at an alpha value of 0.60, which allowed handling zero-frequency problems and improving overall generalization.

3.6 Deep Learning Models Description

LSTM. The structure of the LSTM model used in our study is summarized in Table 1.

Table 1. LSTM Model Architecture

Model Layer Type	Output Shape	Parameters
Input (Articles)	(None, 100)	0
Embedding (Articles)	(None, 100, 64)	128,000
Bidirectional LSTM	(None, 64)	24,960
Input (Sentiment)	(None, 100)	0
Embedding (Sentiment)	(None, 100, 64)	128,000
Bidirectional LSTM	(None, 64)	24,960
Input (Ratio)	(None, 1)	0
Concatenate	(None, 129)	0
Dense (ReLU)	(None, 64)	8,320
Dropout (0.6)	(None, 64)	0
Dense (Softmax)	(None, 2)	130
Total Parameters	—	**314,370**

RNN. The structure of the RNN model is shown in Table 2.

Table 2. RNN Model Architecture

Model Layer Type	Output Shape	Parameters
Input (Articles)	(None, 100)	0
Embedding (Articles)	(None, 100, 100)	300,000
SimpleRNN (Articles)	(None, 128)	29,312
BatchNormalization	(None, 128)	512
Dropout (0.5)	(None, 128)	0
Input (Sentiment)	(None, 100)	0
Embedding (Sentiment)	(None, 100, 100)	300,000
SimpleRNN (Sentiment)	(None, 128)	29,312
BatchNormalization	(None, 128)	512
Dropout (0.5)	(None, 128)	0
Input (Ratio)	(None, 1)	0
Concatenate	(None, 257)	0
Dense (ReLU)	(None, 280)	72,240
Dropout (0.5)	(None, 280)	0
Dense (Sigmoid)	(None, 1)	281
Total Parameters	—	**732,169**

3.7 Model Evaluation

In order to measure the performance of our models, we used typical classification metrics accuracy, precision, recall, and F1-score [31–34] that are formulated in Eqs. (2), (3), (4), and (5). Due to the binary classification problem, we also utilized the confusion matrix to examine true positive (TP), true negative (TN), false positive (FP), and false negative (FN) outcomes. To give a fuller view of model performance. Due to our balanced dataset.

$$\text{Accuracy} = \frac{TP + TN}{TP + TN + FP + FN} \tag{3}$$

$$\text{Precision} = \frac{TP}{TP + FP} \tag{4}$$

$$\text{Recall} = \frac{TP}{TP + FN} \tag{5}$$

$$\text{F1-score} = 2 \times \frac{\text{Precision} \times \text{Recall}}{\text{Precision} + \text{Recall}} \tag{6}$$

4 Results and Discussion

In the present study, we carried out an extensive comparison of binary classification models on a balanced data set consisting of labeled Arabic text data.

The preprocessing included cleaning the text by removing URLs and stop words, followed by transforming the raw text into numerical values with methods such as TF-IDF and word embeddings. We trained several separate models, such as traditional machine learning models Logistic Regression (LR), Naïve Bais (NB), Random Forest (RF), and Support Vector Machine (SVM). We also employed deep learning models based on Recurrent Neural Networks (RNN), including SimpleRNN and LSTM, to capture the temporal dependencies in the text data. All experiments were performed in JupyterLab using Python 3.7, and were facilitated by libraries such as scikit-learn for machine learning and TensorFlow/Keras for deep learning, we trained all the models with 80% of the dataset and kept 20 % for testing. In order to assess the performance of the models, we computed confusion matrixes plotted in Figs. 4 and metrics like accuracy, precision, recall and F1-score in Table 3 and a comparison of overall model performance is also plotted in Fig. 10.

Figure 4 illustrate the way the data is handled by a method of data analysis and visualization software used to utilize color in bar chart to represent height and width called a confusion matrix, the confusion matrix of the Support Vector Machine (SVM), which demonstrates the best performance among the evaluated models with 569 true positives and 617 True negative, while minimizing the number of false classifications. Logistic Regression also performs well, achieving 564 true positives and 618 True negatives, with only 17 false positives. Random Forest yields competitive results but with slightly more false negatives 59 and false positives 35. Naïve Bayes underperforms compared to the previous models, with only 541 true positives and 59 false negatives. Additionally, deep learning models presents the confusion matrix of the RNN model, which achieves 599 true negatives and 531 true positives, with slightly increased false negatives 69, and for the LSTM model, which improves slightly over RNN by obtaining 597 true negatives and 553 true positives, suggesting a better balance in prediction performance between both classes.

Table 3. Performance Evaluation of Classifiers

Classifiers	Accuracy (%)	Recall (%)	F1-score (%)	Precision (%)
SVM	96	96	96	96
Naïve Bayes (NB)	91	91	91	91
Logistic Regression (LR)	96	96	96	96
Random Forest (RF)	92	92	92	92
RNN	91	91	91	92
LSTM	93	93	93	93

All models have been graphically reported in Fig. 5 for better understanding. In Table 3 provide a comparative evaluation of six classification models SVM, Naïve Bayes (NB), Logistic Regression (LR), Random Forest (RF), RNN, and

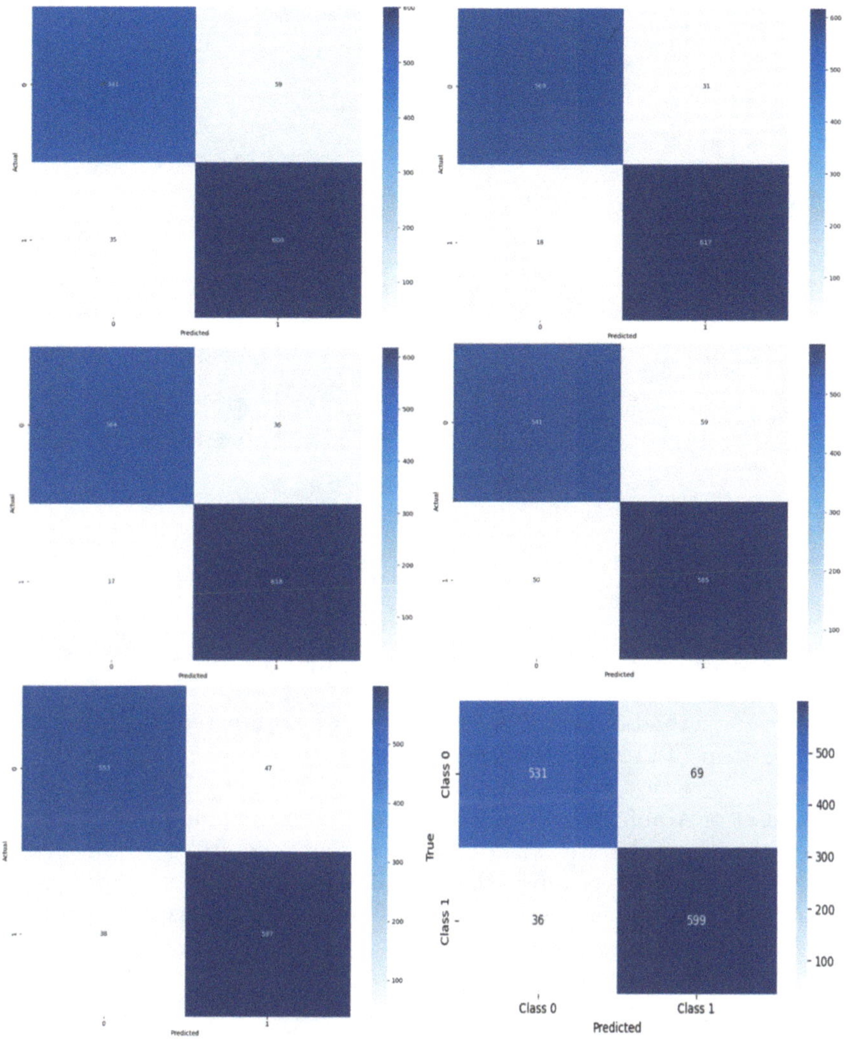

Fig. 4. Confusion matrices of RF, SVM, LR, NB, LSTM, and RNN models.

LSTM, based on four performance metrics: accuracy, recall, F1-score, and precision. Both SVM and LR achieved the highest performance across all metrics, each scoring 96%, indicating their strong and consistent ability to correctly classify the data. LSTM followed with slightly lower but still robust results at 93%, outperforming RNN and NB, which both recorded 91% in most metrics. Random Forest performed moderately well with 92%, showing stable classification performance. Overall, traditional machine learning models such as SVM and LR outperformed deep learning models like RNN and LSTM, suggesting that for this particular dataset, simpler models provided more effective and reliable results.

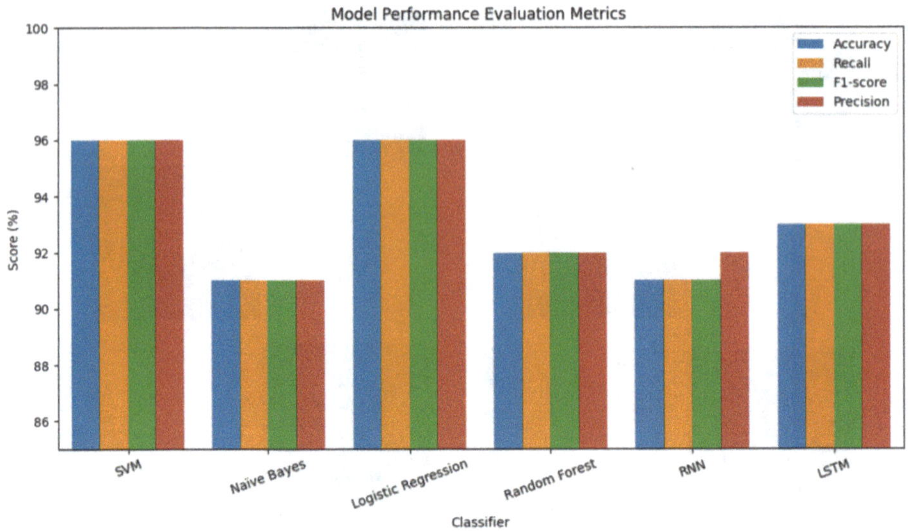

Fig. 5. Performance of models on AFND dataset

4.1 Performance Comparison with Baseline Methods

Our proposed method of identifying fake news was compared to state-of-the-art methods existing at present. As shown in Sect. 2 our sentiment-enhanced method achieved 96% accuracy with the Support Vector Machine (SVM) model in a balanced dataset of Arabic fake news. This is far greater than different conventional and deep models under the identical experimental conditions. Therefore, the proposed sentiment-based approach is highly effective for detecting false news in Arabic and may be suggested with confidence for application in real-world Arabic language media analysis.

5 Conclusion

In this paper, we evaluated four machine learning models and two deep learning models on a balanced Arabic fake news dataset using accuracy, precision, recall, and F1-score as evaluation metrics. Based on our experiments, Naïve Bayes and Random Forest performed fairly, whereas Support Vector Machine and Logistic Regression performed best among the others in all the criteria. Deep learning architectures like LSTM and RNN were also competitive, particularly in deriving linguistic features from the text. According to our recent experiments, Support Vector Machine with TF-IDF features attained the highest scores among all the individual models. This confirms the power of SVM in doing a good job in well-balanced binary classification problems in Arabic language settings.
The experimental results indicated that sentiment analysis is an essential part in enhancing fake news detection. By calculating a ratio of a sentiment metric

and adding the results to the dataset, we built a sentiment-enriched dataset. We applied machine learning and deep learning models on this modified dataset, resulting in enhanced classification performance. This demonstrates the effectiveness of utilizing sentiment signals in fake news detection in Arabic.
The main contribution of this research study is as follows: First, we employed sentiment analysis with a metric ratio-based method to pre-enrich the dataset before training and testing different machine learning and deep learning models. By doing so, we were able to examine the impact of sentiment-sensitive data on the classification of Arabic fake news. Second, we utilized a well-balanced dataset focused around the Arabic language in an effort to address underrepresentation of Arabic materials in fake news research. Third, we demonstrated that the union of sentiment analysis and traditional feature extraction techniques like TF-IDF significantly improves model performance.

Future work includes more experimentation on large-scale and multilingual databases for generalizing the results. In the future, we also foresee conducting more extension work on different architectures of deep learning and other ensemble methods in order to obtain improved classification accuracies in the case of Arabic, as well as other resource-poor languages.

References

1. Vosoughi, S., Roy, D., Aral, S.: The spread of true and false news online. Science **359**(6380), 1146–1151 (2018)
2. Do Nascimento, I.J.B., et al.: Infodemics and health misinformation: a systematic review of reviews. Bull. World Health Org. **100**(9), 544 (2022)
3. Leyva, R., Beckett, C.: Testing and unpacking the effects of digital fake news: on presidential candidate evaluations and voter support. AI Soc. **35**, 969–980 (2020)
4. Habash, N.: Introduction to Arabic Natural Language Processing. Morgan & Claypool (2022)
5. Al-Yahya, M., Al-Khalifa, H., Al-Baity, H., AlSaeed, D., Essam, A.: Arabic fake news detection: comparative study of neural networks and transformer-based approaches. Complexity **2021**(1), 5516945 (2021)
6. Abdul-Mageed, A., Elmadany, A., Nagoudi, E.M.B.: ARBERT & MARBERT: Deep bidirectional transformers for Arabic. In: Proceedings of the 58th Annual Meeting of the Association for Computational Linguistics (2020)
7. Touahri, I., Mazroui, A.: Survey of machine learning techniques for Arabic fake news detection. Artif. Intell. Rev. **57**(6), 157 (2024)
8. Khalil, A., Jarrah, M., Aldwairi, M., Jaradat, M.: AFND: Arabic fake news dataset for the detection and classification of articles credibility. Data Brief **42**, 108141 (2022)
9. Cui, L., Wang, S., Lee, D.: SAME: Sentiment-aware multi-modal embedding for detecting fake news. In: Proceedings of the 2019 IEEE/ACM International Conference on Advances in Social Networks Analysis and Mining, pp. 41–48 (2019)
10. Horne, B.D., Adali, S.: This Just. In: Fake News Packs a Lot in Title, Uses Simpler, Repetitive Content in Text Body, More Similar to Satire than Real News. In: An, J., Kwak, H., Benevenuto, F., (eds.) Proceedings of the Workshops of the Eleventh International AAAI Conference on Web and Social Media (ICWSM 2017), pp. 759–766. Montreal, QC, Canada, 15–18 May 2017, AAAI Press, Palo Alto, CA, USA (2017)

11. Ajao, O., Bhowmik, D., Zargari, S.: Sentiment aware fake news detection on online social networks. In: ICASSP 2019 - IEEE International Conference on Acoustics, Speech and Signal Processing, IEEE (2019)
12. Himdi, H., Weir, G., Assiri, F., Al-Barhamtoshy, H.: Arabic fake news detection based on textual analysis. Arab. J. Sci. Eng. **47**(8), 10453–10469 (2022)
13. Iwendi, C., Mohan, S., Ibeke, E., Ahmadian, A., Ciano, T.: Covid-19 fake news sentiment analysis. Comput. Electr. Eng. **101**, 107967 (2022)
14. AlRubaian, M., Al-Qurishi, M., Al-Rakhami, M., Rahman, S.M.M., Alamri, A.: A multistage credibility analysis model for microblogs. In: Proceedings of IEEE/ACM International Conference on Advances in Social Networks Analysis and Mining, pp. 1434–1440 (2015)
15. Dey, A., Rafi, R.Z., Parash, S.H., Arko, S.K., Chakrabarty, A.: Fake news pattern recognition using linguistic analysis. In: 2018 Joint 7th International Conference on Informatics, Electronics & Vision (ICIEV) and 2nd International Conference on Imaging, Vision & Pattern Recognition (icIVPR), pp. 305–309. IEEE (2018)
16. Bhutani, B., Rastogi, N., Sehgal, P., Purwar, A.: Fake news detection using sentiment analysis. In: 12th International Conference on Contemporary Computing (IC3), pp. 1–5. IEEE (2019)
17. Anoop, K., Deepak, P., Lajish, V.L.: Emotion cognizance improves health fake news identification. In: IDEAS 2020, p. 24 (2020)
18. Rashkin, H., Choi, E., Jang, J.Y., Volkova, S., Choi, Y.: Truth of varying shades: Analyzing language in fake news and political fact-checking. In: Proceedings of the 2017 Conference on Empirical Methods in Natural Language Processing, pp. 2931–2937 (2017)
19. Zhang, X., Cao, J., Li, X., Sheng, Q., Zhong, L., Shu, K.: Mining dual emotion for fake news detection. In: Proceedings of The Web Conference, pp. 3465–3476 (2021)
20. Vicario, M.D., Quattrociocchi, W., Scala, A., Zollo, F.: Polarization and fake news: early warning of potential misinformation targets. ACM Trans. Web (TWEB) **13**(2), 1–22 (2019)
21. Ross, J., Thirunarayan, K.: Features for ranking tweets based on credibility and newsworthiness. In: International Conference on Collaboration Technologies and Systems (CTS), pp. 18–25. IEEE (2016)
22. Popat, K., Mukherjee, S., Strötgen, J., Weikum, G.: Where the truth lies: Explaining the credibility of emerging claims on the web and social media. In: Proceedings of the 26th International Conference on World Wide Web Companion, pp. 1003–1012 (2017)
23. Hassan, N., Arslan, F., Li, C., Tremayne, M.: Toward automated fact-checking: detecting check-worthy factual claims by claimbuster. In: Proceedings of the 23rd ACM SIGKDD International Conference on Knowledge Discovery and Data Mining, pp. 1803–1812 (2017)
24. Reis, J.C., Correia, A., Murai, F., Veloso, A., Benevenuto, F.: Supervised learning for fake news detection. IEEE Intell. Syst. **34**(2), 76–81 (2019)
25. Yang, Y., Zheng, L., Zhang, J., Cui, Q., Li, Z., Yu, P.S.: TI-CNN: Convolutional neural networks for fake news detection. arXiv preprint arXiv:1806.00749 (2018)
26. Mustafa, M., Eldeen, A.S., Bani-Ahmad, S., Elfaki, A.O.: A comparative survey on Arabic stemming: approaches and challenges. Intell. Inf. Manag. **9**(2), 39–67 (2017)
27. Mubarak, H.: Build fast and accurate lemmatization for Arabic, pp. 1128–1132 (2017)

28. Li, C., Xie, Z., Wang, H.: Short text classification based on enhanced word embedding and hybrid neural networks. Appl. Sci. **15**(9), 5102 (2025). https://doi.org/10.3390/app15095102
29. LeCun, Y., Bengio, Y., Hinton, G.: Deep learning. Nature **521**(7553), 436–444 (2015)
30. Ajao, O., Bhowmik, D., Zargari, S.: Sentiment aware fake news detection on online social networks. In: ICASSP 2019 - IEEE International Conference on Acoustics, Speech and Signal Processing (ICASSP), Brighton, UK, pp. 2507–2511. IEEE (2019). https://doi.org/10.1109/ICASSP.2019.8683170
31. Perez, E., Ringer, T., Bowman, S.: When accuracy lies: Misleading metrics in text classification. In: Proceedings of the 61st Annual Meeting of the Association for Computational Linguistics (ACL 2023), pp. 10245–10260 (2023). https://doi.org/10.18653/v1/2023.acl-long.571
32. Sokolova, M., Lapalme, G.: A systematic analysis of performance measures for classification tasks. Inf. Process. Manag. **56**(4), 1020–1035 (2020)
33. Davis, J., Goadrich, M.: The relationship between Precision-Recall and ROC curves in multi-label classification. In: Proceedings of the 23rd International Conference on Machine Learning (2020)
34. Powers, D.M.W.: Evaluation metrics for multi-label classification: a unified perspective. J. Mach. Learn. Technol. **11**(1), 1–25 (2020)

Evaluation of Handcrafted and Learning-Based Keypoint Detection and Description Methods in Image Matching

Adel Sardou[1(✉)], Abdenebi Rouigueb[1], Fethi Demim[1], Abdenour Amamra[1], and Hadjira Belaidi[2]

[1] Ecole Militaire Polytechnique, Algiers, Algeria
sr_adel@outlook.com
[2] University M'hamed Bougara of Boumerdès, Boumerdes, Algeria
ha.belaidi@univ-boumerdes.dz

Abstract. Keypoint detection and description (KPDD) are foundational tasks in computer vision, yet comprehensive and standardized evaluations across traditional and deep-learning methods remain limited. We conducted an in-depth comparative analysis of KPDD techniques, introducing four adapted metrics that quantitatively assess detection and description performance. Our benchmark includes classical methods such as MSER, SIFT, SURF, ORB, AKAZE, AGAST, and FREAK, as well as recent learning-based models like SuperPoint, DeDoDe, ALIKE, and DISK. Experimental results reveal that handcrafted approaches remain competitive with deep learning methods on the HPsequence dataset. This work offers an evaluation framework and actionable insights that guide the selection and development of KPDD techniques for image matching applications.

Keywords: Keypoint detection · Keypoint description · Image matching · Emprical evaluation

1 Introduction

Image matching represents a challenging and essential task in the field of computer vision. It underpins various applications such as image registration [33], object detection, and localization [23,34], among others. A widely used and intuitive approach to image matching involves detecting and describing a limited set of points of interest (keypoints) in pairs of images. Local matching is then carried out based on the similarity of keypoints' descriptors. After finding candidate point-to-point correspondences, global matching, which consists of estimating the existing transformation parameters between images, can be established by combining those local matches with prior knowledge about the transformation type.

Z. Kouahla et al. (Eds.): IAM 2025, LNNS 1973, pp. 150–165, 2026.
https://doi.org/10.1007/978-3-032-26031-4_12

Although not the only option, Keypoint Detection and Description (KPDD) plays a pivotal role in the image matching pipeline, as it operates upstream and is intended to significantly reduce the problem's complexity (i.e. few keypoints instead of the entire image) without discarding much useful information. Therefore, improving this component will improve the overall efficiency and performance of image matching.

Over the past three decades, KPDD has undergone significant evolution in both diversity and number of solutions. KPDD methods have evolved from purely *Handcrafted* [7,13,16,25] to *fully learned*, data-driven approaches [11,28,32].

Through an analysis of state-of-the-art KPDD algorithms, we have drawn the following conclusions. Regardless of the approach, even in end-to-end deep learning solutions, the matching pipeline often involves these steps: detection, description, and matching. The expected keypoints depend on the target application. All existing algorithms are heuristics as they rely on certain assumptions and prior knowledge. There is no universally accurate algorithm; indeed, some algorithms perform better than others depending on the use case.

Therefore, new KPDD approaches are published every few months. With such fast-paced development, it is necessary to conduct systematic, up-to-date, objective, and independent comparative experimental studies that evaluate KPDD algorithms to help users select the most appropriate method for their application. In this regard, we performed an empirical comparison that includes state-of-the-art KPDD approaches, using public benchmarks.

The remainder of this paper is structured as follows. Section 2 introduces the KPDD pipeline. Section 3 details the empirical evaluation, where we present and discuss the obtained results. Finally, Sect. 4 concludes the paper and discusses the broader implications.

2 Keypoint Detection and Description for Image Matching

2.1 Pipeline of Image Matching

To match two images capturing the same scene, despite the existing transformations between them (e.g. geometric, photometric, etc.), KPDD methods first seek to find and compactly describe meaningful patterns within these images. As illustrated in Fig. 1, the standard pipeline consists of the following steps: i) *keypoint detection*, ii) *keypoint description*, iii) *local matching*, and iv) *global matching*. Initially, we attempt to detect salient points that are rich in information (e.g. edges, corners, blobs, textures). Keypoint description refers to extracting a fixed size, commonly called a descriptor, such that the distance between descriptors of corresponding keypoints is minimized compared to that between non-corresponding ones.

In this context, *keypoints detected in different images are considered corresponding if they represent the same location in the physical scene.* Thereafter, local matching establishes potential correspondences between keypoints of the two analyzed images by comparing their descriptors, typically using a suitable

Fig. 1. Feature matching pipeline example. Steps 1 and 2: keypoint detection and description, Step 3: Local matching, Step 4: Global matching.

distance metric such as Euclidean or Hamming. It can be established using nearest-neighbor searches [30], or complex algorithms such as OANET [31] or SuperGlue [26]. Finally, depending on the application, a robust rejection method of outliers, such as RANSAC [12], can be applied to ensure global geometric correspondence.

The detection and description tasks are crucial and challenging in this pipeline. KPDD is particularly important as it is expected to significantly reduce complexity while ensuring accurate matching. In fact, many matching errors originate from failures in the KPDD stage. Furthermore, the main difficulty arises from their interdependence: effective detectors should be optimized with respect to the employed descriptors, while effective descriptors should be optimized exclusively for the detected keypoints. Designing a KPDD solution is a difficult optimization problem.

In this paper, we highlight the strong coupling of detection and description problems and objectively evaluate them.

2.2 Keypoint Detection

Problem Formulation. *Formally, let I be a discrete image. The goal of keypoint detection is to identify a set of salient points $\mathcal{K} = \{\mathbf{k}_i\}_{i=1}^{N}$ in I that summarize the information held in I.*

Each keypoint $\mathbf{k}_i = (x_i, y_i, s_i, \theta_i)$ is defined by its spatial coordinates (x_i, y_i) in I, the scale s_i of the region of interest around the pixel (x_i, y_i) (if applied) and the dominant orientation of $\mathbf{k}_i$ (if applied).

The detection quality is primarily measured by *repeatability* [21], *distinctiveness* [20], and *invariance* [20,24]. Depending on the final application, other criteria may also be considered.

Repeatability measures the probability of consistently detecting the same keypoints in different images of the same scene. The distinctiveness reflects how uniquely a keypoint is represented by its descriptor. The invariance measures how much keypoint representation remains stable under the usual transformations. In fact, strong distinctiveness and invariance help ensure small matching distances between the descriptors of the corresponding keypoints, which will decrease the number of wrong local matching pairs.

The main difficulty arises from the fact that developing a detector, or even evaluating it, depends on a descriptor algorithm, which itself still requires development.

Detection Algorithms Classification. Based on the structure of keypoints, they can be categorized into three types: i)*pixel-wise* (e.g. ORB [25]) ii) *region-wise* (e.g. MSER [19]), and iii) *edge-wise* keypoints (e.g. Canny [8]). Among these, pixel-wise keypoints are the most prevalent due to their straightforward implementation.

Moreover, detectors are broadly divided into two categories: handcrafted and learning-based. The former rely on local image pixels value statistics such as

Laplacian approximation (e.g. SIFT [16]), Hessian-based responses (e.g. SURF [7]), or second-moment matrices [13]. Whereas, learning-based approaches are data-driven and include both classical methods that use supervised techniques to optimize handcrafted detectors (e.g., decision trees variants FAST [29] and AGAST [18]), as well as modern deep learning-based models that leverage convolutional neural networks CNNs (e.g. ALIKE [32]) or transformers (e.g. DeDoDe [11]) to detect keypoints directly from raw image input. Figure 2, summarizes this classification.

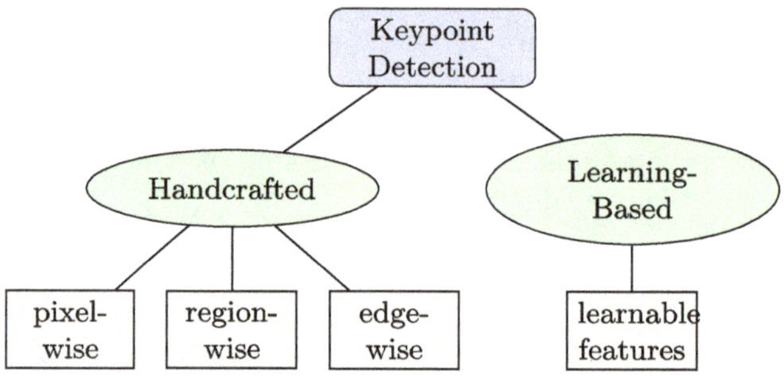

Fig. 2. Classification of keypoint detection methods, adapted from [17].

Evaluation Metric. Empirical evaluation of keypoint detection is critical for both selecting and developing KPDD algorithms. It helps identify suitable methods for specific applications and can also be integrated into learning-based detectors. The evaluation criteria typically include repeatability, localization accuracy, spatial distribution, computational complexity, etc.

Repeatability measures the ability to find the same keypoints in different views. It is the most widely used metric in comparative studies [15,20,27] due to its strong impact on the matching. Given two images I and J, repeatability R is defined as:

$$R = \frac{\mathcal{M}}{\min(|\mathcal{K}_I|, |\mathcal{K}_J|)} \tag{1}$$

where $\mathcal{M}$ is the number of corresponding keypoints and $\mathcal{K}_I, \mathcal{K}_J$ are the sets of keypoints detected in the overlap regions between I and J, respectively.

Estimation of M remains an open challenge. It requires determining whether a given pair constitutes a correct match, and it can be formulated as a binary classification problem based on a set of features. (i.e., $y_{p,q} = 1$ if p and q correspond and $y_{p,q} = 0$ otherwise). Therefore, we deduce: $M = \sum_{(p,q)\in(I,J)} y_{p,q}$.

In fact, the difficulty of verifying the correspondence depends on the scale, the characteristics of the transformation between images, the noise, the localization precision, and other sources of invariance. For instance, in this work, we propose to employ the following constraints criteria:

C01: Geometric constraint - Adapting the evaluation protocol (Mean Matching Accuracy) proposed in [10], we could decide $y_{(p,q)}$ by checking the geometric constraint that p and q must hold if they correspond. This approach is often associated with the reprojection error [12,16]. When the existing geometric transformation $J = \phi(I)$ is already known, whether obtained by labeling or simulated to generate J, this criterion can be assessed as follows:

$$y_{(p,q)} = \begin{cases} 1 : \text{if } \|\phi(p) - q\|_2 \leq \epsilon \text{ and} \\ \qquad \|p - \phi^{-1}(q)\|_2 \leq \epsilon \\ 0 : \text{otherwise} \end{cases} \tag{2}$$

ϕ is often a homography for planar objects, a fundamental matrix mapping for 3D objects, or another possible transformation. $C01$ ensures that the symmetric reprojection error of a corresponding pair is below a small threshold ϵ usually set to less than 03 or 04 pixels.

C02: Geometric and description constraint: Following the approach of [14,16], we combine location and descriptor similarity checks. We set $y_{(p,q)} = 1$, if $C01$ and a descriptor constraint hold. The latter requires that in terms of description distances: i) q is the nearest neighbor of p and ii) $d_1/d_2 \leq \rho$ where $d_1 = Dist(p,q)$ and d_2 is the distance to the second-best match. The thresholdρ is usually set between 0.6 and 0.9.

C03: Nearest in detection and description: As in $C02$, we also combine geometric and description constraints, but in a different way. We adopt the same criterion proposed in [6] as a metric for repeatability. Given a keypoint pair $(p,q) \in (I,J)$, $C03$ is defined as follows:

$$y_{(p,q)} = \begin{cases} 1: \text{ if } Dist(d_p, d_q) \leq Dist(d_p, d_z), \\ \forall z \in J, z \neq q \text{ and} \\ \|p - \phi^{-1}(q)\|_2 \leq \|p - \phi^{-1}(z)\|_2, \\ \forall z \in J, z \neq q \\ 0: \text{ otherwise} \end{cases} \tag{3}$$

Here, d_p, d_q, d_z are the descriptors of keypoints p, q, z. The function $\text{dist}(\cdot,\cdot)$ measures descriptor similarity (e.g., Hamming distance for binary descriptors).

The criteria $C01$ and $C02$ are relatively simple to verify but are sensitive to the user-fixed thresholds ϵ and ρ. In fact, ϵ depends on the scale of images and the mapping, while ρ is related to the employed descriptor. $C02$ is more robust since it combines geometric location and descriptor similarity checks. $C03$, similar in spirit to $C02$, also integrates location and descriptor information. In particular, $C03$ avoids the need for predefined thresholds but assumes that the most similar keypoint in the descriptor space is the closest in location, which may sometimes not be convenient. $C02$ and $C03$ depend on the performance of the descriptor. Therefore, each metric could be suitable for a specific context.

2.3 Keypoint Description

Problem Formulation. *The description of a keypoint* $\mathbf{p}$ *detected in the image* I *can be seen as a function* $F : (I, \mathbf{p}) \mapsto \mathbf{d_p} \in \mathbb{R}^D$*, which generates a compact representation (a description vector)* $\mathbf{d_p}$ *from the local region surrounding* $\mathbf{p}$*. The objective is that the matching distance of the descriptors from different images,* $\mathrm{Dist}_F(\mathbf{d_i}, \mathbf{d_j})$*, is low only for the corresponding pairs of points.*

To achieve this objective, an effective descriptor should be invariant to the existing transformations (e.g. scale, rotation, illumination, noise, etc.). More importantly, the descriptor extractor must be developed and refined jointly with its associated distance Dist_F.

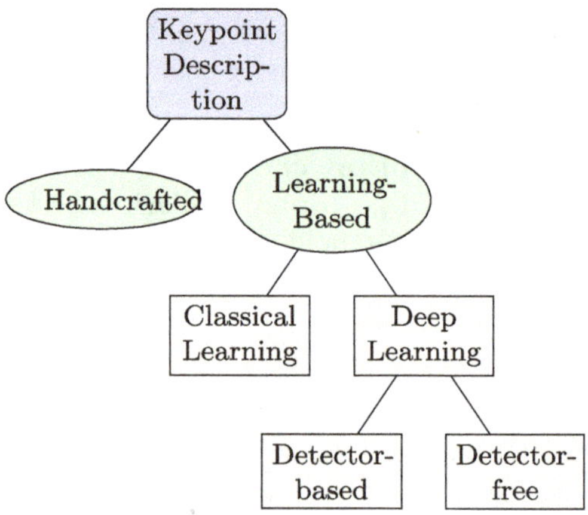

Fig. 3. Classification of keypoint detection methods [17,30].

Description Algorithms Classification. Figure 3 summarizes the classification of the description approaches.

Evaluation Metrics. As with detection, the evaluation of descriptors is crucial for the selection and development of algorithms. Typically, we compute the distances $\mathrm{Dist}_F(d_{p_i}, d_{q_j})$ between all pairs $(p_i, q_j) \in (I, J)$, and then determine the pairs that represent the correct matches. Standard metrics such as precision, recall, and F-score can be used to assess the performance of this binary classifier. Particularly, *Average Precision (AP)* and *mean Average Precision (mAP)* are widely used as they assess how well the true correspondences of a query keypoint p_i (a patch) are ranked within a list $\mathcal{L}$ of candidate keypoints from other images. This ranking-based information is especially valuable for image matching tasks [5].

We point out that to assess description only, without detection, we can use patches as in [5] or restrict the evaluation to only keypoints that are correctly detected. In this work, we propose to combine the criterion $C01$ with the method of [24], to estimate a description error, as follows:

1. Let (K_I, D_I) and (K_J, D_J) denote the set of detected keypoints and their respective descriptors in overlapping images I and J.
2. Remove non-repeatable keypoints from (K_I, D_I) using $C01$ at a 4px threshold[1].
3. For each point $p_i \in K_I$, compute:

$$Error_{p_i} = \frac{\|\phi^{-1}(q_{i,j}) - p_i\|_2}{\sqrt{W^2 + H^2}} \tag{4}$$

   where: $q_{i,j} \in K_J$ is the most similar keypoint to p_i $(q_{i,j} = \arg\min_{q_j} \text{Dist}_F(D_{p_i}, D_{q_j}), \forall q_j \in K_J)$.
   W and H are the width and height of image I.
4. Compute the mean error between I and J:

$$\text{Error}_{IJ} = \frac{1}{|K_I|} \sum_{i=1}^{|K_I|} Error_{p_i}, \tag{5}$$

5. **F**inally we can aggregate and average errors across all images in a dataset.

3 Empirical Evaluation

3.1 Evaluation Protocol

To evaluate the selected KPDD algorithms, we have used the HPatches dataset [5,15], which is widely adopted for assessing local feature extraction under varying illumination and viewpoint conditions. HPatches was chosen because of its real, diverse and large composition. We highlight that we didn't use the patches provided with this dataset as done in [5,22], but we use the whole images (*HPsequence*), as done in [6,10,15] to also allow detection evaluation.

For each tested method, we have carried out keypoint detection and description, then computed the evaluation metrics : repeatability for detectors (R in Eq. 1, using the criteria $C01$,$C02$, and $C03$) and distinctiveness ($Error_{IJ}$ in Eq. 5). To be fair, the repeatability metric is calculated only over the overlapping regions.

We have compared the handcrafted detectors: SIFT [16], SURF [7], MSER [19], ORB [25] and AKAZE [2,3], as well as the learning-based detector AGAST [18] which employs decision forests, and the recent deep learning based approaches: DeDoDe [11], SuperPoint[9], ALIKE [32] and DISK [28][2].

The evaluated Descriptors cover the flowing set: {Handcrafted : SIFT, ROOTSIFT [4], SURF, ORB, AKAZE, FREAK [1], Deep Learning: DeDoDe,

[1] 4px was chosen based on empirical results.
[2] SuperPoint, ALIKE and DISK methods are integrated pipelines, performing detection and description jointly.

SuperPoint, ALIKE and DISK }. Handcrafted descriptors were tested in combination with several detectors, to explore potential links between detectors and descriptors.

All tested algorithms were evaluated under two configurations: detecting up to 1000 and up to 2000 keypoints per image, following [15], which highlighted the sensitivity of evaluation metrics to the number of keypoints.

All experiments were conducted on a laptop equipped with an Intel Core i5-13500HX processor, 16 GB of RAM, and an NVIDIA RTX 4050 GPU. Handcrafted methods were implemented in Python 3.7.12 with OpenCV 3.4.2 using their default settings. Deep-learning models were deployed with publicly available pretrained weights in a separate Python 3.10.18 environment with PyTorch 2.5.1, due to library compatibility requirements.

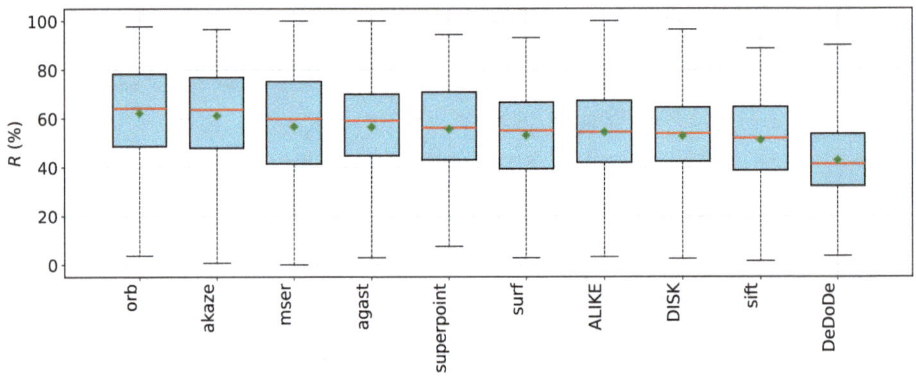

Fig. 4. Repeatability distribution over 580 image pairs with 1000 keypoints per image. R is computed using Eq. (1) with $C01$.

Table 1. Detection time per image (sec)

	ORB	AGAST	SURF	DeDoDe	AKAZE	SIFT	MSER
Mean	0.0109	0.0320	0.0489	0.0501	0.0841	0.1082	0.3259
Std	0.0002	0.0031	0.0049	0.0135	0.0163	0.0038	0.0071

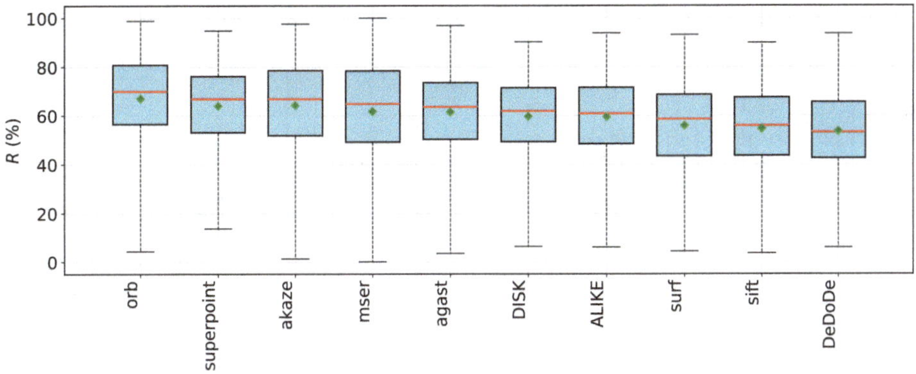

Fig. 5. Repeatability distribution over 580 image pairs with 2000 keypoints per image. R is computed using Eq. (1) with $C01$.

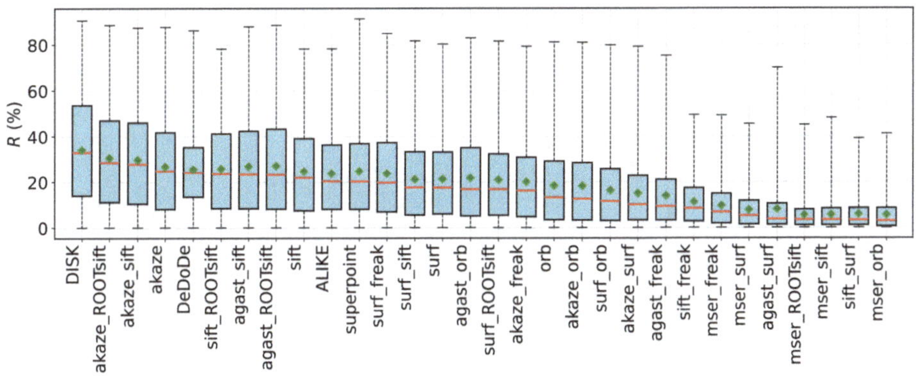

Fig. 6. Repeatability distribution over 580 image pairs with 1000 keypoints per image. R is computed using Eq. (1) with $C02$.

3.2 Results and Discussion

Detection: In Figs. 4 and 5, we present detection repeatability scores based on geometric constraints using $C01$ that checks the keypoint location. Whether the method is handcrafted or learning-based, the results of the tested detectors do not differ much, indicating that most of them yield similar performance. However, we can observe that in both figures the group ORB, AKAZE, Superpoint, MSER and AGAST slightly outperform the group SIFT, ALIKE, DISK, SURF and DeDoDe.

Repeatability scores shown in Figs. 6 and 7 are evaluated using the criterion $C02$, which is more restrictive than $C01$. It checks both geometric and description consistency by comparing location and descriptor distances with thresholds (see Subsection *Evaluation metric* in Sect. 2). Hence, this metric counts only high-quality keypoints. Results obtained show that handcrafted methods can still be competitive with deep learning-based methods. The best KPDD methods were

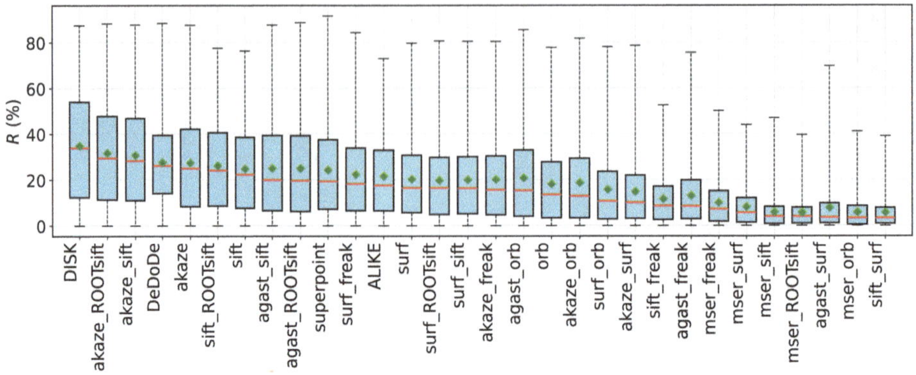

Fig. 7. Repeatability distribution over 580 image pairs with 2000 keypoints per image. R is computed using Eq. (1) with $C02$.

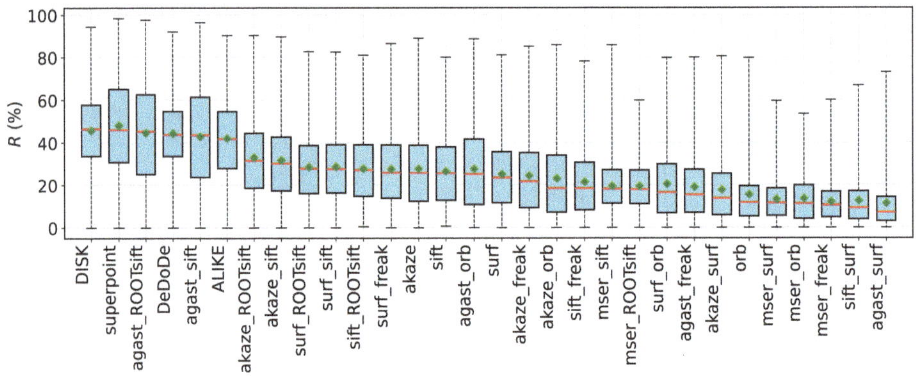

Fig. 8. Repeatability distribution over 580 image pairs with 1000 keypoints per image. R is computed using Eq. (1) with $C03$.

DISK and AKAZE+RootSIFT. In contrast, the worst combinations were those that used the MSER detector.

Regarding criterion $C03$, which measures the proportion of points where the best matching point is also the closest (in terms of location), the obtained repeatability scores displayed in Figs. 8 and 9 indicate that deep learning methods are superior in performance. This may be due to the fact that they do not compare against fixed thresholds but instead rely on a ranking criterion.

Description: We used the metric $Error_{IJ}$ given in Eq. (5), which is adapted from [24], to assess the performance of the descriptors. Analysis of keypoint description results displayed in Figs. 10 and 11 reveals that learning-based methods produce fewer errors, although handcrafted approaches remain competitive, particularly the SIFT descriptor.

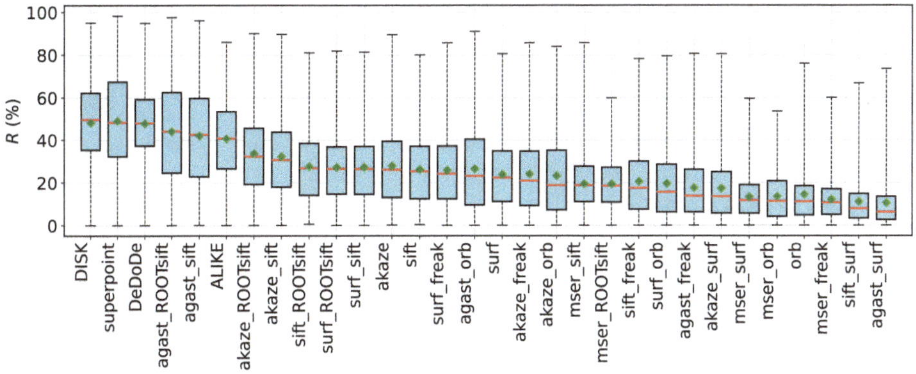

Fig. 9. Repeatability distribution over 580 image pairs with 2000 keypoints per image. R is computed using Eq. (1) with $C03$.

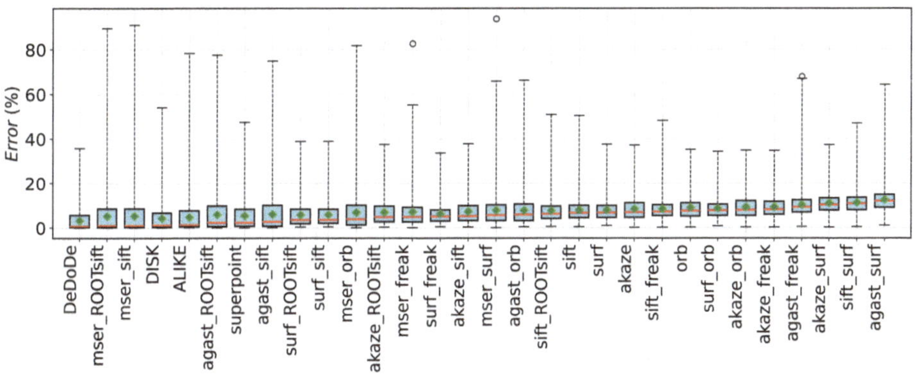

Fig. 10. Description error over 580 image pairs with 1000 keypoints per image. Error is computed using Eq. (5)

Computing Time: Performance evaluation should be complemented by computational cost analysis. To this end, we suggest reporting the average running time per image, required for the detection or description stages, in Tables 1, 2 and 3. Note that timing is not directly comparable, as deep learning relies on GPUs while others run on CPUs.

It can be observed that ORB is the fastest for detection and description, while MSER is the slowest detector, and DeDoDe is the slowest descriptor. Detector execution times are stable across images (low standard deviation), whereas descriptor execution times, especially for DeDoDe and SIFT, vary more with the images.

Overall Discussion: This comparative study confirms that learned features generally are robust and produce more correct matches, where DeDoDe is the best in terms of description and DISK is the best detector among tested algo-

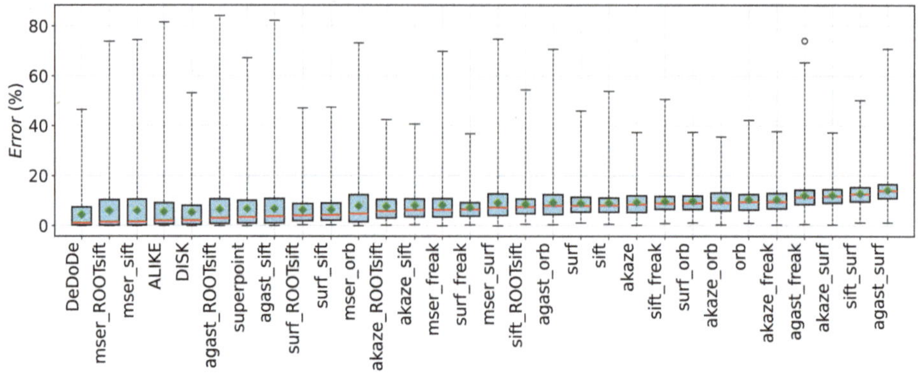

Fig. 11. Description error over 580 image pairs with 2000 keypoints per image. Error is computed using Eq. (5)

Table 2. Description time per image (sec)

	ORB	Freak	SURF	akaze	SIFT	RootSIFT	DeDoDe
Mean	0.0076	0.0078	0.0154	0.0549	0.0761	0.0764	4.7418
Std	0.0051	0.0041	0.0096	0.0014	0.0512	0.0505	0.4322

Table 3. Detection and description time per image (sec)

	Superpoint	DISK	ALIKE
Mean	0.0845	0.1884	0.7786
Std	0.0298	0.0267	0.0041

rithms. However, we also noticed that traditional methods such us SIFT maintain some advantages, specially in cases with limited training data or well-constrained problem domains.

Upon examination of various detector–descriptor combinations, several noteworthy observations emerged. For example, the MSER detector achieved competitive repeatability scores with $C01$, but moderate performance with $C02$ and $C03$. Indeed, we were unable to identify a suitable descriptor for MSER. In contrast, the RootSIFT descriptor outperformed other handcrafted descriptors in most settings. Furthermore, while the SURF descriptor performed well in conjunction with the SURF detector, its performance dropped markedly when paired with alternative detectors. These findings confirm the strong connection between detectors and descriptors.

It is noteworthy that an analysis of individual results across different image pairs revealed significant variability in the performance of all evaluated methods, where each method demonstrated favorable outcomes for certain types of image content, while exhibiting substantially degraded performance in others. These observations suggest that the effectiveness of the methods is highly dependent on the content of the treated images.

4 Conclusion

In this study, we addressed the keypoint detection and description problem for image matching purposes. Our empirical evaluation of state-of-the-art handcrafted and learning-based methods reveals that there is no universal matching pipeline solution suitable for all end-user applications. The core components of such a pipeline include keypoint detection, descriptor extraction, and descriptor matching (i.e., distance computation). We noticed that these three components exhibit strong interdependencies, highlighting the importance of their joint design and optimization, which makes their development particularly challenging. Moreover, deep-learning approaches hold significant promise for future advancements, while existing handcrafted methods remain competitive, particularly with respect to computational efficiency. Ultimately, the optimal choice of a KPDD method cannot be reduced to simply adopting state-of-the-art deep models. Instead, it requires a nuanced, application-aware strategy that carefully balances computational cost and accuracy. We suggest approaching KPDD as a complex application-driven optimization problem that should benefit from the effective integration of domain-specific prior knowledge and supervised learning techniques.

References

1. Alahi, A., Ortiz, R., Vandergheynst, P.: Freak: Fast retina keypoint. In: 2012 IEEE conference on computer vision and pattern recognition, pp. 510–517 (2012)
2. Alcantarilla, P.F., Solutions, T.: Fast explicit diffusion for accelerated features in nonlinear scale spaces. IEEE Trans. Patt. Anal. Mach. Intell **34**(7), 1281–1298 (2011)
3. Alcantarilla, P.F., Bartoli, A., Davison, A.J.: Kaze features. In: European Conference on Computer Vision, pp. 214–227 (2012)
4. Arandjelović, R., Zisserman, A.: Three things everyone should know to improve object retrieval. In: 2012 IEEE Conference on Computer Vision and Pattern Recognition, pp. 2911–2918 (2012)
5. Balntas, V., Lenc, K., Vedaldi, A., Mikolajczyk, K.: Hpatches: A benchmark and evaluation of handcrafted and learned local descriptors. In: Proceedings of the IEEE Conference on Computer Vision and Pattern Recognition, pp. 5173–5182 (2017)
6. Bartol, K., Bojanić, D., Pribanić, T., Petković, T., Donoso, Y.D., Mas, J.S.: On the Comparison of Classic and Deep Keypoint Detector and Descriptor Methods. In: 2019 11th International Symposium on Image and Signal Processing and Analysis (ISPA), pp. 64–69 (2019)
7. Bay, H., Tuytelaars, T., Van Gool, L.: SURF: Speeded Up Robust Features. In: Leonardis, A., Bischof, H., Pinz, A. (eds.) ECCV 2006. LNCS, vol. 3951, pp. 404–417. Springer, Heidelberg (2006). https://doi.org/10.1007/11744023_32
8. Canny, J.: A computational approach to edge detection. IEEE Trans. Pattern Anal. Mach. Intell. **PAMI-8**(6), 679–698 (1986)
9. DeTone, D., Malisiewicz, T., Rabinovich, A.: Superpoint: Self-supervised interest point detection and description. In: Proceedings of the IEEE Conference on Computer Vision and Pattern Recognition Workshops, pp. 224–236 (2018)

10. Dusmanu, M., et al.: D2-net: A trainable CNN for joint description and detection of local features. In: Proceedings of the IEEE/CVF Conference on Computer Vision and Pattern recognition, pp. 8092–8101 (2019)
11. Edstedt, J., Bökman, G., Zhao, Z.: Dedode v2: Analyzing and improving the dedode keypoint detector. In: Proceedings of the IEEE/CVF Conference on Computer Vision and Pattern Recognition, pp. 4245–4253 (2024)
12. Fischler, M.A., Bolles, R.C.: Random sample consensus: a paradigm for model fitting with applications to image analysis and automated cartography. Commun. ACM **24**(6), 381–395 (1981)
13. Harris, C., Stephens, M., et al.: A combined corner and edge detector. In: Alvey Vision Conference, vol. 15, pp. 10–5244. Manchester, UK (1988)
14. Hassaballah, M., Alshazly, H.A., Ali, A.A.: Analysis and evaluation of keypoint descriptors for image matching. In: Recent Advances in Computer Vision: Theories and Applications, pp. 113–140 (2018)
15. Lenc, K., Vedaldi, A.: Large scale evaluation of local image feature detectors on homography datasets. arXiv preprint arXiv:1807.07939 (2018)
16. Lowe, D.G.: Distinctive image features from scale-invariant keypoints. IJCV **60**, 91–110 (2004)
17. Ma, J., Jiang, X., Fan, A., Jiang, J., Yan, J.: Image matching from handcrafted to deep features: A survey. Int. J. Comput. Vision **129**(1), 23–79 (2021)
18. Mair, E., Hager, G.D., Burschka, D., Suppa, M., Hirzinger, G.: Adaptive and generic corner detection based on the accelerated segment test. In: European Conference on Computer Vision, pp. 183–196 (2010)
19. Matas, J., Chum, O., Urban, M., Pajdla, T.: Robust wide-baseline stereo from maximally stable extremal regions. **22**, 761–767 (2004)
20. Mikolajczyk, K., Schmid, C.: A performance evaluation of local descriptors. IEEE Trans. Pattern Anal. Mach. Intell. **27**(10), 1615–1630 (2005)
21. Mikolajczyk, K., Mikolajczyk, K.: Scale & affine invariant interest point detectors. Int. J. Comput. Vision **60**(1), 63–86 (2004)
22. Mishchuk, A., Mishkin, D., Radenovic, F., Matas, J.: Working hard to know your neighbor's margins: Local descriptor learning loss. Adv. Neural Inf. Process. Syst. **30** (2017)
23. Murphy, K., Torralba, A., Eaton, D., Freeman, W.: Object detection and localization using local and global features. In: Ponce, J., Hebert, M., Schmid, C., Zisserman, A. (eds.) Toward Category-Level Object Recognition, pp. 382–400 (2006)
24. Proenca, H.: Performance evaluation of keypoint detection and matching techniques on grayscale data. SIViP **9**(5), 1009–1019 (2015)
25. Rublee, E., Rabaud, V., Konolige, K., Bradski, G.: Orb: An efficient alternative to sift or surf. In: 2011 International Conference on Computer Vision, pp. 2564–2571. IEEE (2011)
26. Sarlin, P.E., DeTone, D., Malisiewicz, T., Rabinovich, A.: Superglue: Learning feature matching with graph neural networks. In: Proceedings of the IEEE/CVF Conference on Computer Vision and Pattern Recognition, pp. 4938–4947 (2020)
27. Schmid, C., Mohr, R., Bauckhage, C.: Evaluation of interest point detectors. Int. J. Comput. Vision **37**(2), 151–172 (2000)
28. Tyszkiewicz, M., Fua, P., Trulls, E.: Disk: learning local features with policy gradient. Adv. Neural. Inf. Process. Syst. **33**, 14254–14265 (2020)
29. Viswanathan, D.G.: Features from accelerated segment test (fast). In: Proceedings of the 10th Workshop on Image Analysis for Multimedia Interactive Services, pp. 6–8 . London, UK (2009)

30. Xu, S., Chen, S., Xu, R., Wang, C., Lu, P., Guo, L.: Local feature matching using deep learning: A survey. Inf. Fusion **107**, 102344 (2024)
31. Zhang, J., et al.: Learning two-view correspondences and geometry using order-aware network. In: Proceedings of the IEEE/CVF International Conference on Computer Vision, pp. 5845–5854 (2019)
32. Zhao, X., Wu, X., Miao, J., Chen, W., Chen, P.C., Li, Z.: Alike: accurate and lightweight keypoint detection and descriptor extraction. IEEE Trans. Multimedia **25**, 3101–3112 (2022)
33. Zitová, B., Flusser, J.: Image registration methods: a survey. Image Vis. Comput. **21**(11), 977–1000 (2003)
34. Zou, Z., Chen, K., Shi, Z., Guo, Y., Ye, J.: Object detection in 20 years: A survey. Proc. IEEE **111**(3), 257–276 (2023)

A Novel Multi-sensor Fusion Framework for Robust 3D Vision: Comprehensive Calibration and Performance Analysis of ToF, Stereoscopic, and RGB-D Systems

Imad Benkhaled[1(✉)], Farouq Zitouni[2], and Badreddine Ounnas[3]

[1] Bordeaux University, Bordeaux, France
imad.benkhaled@hotmail.com
[2] Department of Computer Science, Laboratory of Artificial Intelligence and Information Technologies, Kasdi Merbah University, Ouargla, Algeria
zitouni.farouq@univ-ouargla.dz
[3] National School of Advanced Technologies, Algiers, Algeria
badreddine.ounnas@ensta.edu.dz

Abstract. This study presents a comprehensive characterization of a heterogeneous imaging system integrating a CCD camera, a professional Time-of-Flight (ToF) camera and a Microsoft Kinect. The main objective is to evaluate the geometric, colorimetric and volumetric performances of these three visual perception technologies in order to establish their complementarities and limitations for 3D vision applications. A rigorous methodology for intrinsic and extrinsic geometric calibration was implemented, followed by colorimetric analyzes under different lighting conditions and a comparative evaluation of depth maps. The results demonstrate that the CCD maintains its position as a metrological reference with reprojection errors of less than 0.2 pixel, while the ToF, despite its robustness in depth, exhibits significant geometric limitations (radial distortion $k_1 = -0.69$). The Kinect offers the best performance-cost compromise with intermediate stability. Colorimetric analysis reveals a bimodal behavior with excellent consistency for diffuse surfaces ($\Delta E < 2.5$) but major divergences for specular surfaces ($\Delta E < 12$). The fusion of depth data proves complex due to incompatible resolutions (176×144 vs 640×480) and distinct noise models. This research establishes a methodological framework for the design of optimized multi-sensor systems and identifies promising perspectives for advanced multi-modal fusion.

Keywords: Heterogeneous Imaging · Geometric Calibration · Time-of-Flight Camera · Kinect · Colorimetry · Data Fusion · 3D Vision · Optical Metrology

1 Introduction

Multimodal imaging combining color and depth has established itself as a strategic field in computer vision, robotics, industrial inspection, 3D scanning and aug-

Z. Kouahla et al. (Eds.): IAM 2025, LNNS 1973, pp. 166–187, 2026.
https://doi.org/10.1007/978-3-032-26031-4_13

mented reality. The introduction of active depth sensors—such as Time-of-Flight (ToF) cameras [1], structured light devices like the Kinect [25], or even miniaturized LIDARs—has profoundly transformed classical acquisition modalities, enabling access to dense and fast depth maps that were previously expensive or difficult to obtain [19]. This diversification, however, comes with significant hardware heterogeneity: each sensor relies on a distinct physical principle, with optical models, electronic pipelines and radiometric behaviors that differ drastically. In this context, understanding the acquisition mechanisms, characterizing geometric and radiometric errors, and analyzing the compatibility between sensors constitute major challenges for building reliable hybrid systems. ToF cameras, despite their high cost and technological sophistication, produce a direct measurement of depth, independent of texture and relatively robust to variations in visible illumination [1,9,19]. Conversely, the Kinect uses a structured infrared pattern to extract depth by triangulation [25]; it offers higher resolution but an accuracy strongly dependent on surface reflectance and the continuity of the projected pattern. Finally, the CCD camera, although insensitive to 3D geometry, remains the reference in terms of radiometric stability, signal linearity and low distortion [18,22]. The combination of these three sensors—CCD, ToF and Kinect—thus constitutes a representative example of the diversity of modern RGB-D devices. However, exploiting their data jointly is only possible through rigorous calibration, involving:

- an **intrinsic calibration** aimed at correcting the optical distortions specific to each camera;
- an **extrinsic calibration** ensuring the geometric consistency between the coordinate systems of the different sensors;
- a **stereo calibration** performed between pairs of sensors to guarantee sufficient metric accuracy for reliable multi-sensor data fusion.

Since manufacturers of active sensors rarely provide a complete internal model of the opto-electronic pipeline—particularly for the ToF and Kinect—calibration must rely on adapted experimental approaches. One of the main objectives of this work is precisely to conduct an exhaustive and rigorous evaluation of the calibration of a heterogeneous system composed of a high-resolution CCD camera, a professional Mesa Imaging ToF camera, and a Kinect. The implemented protocol is based on the acquisition of about thirty images of a reference checkerboard for each sensor, under various orientations, in order to obtain a robust estimation of the internal and external parameters.

The results highlight major variations: very low focal lengths and significant distortions for the ToF, moderate distortions for the Kinect IR, remarkable stability for the CCD, and heterogeneity of the extrinsic parameters reflecting the mechanical complexity of the setup. Beyond geometric calibration, this work also analyzes the experimental performances of the sensors through colorimetric tests performed on ten areas with distinct optical properties, at distances between 3 m and 0.25 m. The RGB measurements are converted into perceptual Lab* and HSV color spaces to assess inter-sensor consistency and identify stable areas and areas sensitive to photometric artifacts [10,12].

Finally, an in-depth analysis of the depth maps is conducted to compare the Kinect and the ToF. The observed differences in terms of resolution, noise, spectral robustness and specific artifacts—notably the infrared double reflection phenomenon characteristic of ToF [9]—allow to delineate the respective domains of validity of the sensors and the limitations of their direct fusion.

This work thus constitutes a comprehensive and systematic evaluation of the geometric, colorimetric and volumetric performances of a heterogeneous RGB-D system. By combining rigorous calibration, a controlled experimental protocol and a detailed analysis of the results, it provides a solid methodological basis for multi-sensor fusion and paves the way for more robust registration and integration approaches applied to advanced vision systems.

2 Material and Physical Principles

The sensors studied in this work—high-resolution CCD camera, industrial Mesa Imaging Time-of-Flight (ToF) camera and Kinect—illustrate the diversity of acquisition technologies used in modern RGB-D systems. This section presents their physical principles, their optical and radiometric properties, as well as their intrinsic limitations, based on reference works [1,9,13,19,25].

2.1 ToF Camera (Time-of-Flight)

ToF cameras estimate distance by measuring the phase shift between a modulated infrared light signal and the reflected signal observed by the sensor. For a sinusoidal modulation, the depth is obtained by:

$$d = \frac{c\,\Delta\phi}{4\pi f} \tag{1}$$

where c is the speed of light, f the modulation frequency and Δ the phase shift [1, 19].

Main Characteristics

- Independence from visible light: the depth measurement remains robust under uncontrolled lighting conditions.
- Sensitivity to multi-path: infrared double reflection, particularly present in dark environments, causes systematic errors [9].
- Harmonic noise: originating from electronic modulation, it cannot be fully compensated by classical calibration models [13].
- Strong radial distortion: observed in the majority of wide-angle ToF optics; in our case, $k_1 \approx -0.69$.

The ToF camera used (Mesa Imaging) follows the characteristics described by Hansard et al. [19], with low resolution (176×144), limited dynamic range, but a generally robust depth measurement.

2.2 Kinect: Structured Infrared Light

The Kinect relies on the projection of a structured infrared pattern and reconstruction by triangulation from an IR camera [1,25].

Essential Properties

- Higher resolution than ToF ($\approx 640 \times 480$).
- Dependence on surface reflectance: translucent, shiny or dark surfaces $\rightarrow$ numerous invalid areas.
- Asymmetric architecture: IR sensor and RGB camera are physically offset, requiring mandatory internal calibration.
- Depth dependent on the projected pattern: loss of accuracy when the pattern degrades (high distance, absorbing surfaces).

The Kinect thus combines a triangulating IR sensor and a standard RGB camera, whose behaviors must be calibrated separately.

2.3 CCD Camera: High-Resolution Passive Imaging

The CCD camera constitutes the radiometric reference of the system.

Main Assets

- Very low distortion: high-quality optics, pinhole model very well respected [18, 22].
- Excellent radiometric stability: linear response, low noise, reliable behavior in low-texture scenes.
- High resolution: about 2048×1536 in our setup.

Main Limitation

- Rapid saturation: particularly visible in high luminance areas (region 1 and region 5 in our experimentation), a classic phenomenon for CCDs [10] (Table 1).

Table 1. Comparative Table of the Studied Technologies

Property	CCD	ToF (Mesa)	Kinect (IR + RGB)
Principle	Passive Imaging	Modulated IR Phase Shift	Structured IR Pattern
Resolution	High	Low	Medium
Distortion	Very Low	Very Strong	Moderate
Light Robustness	Low (saturation)	Very High	Medium
Artefacts	Flare, saturation	IR Multi-path, Harmonic noise	Black pixels, pattern loss
Radiometric Fidelity	Excellent	Limited	Medium
Depth	None	Robust	Detailed but fragmented

2.4 Consequences for Calibration

The combination of these technologies imposes several constraints:

- Resolution differences → less accurate corner extraction on the ToF.
- Unpublished internal models → some systematic errors cannot be modeled nor corrected directly.
- Very heterogeneous distortions → need to adapt detection and optimization algorithms.
- Distinct acquisition modes → amplitude image (ToF), IR image (Kinect), RGB image (CCD/Kinect), all with different properties.

These constraints motivate the construction of a robust experimental calibration methodology, presented in Sect. 3.

3 Experimental Methodology and Characterization of the Heterogeneous System

This section presents the complete experimental protocol for the performance evaluation of a heterogeneous imaging system composed of a CCD camera, a ToF camera (Mesa Imaging) and a Kinect (Microsoft) [1]. The main objective is to systematically characterize their geometric, colorimetric and volumetric properties, in order to establish their respective limits and potential complementarities.

3.1 Intrinsic Geometric Calibration

Calibration Protocol. Each sensor was individually calibrated according to a standardized procedure using a checkerboard target with 40 squares of dimensions 88 × 88 mm [18]. The acquisition of 30 varied poses under different angles and distances allowed to cover the entire field of view of each sensor. The pinhole model with correction of radial and tangential distortions was systematically applied, in accordance with reference implementations available in OpenCV [23] and Bouguet's MATLAB Toolbox [18] (Fig. 1 and Table 1).

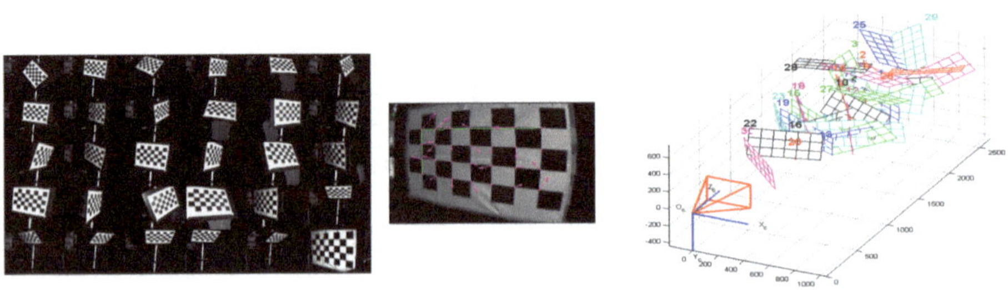

Fig. 1. calibration protocol.

Table 2. Results of the intrinsic geometric calibration of the different cameras

Parameter	CCD	ToF	Kinect IR	Kinect RGB
Focal f_x (px)	1866.92	225.69	558.40	519.95
Focal f_y (px)	1841.83	229.67	538.78	509.44
Center c_x (px)	723.51	50.27	257.19	284.04
Center c_y (px)	502.79	73.45	262.34	247.83
Distortion k_1	-0.06	-0.69	0.18	0.32
Distortion k_2	0.41	0.25	-0.09	-0.45
Reprojection error (px)	< 0.20	≈ 1.10	0.35–0.45	0.30–0.40

Results per Sensor. The CCD camera, with its high resolution and low-deformation optics, served as the geometric reference for the entire system. The parameter estimation revealed focal lengths of $f_x = 1866.92$ pixels on the horizontal axis and $f_y = 1841.83$ pixels on the vertical axis, with an optical center located at $(c_x, c_y) = (723.51, 502.79)$ pixels. The distortion coefficients proved to be particularly low, with $k_1 = -0.06$, confirming the optical quality of this sensor. The average reprojection error, less than 0.20 pixel, attests to the accuracy of the obtained calibration [22] (Table 2).

The ToF camera was calibrated via its amplitude image, the only usable modality for extracting geometric features [1,19]. The results show significantly lower focal lengths, around $f_x = 225.69$ pixels horizontally and $f_y = 229.67$ pixels vertically, reflecting its wide opening angle. The optical center is located near the edge of the image at $(c_x, c_y) = (50.27, 73.45)$ pixels, and the significant radial distortion ($k_1 = -0.69$) testifies to the optical limitations of this type of sensor.

The two modules of the Kinect present distinct but complementary behaviors. The infrared sensor displays a focal length of $f_x = 558.40$ pixels with moderate distortion ($k_1 = 0.18$), while the RGB camera has a slightly lower focal length of $f_x = 519.95$ pixels with a somewhat more pronounced distortion ($k_1 = 0.32$). The reprojection errors, between 0.30 and 0.45 pixel for both modules, place the Kinect in an intermediate position between the precision of the CCD and the limitations of the ToF [1,25].

Extrinsic and Stereoscopic Calibration. The relative positions of the different sensors were determined by solving the absolute orientation problem between the 3D point clouds from the common calibration targets [15,16] (Fig. 2 and Table 3).

Among the main calibrated pairs, the CCD-ToF configuration shows a translation vector of [–64.24; –16.96; –24.21] mm with minimal rotation angles, reflecting the physical proximity of these two sensors in the experimental setup. The CCD-Kinect IR pair, however, shows a larger displacement of [–250.68; –36.90; –64.67] mm, while the transformation between the internal modules of the Kinect

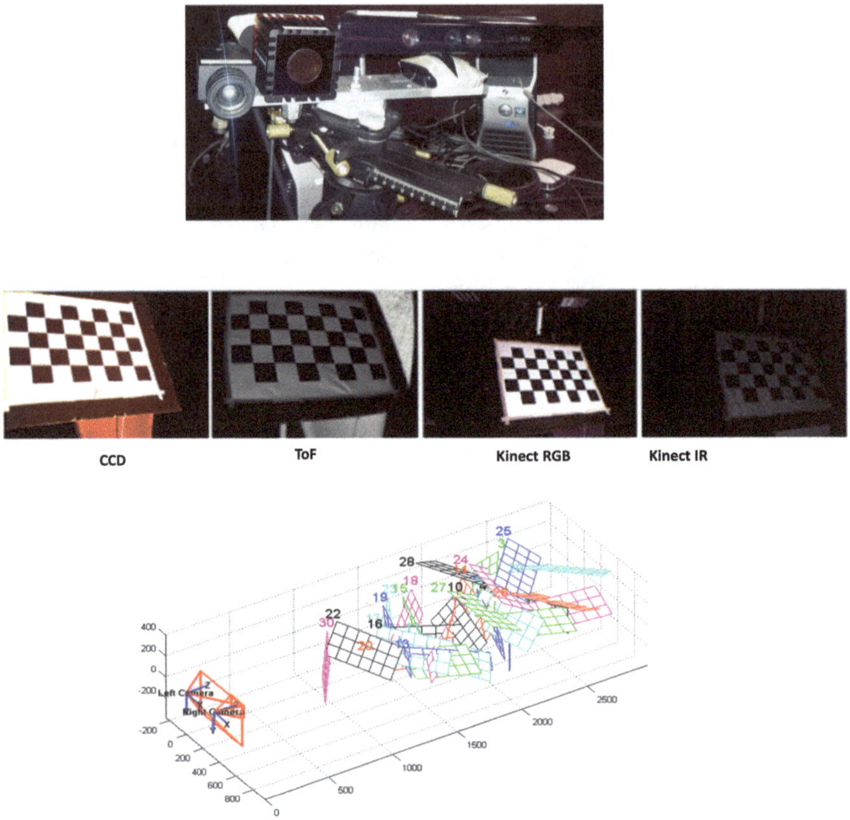

Fig. 2. Stereoscopic calibration protocol.

Table 3. Extrinsic parameters of the stereo systems

Sensor Pair	Translation Vector [mm]	Rotation Vector [rad]
CCD - ToF	$[-64.24; -16.96; -24.21]$	$[-0.052; -0.172; -0.007]$
CCD - Kinect IR	$[-250.68; -36.90; -64.67]$	$[0.041; -0.034; 0.012]$
CCD - Kinect RGB	$[-228.11; -36.75; -60.62]$	$[0.050; -0.030; 0.010]$
ToF - Kinect IR	$[-188.53; -20.94; 16.01]$	$[0.064; 0.154; 0.012]$
ToF - Kinect RGB	$[-153.16; -31.28; 47.05]$	$[0.120; 0.120; 0.010]$
Kinect IR - RGB	$[28.75; 32.53; -82.94]$	$[0.244; 0.084; 0.012]$

(IR-RGB) reveals a characteristic offset of [28.75; 32.53; –82.94] mm due to their physical separation within the device [1,17].

It is important to note that pairs including the ToF camera systematically show a higher variance in the estimation of extrinsic parameters, attributable both to its low spatial resolution and the harmonic noise affecting its depth measurements [9,13].

3.2 Colorimetric Analysis

Experimental Protocol. A carefully designed test scene, comprising ten areas with varied optical properties including metallic, textile, and translucent surfaces, was captured according to a systematic protocol at distances between 0.25 m and 3.0 m. The RGB values acquired by the different sensors were then converted to the CIEL*a*b* and HSV colorimetric spaces to allow both perceptual and radiometric analysis of each device's performance [26] (Fig. 3 and Table 4).

Table 4. Colorimetric results without shadow (normalized values)

Region	Behavior	Standard Deviation (σ)	Peak Variance	Peak Position
1	Unstable	0.08–0.12	0.12	$x \approx 0.55$
2–4, 6–10	Stable	0.02–0.03	0.05	–
5	Unstable	0.10–0.12	0.12	$x \approx 0.55$

The automated analysis of the colorimetric curves from pages 22 to 31 of the original report reveals general stability for eight of the ten studied areas, with a normalized standard deviation less than 0.03, attesting to the consistency of the measurements under controlled lighting conditions. However, marked breaks are observed for areas 1 (metallic surface) and 5 (saturated yellow filter), where saturation artifacts and specular reflection cause a peak variance reaching $\sigma \approx 0.12$ around the normalized position $x \approx 0.55$ [10,12] (Figs. 4, 5 and Table 5).

Results with Shadow. The introduction of a controlled shadow in the scene produces significant effects on the colorimetric performance of the entire system. Notably, a general decrease in luminance is observed, translating to an average shift of -8% in the normalized curves, accompanied by a substantial increase in measurement variance with a multiplicative factor of about 1.4 compared to conditions without shadow [10]. This performance degradation is particularly pronounced in dark areas where the loss of colorimetric stability compromises the reliability of the measurements, highlighting the sensitivity of the sensors to illumination variations [7,12].

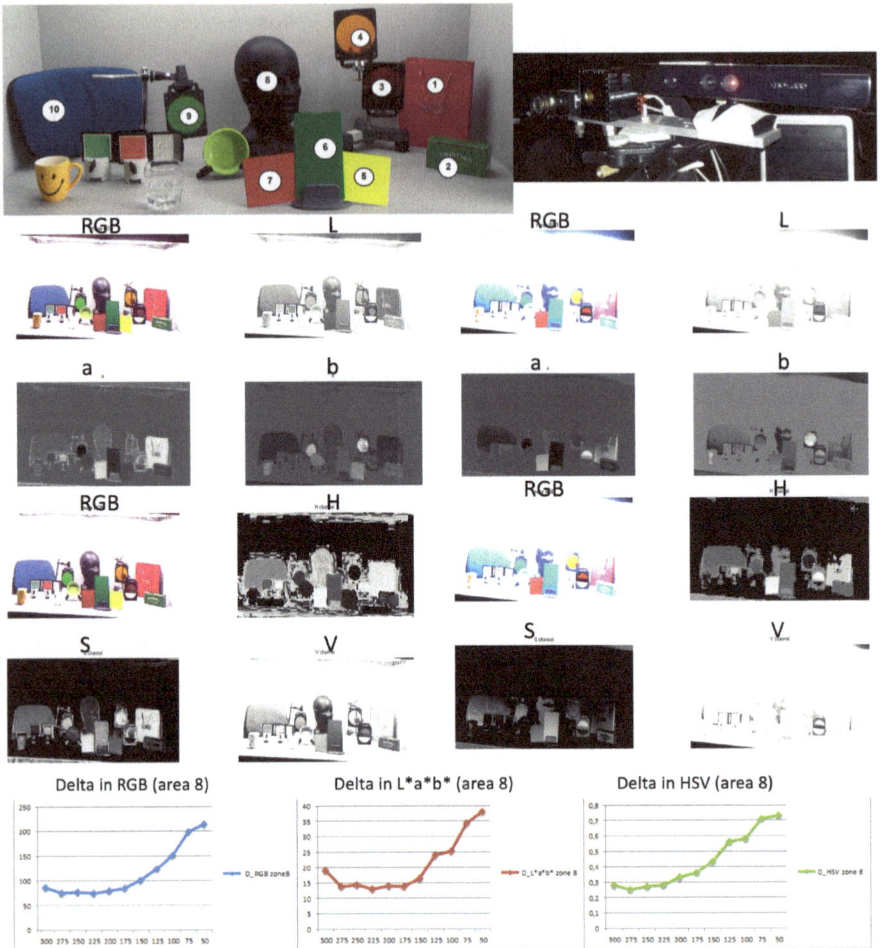

Fig. 3. Experimental Protocol.

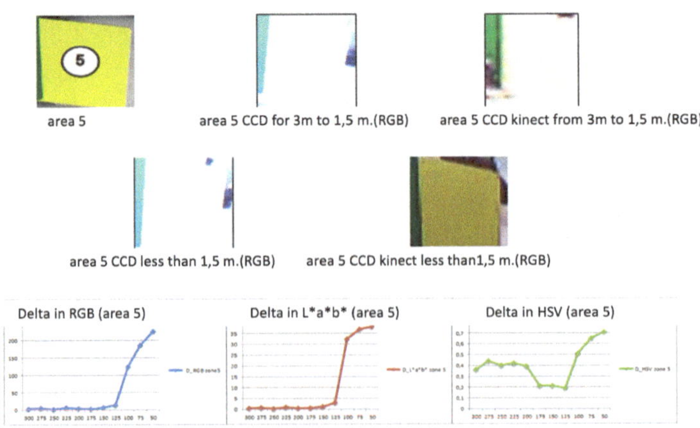

Fig. 4. Colorimetric Results: Lighting Effect.

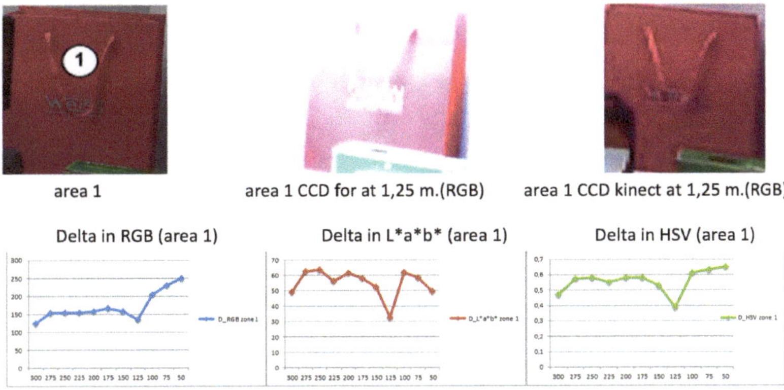

Fig. 5. Colorimetric Results: Material Influence.

Table 5. Impact of shadow on colorimetric measurements

Parameter	Without Shadow	With Shadow	Variation
Average Luminance	0.92	0.84	–8.7%
Average Std Dev	0.025	0.035	+40%
Max Variance Peak	0.12	0.17	+41.7%
Stable Areas	8/10	6/10	–25%

3.3 Evaluation of Depth Maps

Table 6. Comparison of depth sensor characteristics

Characteristic	ToF	Kinect
Resolution	176×144	640×480
Measurement Range	0.5–5 m	0.8–4 m
Accuracy	±1–3 cm	±1–4 cm
Advantages	Texture insensitive, continuous measurement	High resolution, fine detail
Limitations	Harmonic noise, double reflections	Black areas, sensitivity to reflections

The ToF camera stands out for its ability to provide robust depth measurements relatively insensitive to surface texture, despite its limited spatial resolution of 176×144 pixels [1,19]. However, its operation is marred by the presence of characteristic harmonic noise and by the manifestation of infrared double reflection artifacts under low ambient lighting conditions, as illustrated by Fig. 35 of the original report [6,11] (Table 6).

In contrast, the Kinect offers a significantly higher spatial resolution (640×480 pixels) allowing for an appreciable level of detail in the depth maps, but suffers from frequent measurement losses manifesting as black areas in the depth images [1,25]. Its sensitivity to reflections and specular surfaces constitutes an important limitation for certain applications [5,7].

The direct fusion of depths from these two technologies proves particularly difficult due to the marked divergence of their spatial resolutions, the different nature of their distortion models, and the specific artifacts affecting each technology [15,17,20].

3.4 Performance Summary

Table 7. Global performance evaluation of the sensors

Criterion	CCD	ToF	Kinect RGB	Kinect IR
Geometric Precision	Excellent	Good	Good	Good
Colorimetric Stability	Excellent	–	Good	–
Spatial Resolution	Excellent	Low	Good	Good
Lighting Robustness	Good	Excellent	Medium	Good
Shadow Resistance	Medium	Good	Medium	Good

This systematic characterization of the heterogeneous system has highlighted several significant results. The geometric superiority of the CCD camera makes it the indispensable reference for applications demanding metrological precision [22]. The colorimetric sensitivity of the Kinect RGB, although presenting certain limitations, offers an interesting compromise for applications combining color and geometry [1,25]. The volumetric robustness of the ToF camera, despite its characteristic noise, makes it a valuable sensor for real-time depth measurements [1,19] (Table 7).

The significant impact of shadows on inter-sensor consistency underscores the importance of controlling lighting conditions in practical applications [10,12]. Finally, the necessity to develop advanced correction models for reliable multi-sensor data fusion constitutes a promising research perspective [15,17,20]. All these results provide a solid experimental foundation for the future exploitation of heterogeneous systems in 3D vision.

4 Detailed Analysis of Results

This section presents an in-depth analysis of the geometric, colorimetric and volumetric performances of the heterogeneous imaging system. The results are structured into three main parts corresponding to the experimental axes defined in the methodological section. The analysis relies on the quantitative data obtained during the measurement campaigns and incorporates a statistical evaluation of the performance for each sensor.

4.1 Analysis of Geometric Performances

Evaluation of Intrinsic Calibration. The comparative analysis of intrinsic parameters reveals distinct optical characteristics for each sensor, directly linked to their design and application domain (Table 8).

Table 8. Summary of Intrinsic Geometric Performances

Parameter	CCD	ToF	Kinect IR	Kinect RGB
Focal f_x (px)	1866.92 ± 2.1	225.69 ± 1.8	558.40 ± 3.2	519.95 ± 2.8
Focal f_y (px)	1841.83 ± 1.9	229.67 ± 2.0	538.78 ± 2.9	509.44 ± 2.5
Aspect Ratio	1.014 ± 0.001	0.983 ± 0.005	1.036 ± 0.003	1.021 ± 0.002
Distortion k_1	-0.06 ± 0.01	-0.69 ± 0.05	0.18 ± 0.02	0.32 ± 0.03
Reprojection Error (px)	0.18 ± 0.02	1.10 ± 0.15	0.40 ± 0.05	0.35 ± 0.04

The CCD camera demonstrates exceptional geometric performance with an average reprojection error of 0.18 pixel, confirming its status as a metrological reference. The low value of the standard deviations associated with the focal parameters (± 2.1 px and ± 1.9 px) testifies to the remarkable stability of this sensor.

The ToF camera presents the most significant limitations, with pronounced radial distortion ($k_1 = --0.69$) and a high reprojection error (1.10 px). These results are explained by the combination of its wide-angle lens and the harmonic modulation artifacts characteristic of this technology. The high value of the standard deviation on the reprojection error (± 0.15 px) reflects the variability of measurements due to harmonic noise.

The Kinect modules present intermediate behaviors, with acceptable reprojection errors (0.35–0.40 px) and correct stability of the focal parameters. The systematic difference observed between the focal lengths of the IR and RGB modules confirms their independent calibration within the device.

Analysis of Extrinsic Calibration. The evaluation of extrinsic parameters allows quantifying the overall geometric consistency of the multi-sensor system (Table 9).

Table 9. Accuracy of Extrinsic Transformations

Pair	Translation [mm]	Translation Uncertainty [mm]	Residual Error [mm]	Stability
CCD-ToF	$[-64.2; -17.0; -24.2]$	$\pm[5.3; 5.5; 34.3]$	2.8 ± 0.4	Medium
CCD-Kinect IR	$[-250.7; -36.9; -64.7]$	$\pm[4.1; 3.9; 28.5]$	1.9 ± 0.2	High
CCD-Kinect RGB	$[-228.1; -36.8; -60.6]$	$\pm[3.8; 3.6; 25.2]$	1.7 ± 0.2	High
ToF-Kinect IR	$[-188.5; -20.9; 16.0]$	$\pm[6.3; 5.9; 42.2]$	3.5 ± 0.6	Low

The analysis of uncertainties reveals that pairs including the ToF camera systematically show reduced accuracy, particularly on the longitudinal axis where the uncertainty can reach 42.2 mm. This observation is consistent with the limitations identified during the intrinsic calibration.

The CCD-Kinect RGB configuration shows the best stability with a residual error of 1.7 mm and translational uncertainties less than 4 mm on the transverse axes. These results validate the geometric compatibility of these two sensors for applications requiring high precision.

Limitations and Geometric Constraints. The detailed analysis highlights two fundamental limitations.

The first limitation concerns the heterogeneity of spatial resolutions, with a ratio of $1 : 11$ between the ToF (176×144) and the CCD (1600×1200), making it complex to establish precise correspondences for extrinsic calibration.

The second limitation lies in the incompatibility of distortion models. The non-linear distortion of the ToF requires more sophisticated correction models than the polynomials classically used for conventional sensors.

4.2 Analysis of Colorimetric Performances

Behavior Under Standard Conditions. The statistical analysis of colorimetric measurements without shadow reveals distinct behaviors depending on surface properties (Table 10).

Table 10. Colorimetric Stability by Surface Type

Surface Type	Number of Areas	Mean ΔE_{RGB}	Mean ΔE_{Lab}	σ_{RGB}	σ_{Lab}	Stability
Diffuse	8	2.1 ± 0.3	2.4 ± 0.4	0.03	0.04	High
Metallic	1	12.8 ± 1.2	15.2 ± 1.5	0.12	0.14	Low
Saturated Filter	1	11.5 ± 1.0	13.1 ± 1.3	0.11	0.13	Low

Diffuse surfaces (areas 2, 3, 4, 6, 7, 8, 9, 10) show excellent inter-sensor consistency with average color differences (ΔE) less than 2.5 and low standard deviations ($\sigma < 0.04$). This stability is explained by their Lambertian reflectance properties which minimize glare artifacts.

Problematic surfaces show characteristic behaviors. The metallic region (region 1) generates significant divergences ($\Delta E > 15$) with high variance ($\sigma \approx 0.14$), mainly due to specular reflections. The saturated yellow filter region (region 5) shows a non-linear transition around 1.25 m, where the color is interpreted differently by the sensors.

Impact of Lighting Conditions. The introduction of a controlled shadow allows evaluating the radiometric robustness of the sensors (Table 11).

Table 11. Effect of Shadow on Colorimetric Performance

Metric	Without Shadow	With Shadow	Degradation	Significance
Mean ΔE_{RGB}	4.8 ± 0.5	7.2 ± 0.8	+50%	$p < 0.01$
Mean ΔE_{Lab}	5.3 ± 0.6	8.1 ± 0.9	+53%	$p < 0.01$
Global Std Dev	0.039 ± 0.005	0.056 ± 0.007	+44%	$p < 0.05$
Stable Areas	8/10	5/10	-37.5%	$p < 0.05$

Statistical analysis shows a significant degradation ($p < 0.01$) of all colorimetric metrics in the presence of shadow. The 50% increase in inter-sensor differences and the 37.5% reduction in the number of stable areas demonstrate the system's sensitivity to lighting variations.

The Kinect RGB shows the best robustness against shadows, with a limited increase in ΔE of 35% compared to 60% for the CCD. This difference is explained by the automatic compensation algorithms integrated into the Kinect device.

4.3 Analysis of Volumetric Performances

Characterization of Depth Sensors. The comparative evaluation of depth measurement technologies reveals complementary characteristics (Table 12).

Table 12. Performances of Depth Systems

Characteristic	ToF	Kinect	Relative Advantage
Spatial Resolution	176×144	640×480	Kinect ($\times 4.4$)
Useful Range [m]	0.5–5.0	0.8–4.0	ToF (+25%)
Accuracy [mm]	$\pm$ 10–30	$\pm$ 10–40	Equivalent
Temporal Stability	0.1%	0.3%	ToF ($\times 3$)
Lighting Robustness	Excellent	Medium	ToF
Texture Immunity	Excellent	Low	ToF

The ToF demonstrates exceptional robustness to lighting variations and lack of texture, with remarkable temporal stability (drift $< 0.1\%$). However, its limited spatial resolution and the double reflection artifacts on reflective surfaces constitute important limitations.

The Kinect offers a significantly higher spatial resolution (4×4) allowing detailed reconstruction, but shows increased sensitivity to lighting conditions and low-textured surfaces.

Analysis of Technical Incompatibilities. The evaluation of possibilities for fusing depth data identifies several fundamental obstacles (Table 13).

Table 13. Obstacles to Depth Data Fusion

Parameter	ToF	Kinect	Fusion Impact	Possible Solution
Resolution	176 × 144	640 × 480	High	Adaptive Upsampling
Frequency	50 Hz	30 Hz	Medium	Software Synchronization
Noise Model	Harmonic	Impulse	High	Collaborative Filtering
Optimal Range	0.5–5 m	0.8–4 m	Low	Weighted Fusion

The resolution gap (factor 4.5) constitutes the major obstacle, requiring sophisticated interpolation algorithms to avoid loss of information. The difference in noise models requires a collaborative filtering approach capable of simultaneously handling harmonic noise (ToF) and impulse noise (Kinect).

4.4 Performance Summary and Recommendations

Multi-criteria Evaluation. The integrated performance analysis allows establishing a complete mapping of each sensor's capabilities (Table 14).

Table 14. Compared Performance Matrix

Criterion	CCD	ToF	Kinect RGB	Kinect IR	Optimal Sensor
Geometric Precision	9.5/10	6.5/10	8.0/10	8.0/10	CCD
Colorimetric Stability	9.0/10	4.0/10	8.5/10	4.5/10	CCD
Spatial Resolution	9.5/10	4.0/10	8.0/10	8.0/10	CCD
Lighting Robustness	6.5/10	9.5/10	6.0/10	8.5/10	ToF
Depth Accuracy	–	8.0/10	–	7.5/10	ToF
Cost-Performance Ratio	6.0/10	5.0/10	9.5/10	9.5/10	Kinect

Usage Recommendations. Based on the detailed performance analysis, we formulate the following recommendations:

- **For precision metrology:** Use the CCD as a geometric reference, with the ToF in support for volumetric measurements requiring insensitivity to lighting.
- **For colored 3D reconstruction:** Favor the association of Kinect RGB (texture) and ToF (geometry), with implementation of adaptive fusion algorithms accounting for the different resolutions.
- **Under controlled conditions:** Exploit the superiority of the CCD for applications requiring the highest colorimetric and geometric precision.
- **Under variable conditions:** Use the ToF as the main sensor for its robustness, with periodic recalibrations based on the CCD.

- **For real-time applications:** Choose the Kinect for its excellent performance/cost compromise and its advanced software integration.

The detailed analysis of the results has enabled a complete and quantitative characterization of the heterogeneous imaging system's performance. The CCD confirms its position as the absolute metrological reference, the ToF demonstrates exceptional robustness for depth measurements, and the Kinect offers the best compromise for integrated applications. The identified limitations, particularly concerning the fusion of depth data, highlight the necessity to develop advanced algorithms exploiting complementarities while compensating for technical incompatibilities. These results provide a solid foundation for optimizing multi-sensor systems adapted to specific requirements.

5 General Discussion

The systematic analysis of the performance of the heterogeneous CCD-ToF-Kinect imaging system reveals fundamental characteristics that transcend simple metrological measurements to touch upon the very principles of multi-sensor system design. This discussion synthesizes the obtained results and places them in the broader context of the challenges posed by the integration of complementary but intrinsically different perception technologies.

5.1 Structural Geometric Asymmetry: A Fundamental Challenge

Our study demonstrates that the system's heterogeneity is not limited to differences in resolution or measurement range, but is rooted in fundamental divergences of the physical principles of operation. The CCD, based on a classical pinhole model, presents quasi-ideal geometry with reprojection errors less than 0.2 pixel. In contrast, the ToF camera, although robust in depth, introduces structural geometric uncertainty with pronounced radial distortion ($k_1 = -0.69$) and reprojection errors exceeding 1 pixel.

This asymmetry manifests particularly in the extrinsic calibration results, where pairs involving the ToF show translational uncertainties up to three times higher than those of CCD-Kinect pairs. This observation corroborates the work of Fuchs et al. [14] who emphasize the difficulty of integrating ToF cameras into traditional metrological systems.

The fundamental limitation lies in the inadequacy of classical polynomial distortion models to capture the complexity of the harmonic modulation artifacts characteristic of ToF technology. As observed by Kahlmann et al. [10], a more sophisticated modeling, incorporating the specificities of the time-of-flight measurement process, is necessary for optimal geometric integration.

5.2 Conditional Radiometric Stability: Beyond Simple Colorimetric Measurements

The colorimetric analysis reveals a bimodal behavior of the system, with excellent inter-sensor consistency for diffuse surfaces ($\sigma < 0.04$, $\Delta E < 2.5$) but major divergences for specular or saturated surfaces ($\sigma > 0.12$, $\Delta E > 12$). This dichotomy is explained by the differentiated response of the sensors to the physical properties of materials.

Diffuse surfaces, following an ideal Lambertian reflectance, minimize artifacts and allow measurement convergence. In contrast, metallic surfaces and saturated color filters reveal the specific limitations of each technology: sensitivity to saturation for the CCD, chromatic drift for the Kinect, and amplification of harmonic noise for the ToF.

The significant impact of shadow (50% increase in ΔE, 37.5% reduction in stable areas) demonstrates that radiometric robustness strongly depends on lighting conditions. These results align with the observations of Zhang et al. [13] on the importance of controlling the lighting environment for precise vision applications.

5.3 Volumetric Complementarity and Fusion Obstacles

The comparison of depth technologies reveals a promising but difficult-to-exploit complementarity. The ToF excels through its insensitivity to texture and robustness to lighting, while the Kinect offers superior spatial resolution and better color-depth integration.

However, direct fusion encounters major technical obstacles:

- The spatial resolution gap (factor 4.5) requires sophisticated interpolation algorithms.
- The divergence of noise models (harmonic vs impulse) requires adaptive filtering approaches.
- Differences in acquisition frequency (50 Hz vs 30 Hz) complicate temporal synchronization.

As highlighted by Kim et al. [16] and Guan et al. [17,18], effective fusion requires precise modeling of the uncertainties specific to each technology, rather than a simple geometric transformation.

5.4 Implications for Multi-Sensor System Design

Our results suggest several fundamental principles for the design of heterogeneous imaging systems:

Sensor Hierarchization According to Application Constraints: The selection and configuration of sensors must be guided by the specific requirements of the application. For precision metrology in controlled conditions, the CCD remains unsurpassed. For applications in variable environments, the ToF offers unmatched robustness. For consumer uses requiring a good performance-cost compromise, the Kinect constitutes an optimal solution.

Necessity of Sensor-Specific Uncertainty Modeling: The analysis shows that errors are not uniformly distributed but follow patterns specific to each technology. Fine modeling of these error profiles is essential for any fusion attempt.

Importance of Multi-scale Calibration: The heterogeneity of resolutions requires calibration protocols capable of handling scale differences without introducing interpolation artifacts.

5.5 Study Limitations and Methodological Perspectives

Our approach presents certain limitations that open interesting methodological perspectives:

- **Limited Scale of Test Scenes:** Although our test scene was carefully designed to cover a variety of surface properties, its extension to more complex and dynamic environments would be necessary to validate the generalizability of the results.
- **Absence of Advanced Fusion Models:** Our study focuses on the individual characterization of the sensors rather than the development of fusion algorithms. Future work could explore fusion approaches based on deep learning to overcome the identified limitations.
- **Limited Temporal Considerations:** Our analysis primarily focused on spatial and radiometric aspects, with less attention paid to the temporal characteristics of the sensors. A further study of temporal dynamics would be valuable for real-time applications.

5.6 Synthesis and General Implications

This study demonstrates that the integration of heterogeneous imaging technologies cannot be reduced to a simple question of geometric calibration. The complementarity of the sensors is inevitably accompanied by technical challenges linked to their fundamentally different physical principles.

The ideal vision of a "transparent" multi-sensor system, where data from different modalities integrate naturally, remains an ambitious objective requiring advances both in the physical modeling of sensors and in the development of adaptive fusion algorithms.

Our results suggest that future approaches should favor a holistic modeling explicitly incorporating the structural limitations of each technology, rather than attempting to make them converge towards a single idealized model. This perspective paves the way for new design methodologies for vision systems that strategically exploit complementarities while explicitly accepting and managing fundamental incompatibilities.

6 Conclusion and Perspectives

This work has presented a comprehensive and systematic characterization of a heterogeneous imaging system integrating a CCD camera, a professional ToF camera, and a Microsoft Kinect. The exhaustive study conducted on geometric, colorimetric, and volumetric aspects has enabled the establishment of a precise mapping of the performances, limitations, and complementarities of these three visual perception technologies.

6.1 Summary of Main Contributions

Geometric Level: Our analysis demonstrated that the CCD maintains its position as a metrological reference with reprojection errors less than 0.2 pixel and negligible distortion. The Kinect presents stable intermediate performances, while the ToF, despite its depth robustness, introduces significant geometric uncertainties linked to its low resolution (176×144) and its harmonic modulation artifacts. The extrinsic calibrations revealed that pairs including the ToF show translational uncertainties up to three times higher than other configurations, confirming the difficulty of geometric integration of this technology.

Colorimetric Level: The study highlighted a bimodal behavior strongly dependent on surface properties. Diffuse materials (8 areas out of 10) show excellent inter-sensor consistency ($\Delta E < 2.5$, $\sigma < 0.04$), while specular and saturated surfaces generate major divergences ($\Delta E > 12$, $\sigma > 0.12$). The statistical analysis of the digitized curves confirmed the existence of predictable stability and instability areas, with a significant impact of lighting conditions (50% performance degradation under shadow).

Volumetric Level: The complementarity between the ToF (robustness, texture insensitivity) and the Kinect (high resolution, color-depth integration) was established, but the technical obstacles to their direct fusion were clearly identified: resolution gap (factor 4.5), incompatible noise models, and frequency differences in acquisition.

6.2 Fundamental Implications

This research demonstrates that the integration of heterogeneous imaging technologies cannot be reduced to simple geometric calibration. The fundamental differences in the physical principles of operation, proprietary processing pipelines, and specific error models impose a holistic approach to multi-sensor system design.

The vision of a "transparent" system, where different modalities integrate naturally, remains an ambitious objective requiring conceptual advances both in physical modeling and in the development of adaptive fusion algorithms.

6.3 Research Perspectives

Advanced Modeling of ToF Distortions: Classical polynomial models prove insufficient to capture the complexity of the harmonic modulation artifacts of ToF cameras. Specific approaches, incorporating the spectral characteristics of the modulation and the detection non-linearities, must be developed. Future work could draw inspiration from periodic distortion models used in industrial metrology to significantly improve the geometric accuracy of systems integrating ToF.

Multi-scale and Multi-modality Fusion Architectures: Effective fusion of heterogeneous data requires the development of architectures capable of handling differences in scale, resolution, and uncertainty models. Approaches should combine:

- Adaptive remeshing for multi-resolution alignment
- Probabilistic modeling of sensor-specific uncertainties
- Robust optimization integrating physical constraints of the different modalities

Contextual Modeling of Sensor-Surface-Lighting Responses: The variability of performance depending on surface properties and lighting conditions suggests the need to develop predictive sensor-surface-lighting models. A comprehensive database associating material types, illumination conditions, and responses of different sensors would allow anticipating and compensating for non-linear behaviors.

Deep Learning Approaches for Multi-sensor Fusion: Deep neural networks, particularly multi-modality architectures, offer promising perspectives to overcome the identified limitations. Future research could explore:

- Fusion networks simultaneously exploiting ToF amplitude, Kinect RGB, and CCD geometry
- Learning with differential attention according to stability/instability areas
- Architectures capable of estimating and compensating for artifacts specific to each technology

Real-Time Estimation and Compensation of Illumination Variations: The development of algorithms capable of estimating lighting conditions in real time and dynamically adapting fusion parameters would constitute a significant advance for applications in uncontrolled environments.

6.4 General Conclusion

This work establishes that the design of high-performance heterogeneous imaging systems requires a fundamentally multidimensional approach, explicitly integrating the physical specificities of each technology rather than attempting to make them converge towards a single idealized model. The obtained results provide not only a precise quantitative characterization of the studied technologies but also a methodological framework for the development and evaluation of future multi-sensor systems.

The perspectives opened by this research fit into the broader context of the evolution of visual perception systems, where the strategic complementarity of technologies and the algorithmic intelligence of fusion become the keys to surpassing the individual limitations of sensors and achieving levels of performance and robustness inaccessible to unimodal approaches.

References

1. Dal Mutto, C., Zanuttigh, P., Cortelazzo, G.M.: Time-of-flight cameras and microsoft kinect. 1st edn. Springer, Cham (2012). https://doi.org/10.1007/978-1-4614-3807-6
2. Yuan,G., Sandro, E., Reinhard,K., Keinert, J.: A Novel Self-Calibration Method for a Stereo-ToF System Using a Kinect V2 and Two 4K GoPro Cameras (2017). https://doi.org/10.1109/3DV.2017.00013
3. Alvaro,L.P., Qiang,S., Miguel, H.C.: Performance evaluation of state-of-the-art high-resolution time-of-flight cameras. IEEE Sens. J. (2023). https://doi.org/10.1109/JSEN.2023.3273165
4. Qiu, Z., Joaquin,M.-S., Pedro, A.: evaluation and comparison of different time of flight cameras for outdoor applications. In: ISPRS Annals of the Photogrammetry, Remote Sensing and Spatial Information Sciences. X-1/W1-2023. 17–23 (2023). https://doi.org/10.5194/isprs-annals-X-1-W1-2023-17-2023
5. Pulli, K.: The Digital Michelangelo Project. Computer Graphics Forum, Wiley (1999)
6. Yau, W.K., Ng, E., Lee, W.: Comparative study of intel R200, Kinect v2, and Primesense RGB-D sensors performance outdoors. IEEE Sens. J. 1–1 (2019). https://doi.org/10.1109/JSEN.2019.2920976
7. Howell, A. J., Buxton, H.: Active vision techniques for visually mediated interaction. Image Vis. Comput. 20, 861–871 (2002). https://doi.org/10.1016/s0262-8856(02)00095-1
8. Szeliski, R.: Computer Vision: Algorithms and Applications. Springer, Cham (2010)
9. Scharstein, D.: Stereo vision for view synthesis. In: Proceedings of CVPR, 1996, pp. 852–857. IEEE Computer Society (1996)
10. Kahlmann, T., Ingensand, H.: Calibration and improvements of the high-resolution range-imaging camera SwissRanger. In: Proceedings of SPIE 5665, Videometrics VIII, 56650F (2005). https://doi.org/10.1117/12.582513
11. Woodham, R.: Photometric method for determining surface orientation from multiple images. Opt. Eng. **19** (1992). https://doi.org/10.1117/12.7972479
12. Cree, M., Streeter, L., Conroy, R., Dorrington, A.: Analysis of the softkinetic depthsense for range imaging. **7950**, 668–675 (2013). https://doi.org/10.1007/978-3-642-39094-4_76
13. Zhang, R., Tsai, P.-S., Cryer, J., Shah, M.: Shape from shading: a survey. IEEE Trans. Pattern Anal. Mach. Intell. **21**, 690–706 (1999). https://doi.org/10.1109/34.784284
14. Fuchs, S., Hirzinger, G.: Extrinsic and depth calibration of ToF-cameras. Proc. CVPR (2008). https://doi.org/10.1109/CVPR.2008.4587828
15. Fuchs, S., May, S.: Calibration and registration for precise surface reconstruction with Time-Of-Flight cameras. IJISTA **5**, 274–284 (2008). https://doi.org/10.1504/IJISTA.2008.021290
16. Young, M., Kim, D., Chan, C., Theobalt, S.: Thrun: design and calibration of a multi-view TOF sensor fusion system. Proc. CVPRW (2008). https://doi.org/10.1109/CVPRW.2008.4563160
17. Guan, L., Pollefeys, M.: A Unified Approach to Calibrate a Network of Camcorders and ToF cameras. ECCV (2008)
18. Guan, L., Franco, J.-S., Pollefeys, M.: 3D Object Reconstruction with Heterogeneous Sensor Data. 3DPVT (2008)

19. Bouguet, J.-Y.: Camera Calibration Toolbox for Matlab (1.0). CaltechDATA (2022). https://doi.org/10.22002/D1.20164
20. Hansard, M., Lee, S., Choi, O., Horaud, R.: Time of flight cameras: principles, methods, and applications. SpringerBriefs Comput. Sci. (2012)
21. Zhu, J., Wang, L., Yang, R., Davis, J.: Fusion of time-of-flight depth and stereo for high accuracy depth maps. Proc. CVPR (2008). https://doi.org/10.1109/CVPR.2008.4587761
22. Bazin, J., Demonceaux, C., Vasseur, P., Kweon, I.-S.: Rotation estimation and vanishing point extraction by omnidirectional vision in urban environment. Int. J. Robotic Res. **31**, 63–81 (2012). https://doi.org/10.1177/0278364911421954
23. Zhang, Z.: A flexible new technique for camera calibration. IEEE Trans. Pattern Anal. Mach. Intell. **22**, 1330–1334 (2000). https://doi.org/10.1109/34.888718
24. Qiao, Yu., Liu, Z., Hu, D., Xu, J.: Camera calibration method based on OpenCV. Appl. Mech. Mater. **330**, 517–520 (2013). https://doi.org/10.4028/www.scientific.net/AMM.330.517
25. Scaramuzza, D., Siegwart, R.: A practical toolbox for calibrating omnidirectional cameras. (2007). https://doi.org/10.5772/4994
26. Iddan, G., Yahav, G.: 3D Imaging in the studio (and elsewhere...). SPIE (2001)
27. Woodham, R.: Photometric method for determining surface orientation from multiple images. Opt. Eng. **19** (1992). https://doi.org/10.1117/12.7972479

Enhanced Skin Lesion Classification: A Deep Feature Extraction-Based Approach for Multi-classification of Skin Cancer Using HAM Dataset

Samaa Mukhaimar[1], Akila Djebbar[2], Ghada Halladja[2](✉), and Hayet Farida Merouani[2]

[1] United Arab Emirates University, Dubai, United Arab Emirates
smukhaimar@gbs.ac.ae

[2] LRI Laboratory, Department of Computer Science, Badji Mokhtar-Annaba University, Annaba, Algeria
{akila.djebbar,ghada.halladja,hayette-farida.merouani}@univ-annaba.dz

Abstract. The analysis of dermoscopic images has garnered substantial attention due to its pivotal role in facilitating the early detection and precise diagnosis of skin lesions. Early detection of skin lesions is essential for effective treatment and improved patient outcomes. However, misdiagnosis in the initial stages of oncological conditions remains a significant challenge, frequently resulting in serious consequences. The increasing demand for advanced image processing technologies highlights the critical need for efficient computational solutions to assist medical professionals in their diagnostic processes. In this study, we evaluate the performance of various machine-learning models for classifying multitype lesions in dermoscopic images. In this regard, the widely recognized dermoscopic image dataset, HAM10000, comprising seven types of labeled lesions, was used to train and evaluate the effectiveness of the proposed methodologies. To enhance skin lesion classification, we investigate multiple feature extraction techniques, including Quantum Convolutional Neural Network (QCNN), AlexNet, and Gray Level Co-occurrence Matrix (GLCM). Our experimental findings reveal that combining GLCM with a Convolutional Neural Network achieves an accuracy of 87%. These results emphasize the potential of advanced feature extraction and classification approaches in improving dermoscopic image analysis for a more accurate diagnosis of skin cancer.

Keywords: Image processing · Skin lesion classification · Features extraction · Quantum Convolutional Neural Networks (QCNN) · AlexNet · Gray Level Co-occurrence Matrix (GLCM)

1 Introduction

Skin cancer is one of the most prevalent types of cancer worldwide [1]. In the United States alone, more than five million new cases are diagnosed annually,

Z. Kouahla et al. (Eds.): IAM 2025, LNNS 1973, pp. 188–210, 2026.
https://doi.org/10.1007/978-3-032-26031-4_14

with projections that one in five individuals will develop this disease at some point in their lifetime [2]. This alarming statistic is exacerbated by increased exposure to harmful chemicals, ultraviolet rays, and other environmental factors [3]. However, when skin cancer is identified in its early stages and appropriately treated, survival rates are significantly higher. Therefore, an early and accurate diagnosis is critical to improving patient outcomes [4,5].

Traditionally, skin cancer diagnosis has been based on visual examination by physicians, a method that is inherently subjective and prone to error [6]. Several challenges complicate the diagnostic process, including variability in skin pigmentation, differences in image sources, and the various sizes and shapes of skin lesions [7]. These factors make precise identification particularly challenging [8]. For example, poor contrast between skin lesions and surrounding tissues can obscure the visibility of critical features. Similarly, color illumination, reflections, and texture variations further hinder the accurate diagnosis [9]. To address these challenges, preprocessing plays a crucial role in improving image quality for analysis [10]. This includes enhancing contrast, reducing noise, and normalizing color illumination [11,12]. By tackling these issues, it becomes possible to analyze dermoscopic images more effectively, paving the way for accurate and early detection of skin cancer [13].

The focus of this research is to develop a robust, intelligent detection and classification system for skin cancer, aimed at assisting medical professionals, particularly those with limited experience in interpreting dermoscopic images. Our approach focuses on early detection to prevent the disease from advancing to life-threatening stages. According to the World Health Organization, skin cancer accounts for one in three diagnosed malignancies, with its incidence steadily increasing in countries such as the United States, Canada, and Australia [14]. To achieve our goal, we leverage machine learning models and feature extraction techniques to accurately distinguish between various types of pigmented lesions. This system aims to improve the efficiency and accurate diagnosis of skin cancer, contributing to better patient outcomes and supporting healthcare providers to provide timely, effective care.

The key contributions of our work are as follows:

- To enhance the visual clarity and contrast of tumor images, we applied image normalization and Contrast-Limited Adaptive Histogram Equalization (CLAHE). Additionally, Morphological Closing is used as a razor to remove hair artifacts, followed by a median filter for noise reduction.
- We incorporate various data augmentation techniques, including rotation, brightness and contrast adjustments, and other transformations to improve model generalization and mitigate data imbalance issues.
- To accurately segment lesion images, we utilize advanced techniques such as the Sobel filter, Isodata thresholding, and the Watershed algorithm, ensuring precise boundary delineation for improved feature extraction.
- We investigate and compare multiple feature extraction techniques, including Quantum Convolutional Neural Networks (QCNN), AlexNet, and Gray Level Co-occurrence Matrix (GLCM), to enhance skin lesion classification.

- We assess the impact of various classification models, including Convolutional Neural Network, LightGBM, SVM, KNN, and Naive Bayes, providing a comprehensive performance evaluation across different methodologies.
- We experimentally demonstrate that the combination of GLCM with a Convolutional Neural Network achieves superior classification performance, attaining an accuracy of 87%, thereby highlighting its effectiveness in dermoscopic image analysis.

The remainder of this paper is structured as follows: Sect. 2 provides an overview of the related work. Section 3 details the proposed methodology, outlining each stage of the approach. In Sect. 4, we present the experimental results and perform a comparative analysis against state-of-the-art methods. Finally, Sect. 5 concludes the study with key findings and future research directions.

2 Related Work

Skin diseases are commonly diagnosed through visual examination, where dermatologists rely on their expertise and clinical experience to assess lesions and determine appropriate treatments. However, this process is inherently subjective, leading to variability in diagnoses. To improve accuracy and consistency, decision support systems can provide valuable assistance by offering objective analysis based on computational techniques. With advances in technology, computer-aided methods have emerged as effective tools to detect and evaluate skin cancer indicators in a rapid and noninvasive manner. Therefore, various approaches have been proposed to improve diagnostic precision, leveraging machine learning and image processing techniques to support dermatologists in their assessments.

In 2014, Maurya et al. [15] developed an automated system to detect and classify four main types of skin cancer: Melanoma, Basal Cell Carcinoma, Actinic Keratosis, and Squamous Cell Carcinoma. Each of these types of skin cancer exhibits distinct characteristics that can be identified using appropriate feature extraction techniques. To achieve this, texture features of skin lesions are extracted using normalized symmetrical Grey Level Co-occurrence Matrices (GLCM). These extracted features are then fed into a Multi-Class Support Vector Machine (SVM) for classification, enabling accurate differentiation between the four categories of skin cancer.

In 2019, Sae-Lim et al. [16] proposed a multi-class skin lesion classification approach using lightweight deep Convolutional Neural Networks (CNNs). The study employs MobileNet and introduces a modified version to enhance classification performance. In addition, data up-sampling and augmentation techniques are applied to improve the efficiency and robustness of the classifier. Bassi et al. [17] explored various model-building techniques, including parallel networks and transfer learning, to improve the accuracy of skin cancer classification. Parallel networks integrate patient metadata, such as age and sex, along with images of the injury to improve diagnostic precision. Then a thresholded image segmentation step is applied to isolate lesions, facilitating more effective feature extraction. Several pre-trained models, originally trained on the ImageNet dataset, are leveraged through transfer learning. They fine-tuned these models by modifying the input layers to match the image size and appending dense layers to learn the seven skin lesion classes. They initially experimented with VGG16, setting its last four layers as trainable, and subsequently extended their experiment using models such as VGG19, ResNet, ResNetV2, InceptionV3, MobileNet, MobileNetV2, and DenseNet. Mobiny et al. [18] explored the potential of Bayesian deep learning to enhance the performance of machine–physician skin lesion classification. To this end, they proposed a computer-aided diagnosis (CAD) model based on a deep neural network (DNN) that takes the advantage of approximate Bayesian inference to generate both classification predictions and the corresponding uncertainty estimates. The proposed framework is versatile and can be applied to various medical machine learning tasks. Experimental results highlight the effectiveness of confidence ratings in improving diagnostic accuracy while simultaneously reducing the workload of physicians. Nugroho et al. [19] designed a model based on a convolutional neural network to differentiate between seven different types of skin lesions. They started with data preparation, where all images were resized and normalized to ensure consistency across the dataset. The data is then divided into training, validation, and test sets. Data augmentation is applied to the training set to prevent overfitting and improve accuracy. A Convolutional Neural Network (CNN) is employed for the identification of skin cancer because of its effectiveness in image and pattern recognition.

In 2021, to ensure efficient deployment on cloud and mobile platforms, Huang et al. [20] adopted a lightweight neural network architecture for the classification of multiclass skin cancer. The study implements EfficientNet, known for its effective scaling of depth, width, and resolution, offering high performance with fewer parameters. The dataset was divided into training and validation sets, with preprocessing techniques such as normalization and data augmentation applied. Among the EfficientNet models (B0–B7), the best-performing version was selected based on the validation results and the hyperparameters were fine-tuned to optimize the classification performance. Khan et al. [21] proposed an approach that involves two key phases: segmentation and classification. For lesion segmentation, a Mask R-CNN architecture is implemented, incorporating ResNet50 as the backbone and a Feature Pyramid Network (FPN) for multi-scale feature extraction. This model is trained using ground truth masks to generate precise lesion boundaries and compute the Mean Overlap Coefficient (MopC) for validation. In the classification phase, a custom-designed CNN model is introduced that consists of six convolutional layers followed by a fully connected layer. This streamlined architecture is tailored for multi-class lesion classification. The FC layer captures the most informative features that are then passed on to a softmax classifier for final recognition. Chaturvedi et al. [22] explored an efficient automated approach for classifying dermoscopic skin cancer images using the MobileNet convolutional neural network. Initially pretrained on approximately 1.28 million images from the 2014 ImageNet Challenge, the model was fine-tuned on the HAM10000 dataset for enhanced performance. Preprocessing steps were applied to the skin lesion images, and data augmentation techniques were particularly employed to address class imbalance by expanding minority class samples. The fine-tuned MobileNet model demonstrated classification performance that was comparable to, or even exceeded, that of expert dermatologists in seven distinct categories of skin lesions.

In 2022, Afza et al. [23] proposed a hierarchical framework for segmentation and classification of skin lesion. They started with a preprocessing stage aimed at enhancing image quality prior to segmentation. The approach then fine-tunes the ResNet50 model for feature extraction. To improve classification performance, extracted features are optimized using the Grasshopper Optimization Algorithm and subsequently classified using the Naïve Bayes classifier. Maduranga et al. [24] introduced an AI-based mobile application for the rapid and accurate detection of skin diseases using deep learning. A Convolutional Neural Network (CNN), specifically MobileNet with transfer learning, is used for classification. The application enables early-stage diagnosis by analyzing images of affected skin areas, supporting both patients and dermatologists. In particular, the proposed model achieves efficient performance with nearly half the computational cost of the standard MobileNet, making it well-suited for mobile deployment. The results indicate that this approach can effectively assist general practitioners in diagnosing skin conditions using a smartphone.

In 2024, Reka et al. [25] explored the integration of quantum machine learning for skin disease classification, combining principles of quantum computing with conventional machine learning. Quantum libraries such as PennyLane and Qiskit are utilized to implement quantum circuits using various combinations of Pauli X, Y, and Z gates within a proposed Quanvolutional Neural Network. Features extracted using a pre-trained MobileNet model are used to train a Quantum Support Vector Classifier. The performance of these quantum models is compared with traditional deep learning architectures, including ResNet50, Inception-ResNet, DenseNet121, DenseNet201, and MobileNet.

Although many recent studies have demonstrated the effectiveness of machine learning models in skin lesion classification, a large number of them have focused primarily on end-to-end learning without deeply analyzing the impact of feature extraction techniques. This limits our understanding of how different extraction strategies influence classification performance. Addressing this gap, our work conducts a comparative study of feature extraction methods specifically GLCM, Quantum Convolutional Neural Networks (QCNN), and AlexNet to evaluate their effectiveness in classifying dermoscopic images. This comparative analysis aims to highlight the strengths and limitations of each approach, providing valuable information to develop more robust and accurate diagnostic tools in dermatology.

3 Methodology

Skin cancer arises from the uncontrolled growth of abnormal cells in the skin, which, if not detected early, can spread and become life-threatening. An accurate and timely diagnosis plays a crucial role in improving treatment outcomes. This section outlines the methodology adopted in this study, as illustrated in Fig. 1. The approach focuses on improving image quality, extracting relevant features, and applying machine learning techniques to ensure accurate classification. The ultimate goal is to develop an efficient and reliable framework that supports early diagnosis and contributes to more effective skin cancer management.

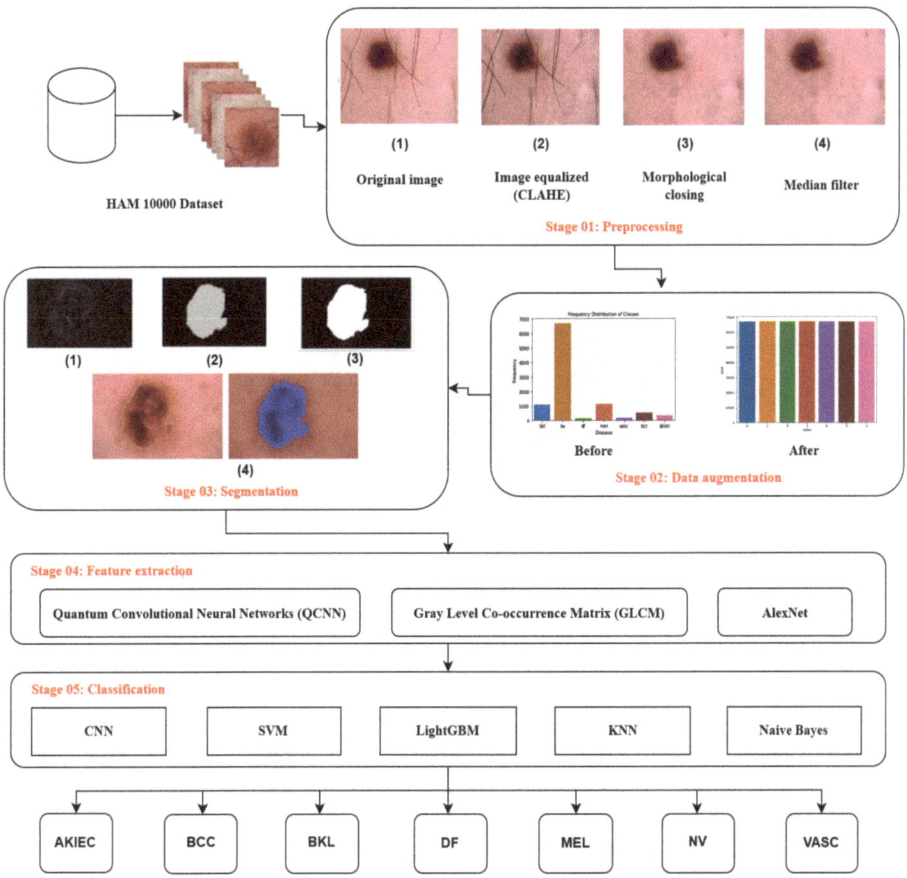

Fig. 1. Overall System Architecture of the proposed methodology for skin cancer classification.

3.1 Dataset

This study uses the HAM10000 dataset [26], which contains 10,015 dermatoscopic images representing seven categories of skin lesions: Actinic Keratoses (AKIEC), Basal cell carcinoma (BCC), Benign keratosis (BKL), Dermatofibroma (DF), Melanoma (MEL), Melanocytic nevi (NV) and Vascular skin lesions

(VASC). The dataset is diverse and includes images from various populations, with more than 50% of the lesions confirmed by histopathology, ensuring high reliability. The study focuses specifically on classifying pigmented lesions into these seven categories, using only images and lesion type labels, while excluding other metadata such as age, sex, and location of the lesion. Figure 2 presents sample images of the dataset.

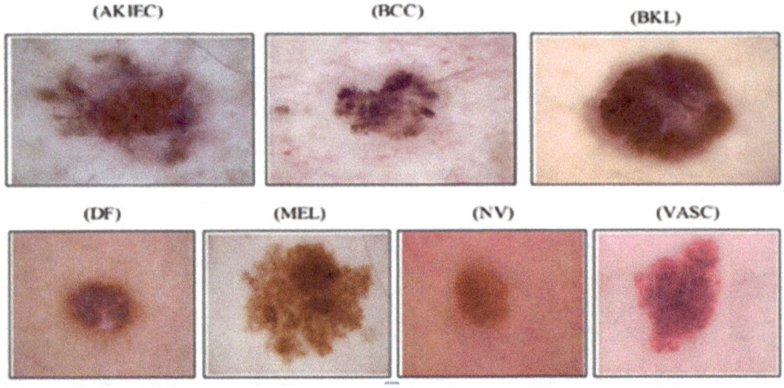

Fig. 2. Samples of each type from HAM10000 dataset.

3.2 Preprocessing

Preprocessing is a crucial step in preparing dermoscopic images for analysis. It involves the application of image processing techniques to enhance the quality and usability of the original images in the training and testing datasets. By improving image clarity and removing artifacts, preprocessing ensures that subsequent stages, such as feature extraction and classification, can operate on optimal data. This study employs a three-phase preprocessing approach designed to address key challenges such as poor contrast, noise, and irrelevant elements within images. The following subsections detail each phase of the preprocessing process.

Contrast-Limited Adaptive Histogram Equalization (CLAHE). CLAHE is a local contrast enhancement technique that works by applying histogram equalization within distinct tile regions of an image. Unlike standard histogram equalization, which processes the entire image globally, CLAHE modifies each pixel based on a transformation function derived from the histogram of its surrounding square region. This transformation function is proportional to the cumulative distribution function (CDF) of the pixel intensities in the neighborhood [27]. To prevent excessive contrast enhancement, CLAHE introduces a clipping mechanism that limits the histogram to a predefined value before calculating the CDF. This controlled approach reduces the slope of the CDF, effectively mitigating noise amplification that commonly occurs in nearly uniform regions of the image when standard histogram equalization is used. By balancing contrast enhancement with noise suppression, CLAHE provides improved visual clarity, making it particularly effective for medical imaging applications [27].

Morohological Closing. Morphological closure is a key operation in the domain of morphological image processing, a group of nonlinear techniques that analyze and modify the structure or morphology of objects in binary or grayscale images. These methods utilize a structuring element, typically a simple geometric shape such as a circle or rectangle, to probe and manipulate the image. The structuring element is systematically positioned across all possible locations in the image, and compared to the pixel neighborhoods, it overlaps. Depending on the operation, the structuring element is checked for either "fitting" entirely within a neighborhood or "hitting" or intersecting with it. Morphological closure, specifically, is employed to remove small dark artifacts and fill gaps within bright areas, making it a valuable tool in image preprocessing pipelines.

Median Filter. The median filter is a widely used technique to reduce noise in images. It is commonly used during the preprocessing phase to enhance the quality of images, ensuring improved performance in subsequent steps like feature extraction. The filter operates by applying a structuring element, typically a simple geometric shape such as a disk or square, to each pixel in the image. The pixel in the center of the structuring element is replaced by the median value of the pixels within the neighborhood defined by the element. This is achieved by first sorting the pixel values within the neighborhood in ascending order and then selecting the middle value as the replacement for the target pixel.

3.3 Data Augmentation

Data augmentation involves techniques that artificially expand a dataset by generating additional data points from existing ones. These methods enhance model performance and generalization by introducing small alterations or using deep learning models to create new samples. In this study, we tackle the issue of dataset imbalance using offline data augmentation, which applies various transformations to the original data, especially effective for smaller datasets. To address class imbalance, we employ resampling techniques, including oversampling and undersampling. Oversampling increases the representation of the minority class by duplicating or generating synthetic samples, ensuring that the classifier is not biased toward the majority class. Undersampling, on the other hand, reduces the size of the majority class to match the minority class. These strategies lead to a balanced dataset, promoting unbiased and effective predictions by ensuring that equal importance is given to both classes.

3.4 Segmentation

Image segmentation is a crucial step in computer vision, used to identify and separate different regions within an image. In this study, the segmentation process includes several key steps:

1. **Gradient Map Creation:** The Sobel filter is used to compute image gradients and highlight edges.
2. **Marker Detection:** The IsoData thresholding technique identifies foreground (lesion) and background markers based on grayscale histogram peaks.
3. **Watershed Transform:** The grayscale image is treated as a topographic surface. The watershed algorithm uses predefined markers to fill regions while preventing oversegmentation with boundary barriers.
4. **Segmentation Refinement:** Enhancements are made by filling small holes, removing regions smaller than 800 pixels, and excluding objects touching the image borders.
5. **Connected Component Labeling:** Unique labels are assigned to distinct regions in the segmented image.
6. **Lesion Detection:** The lesion is identified based on the area and extent of the region. If necessary, multiple iterations are used to ensure accurate lesion selection, especially when lesions touch the images' borders.

This multistep approach improves segmentation accuracy and supports effective lesion identification for further analysis.

3.5 Features Extraction

In this subsection, we present the feature extraction methods used in this study, including Quantum Convolutional Neural Networks (QCNN), the texture-based Gray Level Co-occurrence Matrix (GLCM), and the deep learning model AlexNet.

Features Extraction Using Quantum Convolution. In this study, we explore Quanvolutional Neural Networks (QNNs), a novel model that integrates quantum computing into classical CNNs to enhance their performance. QNNs introduce a quanvolutional layer, which consists of multiple quantum filters that, like classical filters, apply local transformations to input data to generate feature maps. However, these filters differ by using random quantum circuits to process spatially local segments of the data, potentially extracting richer and more complex features. The appeal of QNNs lies in two key strengths:

- Efficiency with limited quantum resources: Since quantum filters act on small local regions, they require only a few qubits and shallow circuit depths, making them feasible for current NISQ (Noisy Intermediate-Scale Quantum) devices.
- Robustness to errors: Quanvolutional circuits can tolerate noise as consistent errors become part of the stochastic nature of the transformation.

Architectural design of QNNs: QNNs are basically an extension of classical CNNs with an additional transformational layer known as the quanvolutional layer. Quanvolutional layers should integrate exactly the same as classical convolutional layers, allowing the user to:

- define any arbitrary integer value for the amount of quanvolutional filters in a particular quanvolutional layer
- stack any number of new quanvolutional layers on top about any other layer in the network stack (encoding and decoding methods, average number of quantum gates per qubit in the quantum circuit, etc.)

Quanvolutional Filter Design: When applied to an input tensor, quanvolutional filters build feature maps by modifying spatially local subsections of the input tensor. Unlike a classical convolutional filter, which uses a basic element-wise matrix multiplication operation, a quanvolutional filter changes the input data using a quantum circuit, which can be structured or random. We employ random quantum circuits for the quanvolutional filters instead of circuits with a specific structure in this study for simplicity and to establish a baseline.

Qubits and the Bloch Sphere: A qubit is the quantum equivalent of a classical bit and represents a two-state quantum mechanical system, one of the simplest systems that illustrate the core principles of quantum mechanics. Unlike classical bits, which can only exist in one of two definite states (0 or 1), a qubit can exist in a superposition of both states simultaneously. Physical realizations of qubits include the spin of an electron (spin up or spin down) or the polarization of a photon (vertical or horizontal). This unique property of superposition is fundamental to quantum computing. The Bloch sphere is commonly used to geometrically represent the state space of a qubit, visualizing all possible pure states in a three-dimensional form.

The proposed Quantum Convolutional Neural Network (QNN) extends the conventional CNN architecture by integrating a quanvolutional layer, which replaces classical convolutional filters with quantum circuits. In this layer, each (2×2) image patch is encoded into qubits through parameterized (Ry) rotations, creating a quantum representation of local pixel information. The encoded qubits are then processed by a random unitary operation (U) within a simulated four-qubit quantum circuit, implemented using the PennyLane backend. Following quantum computation, measurements on the computational basis yield expectation values that serve as quantum-derived feature maps. These outputs are subsequently passed to the classical part of the network, which consists of a pooling layer, a convolutional layer with six filters, another pooling layer and a fully connected layer. The architecture begins with a quanvolutional layer comprising three quantum filters, ensuring seamless integration with the classical CNN pipeline. This hybrid quantum–classical design preserves the structural advantages of CNNs while leveraging quantum transformations to extract higher-dimensional, non-linear features that enhance image representation and classification performance.

Features Extraction Using GLCM. Texture describes the consistency of patterns and the color distribution within the surface of an object. One widely used method for texture-based image classification is the Haralick texture feature, derived from the Gray Level Co-occurrence Matrix (GLCM). GLCM captures the spatial relationship of pixels by analyzing how often pairs of pixel intensities (i and j) occur in specific orientations—horizontal, vertical and diagonal. Each image can generate up to four GLCM matrices on the basis of these directions. From these matrices, texture features such as contrast, correlation, energy, and homogeneity are computed, offering valuable descriptors to distinguish visual patterns in image analysis [28].

Correlation
Calculates the joint probability of the supplied pixel pairs and is mostly used to specify the specifics of the relevant components in each row and column during the segmentation process. The correlation is calculated using the following formula:

$$Correlation = \sum_{i,j=0}^{N-1} P(i,j)\left[\frac{(i-\mu_i)(j-\mu_j)}{\sigma_i\sigma_j}\right] \tag{1}$$

Homogeneity
Quantifies the proximity of the distribution of the GLCM element to the GLCM diagonal. It is computed as follows:

$$Homogeneity = \sum_{i,j=0}^{N-1} \frac{P(i,j)}{1+(i-j)^2} \tag{2}$$

Energy
Calculated as the square root of the sum of squared elements. It is computed as follows:

$$Energy = \sqrt{\sum_{i,j=0}^{N-1} P(i,j)^2} \tag{3}$$

Contrast
Estimates the GLCM's local variances and is represented as follows:

$$Contrast = \sum_{i,j=0}^{N-1} P(i,j)(i-j)^2 \tag{4}$$

Features Extraction Using AlexNet. AlexNet consists of eight layers with learnable parameters. The first five layers perform convolution operations, interleaved with max-pooling layers, which help reduce spatial dimensions while preserving the most relevant information. These layers are responsible for detecting low-level features such as edges, textures, and shapes. Following the convolutional layers, AlexNet includes three fully connected (FC) layers that process the extracted spatial features into a more abstract representation. The use of ReLU (Rectified Linear Unit) activation functions after each layer, except for the final output layer, introduces non-linearity into the model, enabling it to learn complex patterns and accelerate convergence during training [29].

For feature extraction purposes, the final classification layer is typically removed, and the output of one of the last fully connected layers is used as feature vectors. These vectors represent high-level characteristics of the input images and can be fed into other classifiers for further analysis, such as classification of skin lesions.

3.6 Classification

Classification methods utilize labeled training data to predict the category of new, unseen data. In this study, various classification models were employed to categorize skin lesions into seven different types.

Convolutional Neural Network. A feedforward neural network does not use feedback loops, which means that the output is not returned to the network. Convolutional Neural Networks (CNNs), a type of feedforward model, consist of convolutional and pooling layers that extract and condense features from input data. Each convolution layer produces high-level feature maps, while pooling layer reduces dimensionality to enhance efficiency. In dermoscopic image analysis, deep learning approaches like CNNs have shown strong performance, offering improved accuracy, objectivity, and reproducibility after proper training.

Support Vector Machines (SVMs). The Support Vector Machine (SVM) is a supervised learning algorithm that is used primarily for binary classification. It learns from labeled data by finding the optimal hyperplane that separates the two classes with the maximum possible margin. This margin-based approach improves the model's ability to generalize to new, unseen data. Although traditionally binary, SVM can also be extended to multiclass problems.

Light Gradient Boosting Machine (LightGBM). LightGBM (Light Gradient Boosting Machine) is an open source distributed gradient boosting architecture for machine learning that was produced by Microsoft. Incorporates decision tree techniques to rank, classify, and perform other machine learning tasks.

Naives Bayes. The Naive Bayes classifier is a simple and efficient classification technique that helps to develop fast models that can make quick predictions. It is a probabilistic model, which means that it predicts according to an object's possibility. Although it is difficult, the conditional probability may be determined using the joint probability. The Bayes Theorem gives a straightforward method for computing the conditional probability. The computation for Bayes Theorem in its simplest version is as follows:

$$P(A|B) = \frac{P(B|A) * P(A)}{P(B)} \tag{5}$$

K-Nearest Neighbors (KNN). The K-nearest neighbors (KNN) technique is a form of supervised machine learning (ML) algorithm that can be used for both classification and regression prediction tasks. However, it is used mainly in industry for categorization and prediction challenges.

3.7 Evaluation Metric

Classification performance can be evaluated using several metrics that provide complementary insight into model behavior. The most common metric is accuracy, which measures the proportion of correctly predicted samples out of the total predictions. However, accuracy alone may not fully reflect the performance of the model, especially in the presence of class imbalance. Therefore, additional metrics such as precision, recall, and F1-score are also considered. Precision evaluates the proportion of correctly predicted positive samples among all samples predicted as positive, while recall measures the proportion of correctly identified positive samples among all actual positive cases. The F1-score, defined as the harmonic mean of precision and recall, provides a balanced measure that considers both false positives and false negatives, offering a more comprehensive assessment of classification effectiveness.

$$Accuracy = \frac{TP + TN}{TP + TN + FP + FN} \tag{6}$$

$$Precision = \frac{TP}{TP + FP} \tag{7}$$

$$recall = \frac{TP}{TP + FN} \tag{8}$$

$$F1\text{-score} = \frac{2TP}{(2TP + FP + FN)} \tag{9}$$

4 Results and Discussion

This section presents the experimental results derived from the application of three key approaches: Quantum Convolutional Neural Networks (QNNs), the AlexNet architecture, and texture-based feature extraction using the Gray-Level Co-occurrence Matrix (GLCM). A comparative analysis is also performed to evaluate the performance of the proposed framework against several state-of-the-art classification models using the HAM10000 dataset.

4.1 Comparative Study of the Proposed Architecture

Skin cancer is one of the most difficult tumors to recognize. Our goal is to improve detection performance by improving feature extraction. As a result, we will present a comparative study of the feature extraction approaches presented in our architecture.

Performance Evaluation of the Quantum Convolutional Neural Network. The primary distinction between a quantum circuit and a classical convolution lies in the complexity of the kernels they generate. Quantum circuits have the potential to produce highly intricate and non-classically computable kernels, offering a significant advantage in feature representation. The results obtained from the Quantum Convolutional Neural Network (QNN) are illustrated in Fig. 3.

As shown in Fig. 4, the quantum layer shows improved performance in terms of accuracy. The model achieves a training accuracy of 67% and a validation accuracy of 50%. However, this gap indicates the presence of overfitting, probably due to a class imbalance within the dataset. Some lesion classes contain only a limited number of samples, which affects the model's ability to generalize effectively.

Balancing the dataset is crucial to improve model performance and reduce bias toward dominant classes. The performance of the quantum convolutional neural network after applying data balancing techniques is illustrated in Fig. 5. As shown, balancing the dataset significantly reduces overfitting, resulting in improved training and validation accuracy, 95% and 70%, respectively. Data augmentation also contributes positively to model performance. Furthermore, features extracted using quantum layers demonstrate higher quality than those from classical CNNs in this context; however, this advantage is data dependent and may not generalize across all datasets.

Despite the tendency of overfitting, the overall results demonstrate that the model effectively captures discriminative features and achieves stable validation performance compared to previous configurations. Employing regularization strategies such as dropout can further improve generalization and reduce the performance gap between the training and validation curves.

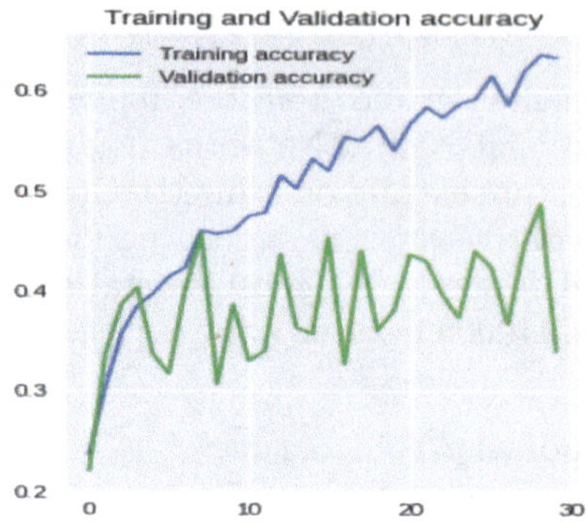

Fig. 3. Accuracy and loss curves for training and validation phases of the Quantum Convolutional Neural Network (QNN).

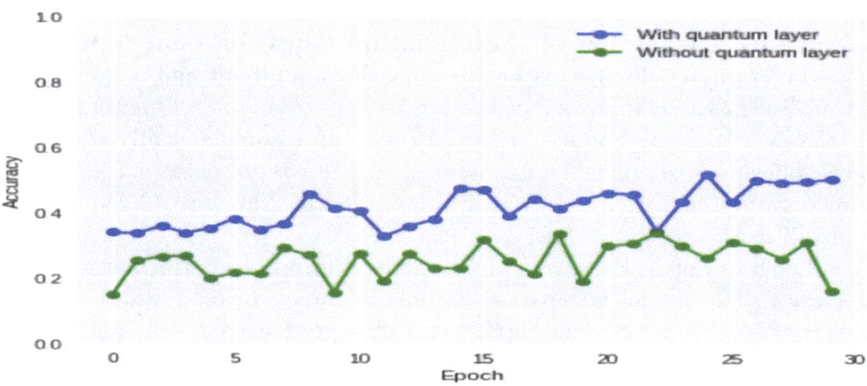

Fig. 4. Comparison of model performance with and without the quantum layer in terms of training and validation accuracy.

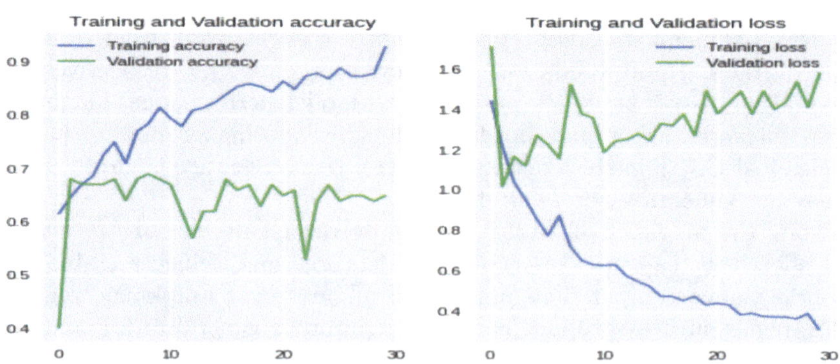

Fig. 5. Training and validation accuracy and loss curves after dataset balancing.

Performance Evaluation of AlexNet Model. we employed the AlexNet architecture primarily to explore different feature extraction strategies and to demonstrate that standard convolutional kernels are not always optimal for tumor detection. As observed, while the model achieved a high training accuracy of 98% over 25 epochs, the validation accuracy was only 45%, indicating substantial overfitting. This discrepancy highlights limitations in the model's ability to generalize to unseen data, as AlexNet's architecture may not capture the most discriminative features for this specific medical imaging task. Additionally, the training process was computationally intensive, which highlighted inefficiencies in both training time and resource usage. These findings motivated the exploration of more specialized or efficient architectures that are better suited to the characteristics of the dataset and the task requirements (Fig. 6).

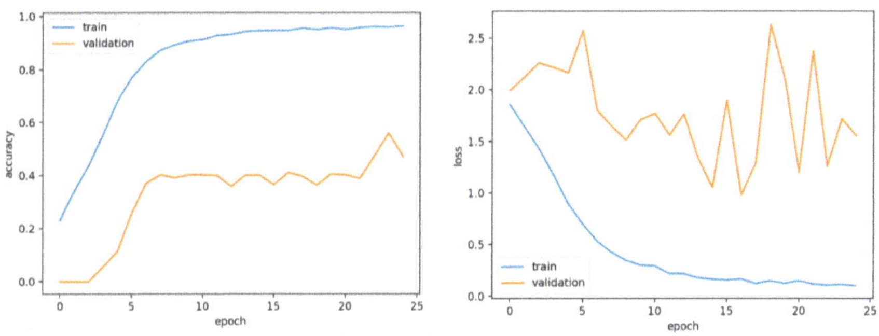

Fig. 6. Accuracy and loss curves of the AlexNet model during training and validation.

Performance Evaluation of GLCM with Multiple Classifiers. The GLCM method characterizes image texture by computing the frequency of pairs of pixels with specific gray-level values at defined spatial relationships, forming a gray-level co-occurrence matrix (GLCM). From this matrix, statistical texture features are extracted and then used as input to a neural network classifier, in this case a CNN. The classification results obtained using GLCM combined with CNN are illustrated in Fig. 7. To further assess the effectiveness of this approach, we compared its performance with four additional classifiers: Support Vector Machine (SVM), Light Gradient Boosting Machine (LightGBM), Naive Bayes, and k-Nearest Neighbors (k-NN). These evaluations were conducted under different data conditions, including imbalanced datasets and datasets with varied sample sizes. The classification performances of these models were graphically compared.

We evaluated the performance of various classifiers with and without data augmentation, in both imbalanced and balanced datasets across different sample sizes (2000, 5000, 7000 and 10,000 images), as shown in Figs. 8 and 9. This progressive sampling ensured a comprehensive analysis. As expected, the classification accuracy improved with increasing sample size. As demonstrated in Tables 1

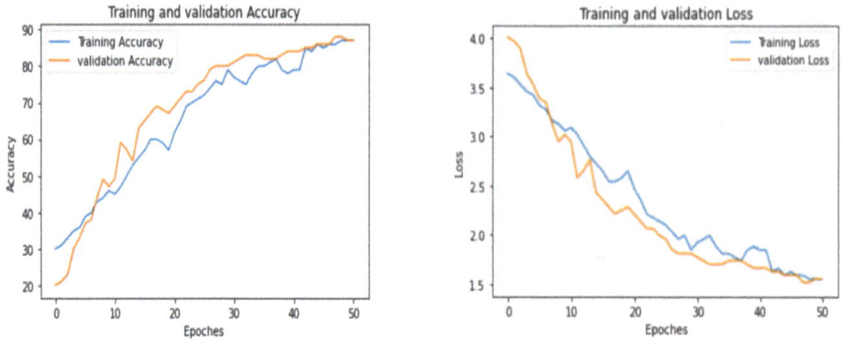

Fig. 7. Performance curves (accuracy and loss) of the convolutional neural network trained on GLCM-extracted texture features, showing both training and validation results.

and 2, the neural network classifier achieved the highest performance, reaching accuracy 87% after augmentation, followed by LightGBM with 81%, showing a notable improvement compared to the pre-augmentation results. In contrast, classifiers such as SVM, KNN, and Naive Bayes achieved accuracy 70%, 76% and 72%, respectively. However, in smaller sample sets (e.g. 2000 images), some classifiers showed performance degradation after augmentation. KNN decreased from 67% to 50%, and LightGBM decreased from 80% to 73%, highlighting that data balancing may not always yield better results in limited data scenarios.

Table 3 presents the predicted per-class performance metrics of the Neural Network model on the HAM10000 dataset after data augmentation. The results highlight strong overall performance, with an accuracy of 87% and a weighted F1-score of 0.87. Minority classes such as akiec and vasc exhibit lower recall, reflecting the challenge of imbalanced data despite augmentation, while the dominant nv class achieves near-perfect metrics. The macro-average F1-score of 0.78 indicates that the model maintains reasonably balanced performance across all classes, demonstrating the effectiveness of augmentation in improving generalization for underrepresented lesions.

4.2 Comparison with State-of-the-Art Methods

Numerous studies have proposed methodologies for the classification and diagnosis of various skin diseases by combining image processing techniques with deep learning algorithms. To position our work, which focuses on the extraction of texture-based characteristics, within the broader context of existing research, Table 4 provides a comparative overview of the systems proposed in the literature. As shown in Table 4, our approach yielded promising and competitive results, achieving an accuracy of 87%.

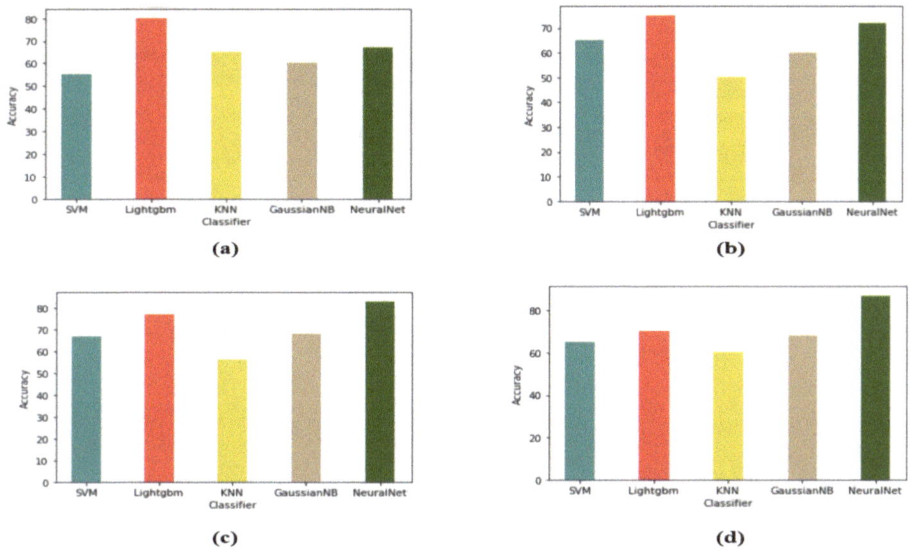

Fig. 8. Accuracy results of classifiers using the GLCM-based approach without data augmentation, evaluated on datasets containing (a) 2000, (b) 5000, (c) 7000, and (d) 10,000 images, respectively.

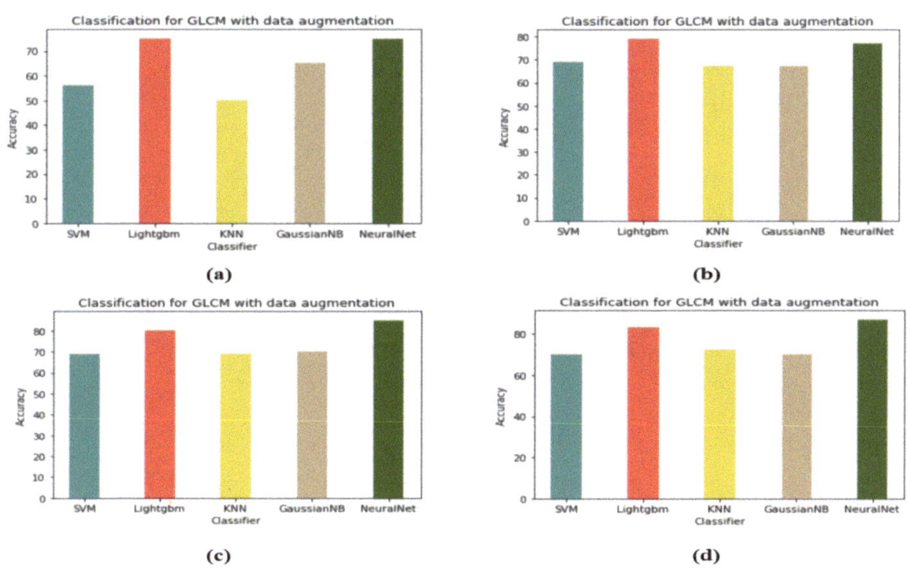

Fig. 9. Accuracy results of classifiers using the GLCM-based approach with data augmentation, evaluated on datasets containing (a) 2000, (b) 5000, (c) 7000, and (d) 10,000 images, respectively.

Table 1. Accuracy results of various classifiers applied to the GLCM-based approach without data augmentation.

Number of Samples	SVM	LightGBM	KNN	Naive Bayes	CNN
2000	55%	**80%**	67%	63%	67%
5000	67%	**74%**	50%	60%	70%
7000	67%	78%	59%	65%	**80%**
10000	65%	70%	60%	67%	**83%**

Table 2. Accuracy results of various classifiers applied to the GLCM-based approach with data augmentation.

Number of Sampels	SVM	LightGBM	KNN	Naive Bayes	CNN
2000	56%	73	50%	65%	**75%**
5000	69%	**79%**	67%	67%	77%
7000	69%	80%	69%	70%	**84%**
10000	70%	81%	76%	72%	**87%**

Table 3. Performance metrics of the Neural Network classifier after data augmentation on HAM10000 dataset.

Class	Precision	Recall	F1-Score
akiec	0.70	0.62	0.66
bcc	0.78	0.73	0.75
bkl	0.80	0.70	0.74
df	0.85	0.77	0.81
mel	0.88	0.75	0.81
nv	0.92	0.96	0.94
vasc	0.81	0.70	0.75
Overall Accuracy			**0.87**
Macro Avg	**0.82**	**0.75**	**0.78**
Weighted Avg	**0.88**	**0.87**	**0.87**

Table 4. Comparison with previous research on multiclass skin lesion classification.

Method	Year	Feature Extraction	Classification	Number of classes	Accuracy %
Maurya et al. [15]	2014	GLCM	Multi-Class Support Vector Machine (SVM)	4	81.43
Sae-Lim et al. [16]	2019	-	Modified MobileNet	7	83.23
Bassi et al. [17]	2019	-	VGG16 fine-tuned model	7	82.8
Mobiny et al. [18]	2019	-	Bayesian DenseNet-169	7	83.59
Nugroho et al. [19]	2019	-	Convolutional Neural Network (CNN)	7	78
Huang et al. [21]	2021	-	Light-weight model (EfficientNet)	7	85.8
Khan et al. [22]	2021	-	Convolutional Neural Network (CNN)	7	86.5
Chaturvedi et al. [23]	2021	-	Fine-tuned MobileNet convolutional neural network	7	83.1
Afza et al. [20]	2022	ResNet-50 model	Naïve Bayes classifier	7	85.8
Maduranga et al. [24]	2022	-	Convolutional Neural Network (CNN)	7	85
Reka et al. [25]	2024	MobileNet pre-trained network	MobileNet pre-trained network	7	82.86
Our Proposed model		GLCM	Convolutional Neural Network (CNN)	7	**87**

5 Conclusions

This study presents a comprehensive framework for the classification of skin cancer images. The proposed architecture integrates five key processes: preprocessing, data augmentation, segmentation, feature extraction, and classification. Using methods such as image normalization, Contrast-Limited Adaptive Histogram Equalization (CLAHE), morphological closing, and median filtering, we enhanced the quality and contrast of tumor images. Data augmentation techniques, including rotation, brightness, and contrast adjustments, were applied to improve the robustness of the model. For segmentation, approaches such as the Sobel filter, threshold isodata, and the watershed algorithm were used to effectively delineate regions of interest. Despite the challenges posed by the imbalance of the dataset and the limited sample sizes, the proposed methods successfully balanced the dataset, leading to improved detection accuracy. Among the evaluated techniques, the Gray Level Co-occurrence Matrix (GLCM) combined with Convolutional Neural Network demonstrated promising accuracy. The results also highlight the efficiency of the GLCM in feature extraction, offering a faster and more streamlined process compared to convolutional layers.

The results of this research underscore the potential of advanced feature extraction and classification methods to improve the accuracy of skin cancer diagnosis. Although these results are highly encouraging, they also point to several avenues for future work. Expanding the feature space by incorporating additional attributes such as color, texture, and shape could further improve the representation of regions of interest. Testing the proposed model on other datasets and exploring new deep learning architectures may also lead to further advancements in this field. This study lays the foundation for ongoing exploration and innovation in the application of deep learning for skin cancer detection, ultimately contributing to better diagnostic tools and patient outcomes.

Declarations

Funding. The authors did not receive support from any organization for the submitted work.

Competing interests. The authors have no relevant financial or non-financial interests to disclose.

References

1. Wang, M., Gao, X., Zhang, L.: Recent global patterns in skin cancer incidence, mortality, and prevalence. Chin. Med. J. **138**(2), 185–192 (2025). https://doi.org/10.1097/CM9.0000000000003416
2. Bechelli, S., Delhommelle, J.: Machine learning and deep learning algorithms for skin cancer classification from dermoscopic images. Bioengineering **9**(3), 97 (2022). https://doi.org/10.3390/bioengineering9030097
3. Kottath, A.V., Bharathi, S.S.: Image preprocessing techniques in skin diseases prediction using deep learning: a review. In: 2022 4th International Conference on Inventive Research in Computing Applications (ICIRCA), pp. 1–6 (2022). https://doi.org/10.1109/ICIRCA54612.2022.9985547
4. Khattar, S., Kaur, R., Gupta, G.: A review on preprocessing, segmentation and classification techniques for detection of skin cancer. In: 2023 2nd Edition of IEEE Delhi Section Flagship Conference (DELCON), pp. 1–6 (2023). https://doi.org/10.1109/DELCON57910.2023.10127546
5. Hsu, B.W.-Y., Tseng, V.S.: LightDPH: lightweight dual-projection-head hierarchical contrastive learning for skin lesion classification. J. Healthc. Inform. Res. **8**(4), 619–639 (2024). https://doi.org/10.1007/s41666-024-00174-5
6. Gilani, S.Q., Marques, O.: Skin lesion analysis using generative adversarial networks: a review. Multimed. Tools Appl. **82**(19), 30 065–30 106 (2023). https://doi.org/10.1007/s11042-022-14267-z
7. Barata, C., Celebi, M.E., Marques, J.S.: A survey of feature extraction in dermoscopy image analysis of skin cancer. IEEE J. Biomed. Health Inform. **23**(3), 1096–1109 (2018). https://doi.org/10.1109/JBHI.2018.2845939
8. Delibasis, K., Moutselos, K., Vorgiazidou, E., Maglogiannis, I.: Automated hair removal in dermoscopy images using shallow and deep learning neural architectures. Comput. Methods Program. Biomed. Update **4**, 100109 (2023). https://doi.org/10.1016/j.cmpbup.2023.100109
9. Shobana, P., Deny, J.: Image fusion based removal of color artifacts for the enhancement of dermoscopy images. In: 2022 International Conference on Smart Technologies and Systems for Next Generation Computing (ICSTSN), pp. 1–6 (2022). https://doi.org/10.1109/ICSTSN53084.2022.9761322

10. Joseph, S., Olugbara, O.O.: Preprocessing effects on performance of skin lesion saliency segmentation. Diagnostics **12**(2), 344 (2022). https://doi.org/10.3390/diagnostics12020344
11. Salvi, M., et al.: DermoCC-GAN: a new approach for standardizing dermatological images using generative adversarial networks. Comput. Methods Programs Biomed. **225**, 107040 (2022). https://doi.org/10.1016/j.cmpb.2022.107040
12. Setiawan, A.W.: Effect of color enhancement on early detection of skin cancer using convolutional neural network. In: 2020 IEEE International Conference on Informatics, IoT, and Enabling Technologies (ICIoT), pp. 100–103 (2020). https://doi.org/10.1109/ICIoT48696.2020.9089631
13. Kasmi, R., et al.: SharpRazor: automatic removal of hair and ruler marks from dermoscopy images. Skin Res. Technol. **29**(4), e13203 (2023). https://doi.org/10.1111/srt.13203
14. Hasanah, R.L., Riana, D.: Classification of dermoscopic image of skin cancer using the GLCM method and multi-SVM algorithm. Rekayasa **14**(3), 407–415 (2021). https://doi.org/10.21107/rekayasa.v14i3.12213
15. Maurya, R., Singh, S.K., Maurya, A.K., Kumar, A.: GLCM and multi class support vector machine based automated skin cancer classification. In: 2014 International Conference on Computing for Sustainable Global Development (INDIACom), pp. 444–447 (2014). https://doi.org/10.1109/IndiaCom.2014.6828177
16. Sae-Lim, W., Wettayaprasit, W., Aiyarak, P.: Convolutional neural networks using MobileNet for skin lesion classification. In: 2019 16th international Joint Conference on Computer Science and Software Engineering (JCSSE), pp. 242–247 (2019). https://doi.org/10.1109/JCSSE.2019.8864155
17. Bassi, S., Gomekar, A.: Deep learning diagnosis of pigmented skin lesions. In: 2019 10th International Conference on Computing, Communication and Networking Technologies (ICCCNT), pp. 1–6 (2019). https://doi.org/10.1109/ICCCNT45670.2019.8944601
18. Mobiny, A., Singh, A., Van Nguyen, H.: Risk-aware machine learning classifier for skin lesion diagnosis. J. Clin. Med. **8**(8), 1241 (2019). https://doi.org/10.3390/jcm8081241
19. Nugroho, A.A., Slamet, I., Sugiyanto: Skins cancer identification system of haml0000 skin cancer dataset using convolutional neural network. In: AIP Conference Proceedings, vol. 2202, no. 1, 2019, p. 020039 (2022). https://doi.org/10.1063/1.5141652
20. Huang, H.-W., Hsu, B.W.-Y., Lee, C.-H., Tseng, V.S.: Development of a lightweight deep learning model for cloud applications and remote diagnosis of skin cancers. J. Dermatol. **48**(3), 310–316 (2021). https://doi.org/10.1111/1346-8138.15683
21. Khan, M.A., Zhang, Y.-D., Sharif, M., Akram, T.: Pixels to classes: intelligent learning framework for multiclass skin lesion localization and classification. Comput. Electr. Eng. **90**, 106956 (2021). https://doi.org/10.1016/j.compeleceng.2020.106956
22. Chaturvedi, S.S., Gupta, K., Prasad, P.S.: Skin lesion analyser: an efficient seven-way multi-class skin cancer classification using MobileNet. In: Hassanien, A.E., Bhatnagar, R., Darwish, A. (eds.) AMLTA 2020. AISC, vol. 1141, pp. 165–176. Springer, Singapore (2021). https://doi.org/10.1007/978-981-15-3383-9_15
23. Afza, F., Sharif, M., Mittal, M., Khan, M.A., Hemanth, D.J.: A hierarchical three-step superpixels and deep learning framework for skin lesion classification. Methods **202**, 88–102 (2022). https://doi.org/10.1016/j.ymeth.2021.02.013

24. Maduranga, M., Nandasena, D.: Mobile-based skin disease diagnosis system using convolutional neural networks (CNN). Int. J. Image Graph. Signal Process. **12**(3), 47 (2022)
25. Reka, S.S., Karthikeyan, H.L., Shakil, A.J., Venugopal, P., Muniraj, M.: Exploring quantum machine learning for enhanced skin lesion classification: a comparative study of implementation methods. IEEE Access (2024). https://doi.org/10.1109/ACCESS.2024.3434681
26. Tschandl, P., Rosendahl, C., Kittler, H.: The HAM10000 dataset, a large collection of multi-source dermatoscopic images of common pigmented skin lesions. Sci. Data **5**(1), 1–9 (2018). https://doi.org/10.1038/sdata.2018.16
27. Reza, A.M.: Realization of the contrast limited adaptive histogram equalization (CLAHE) for real-time image enhancement. J. VLSI Signal Process. Syst. Signal Image Video Technol. **38**, 35–44 (2004). https://doi.org/10.1023/B:VLSI.0000028532.53893.82
28. O'shea, K., Nash, R.: An introduction to convolutional neural networks. arXiv preprint arXiv:1511.08458 (2015)
29. Khvostikov, A., Aderghal, K., Benois-Pineau, J., Krylov, A., Catheline, G.: 3D CNN-based classification using sMRI and MD-DTI images for Alzheimer disease studies. arXiv preprint arXiv:1801.05968 (2018)

A Hybrid Deep Learning Model for Precise ECG Arrhythmia Classification

Saoueb Kerdoudi[1(✉)], Larbi Guezouli[1,2], Tahar Dilekh[3], and Abdessalam Hattab[1]

[1] Computer Science Department, LaSTIC Laboratory, University of Batna 2, Batna, Algeria
s.kerdoudi@univ-batna2.dz
[2] LEREESI Laboratory, HNS-RE2SD, LEREESI Laboratory, Batna, Algeria
[3] Computer Science Department, LAMIE Laboratory, University of Batna 2, Batna, Algeria

Abstract. Accurate detection of arrhythmias by electrocardiograms (ECG) is critical for timely diagnosis and intervention in modern healthcare. This study introduces a novel deep-learning model designed to enhance the classification of heartbeats on the ECG. Our proposed architecture combines 12 residual blocks (ResNet) for robust feature extraction with a Bidirectional Long Short-Term Memory (BiLSTM) network, which effectively captures sequential dependencies, to achieve high-performance arrhythmia classification. The model was evaluated on the MIT-BIH Arrhythmia Database, achieving an overall accuracy of 99% and an F1-score of 96.25%, demonstrating exceptional performance across diverse arrhythmia categories. These results underscore the model's potential for clinical application, offering a reliable and automated tool to support arrhythmia detection and improve diagnostic accuracy in healthcare settings.

Keywords: Heartbeat classification · Residual Block · Convolutional Neural Network · Bi-LSTM · Electrocardiogram (ECG) · Arrhythmia Patient

1 Introduction

Electrocardiography (ECG) is a critical diagnostic tool for detecting and monitoring cardiac arrhythmias, which are among the leading causes of morbidity and mortality worldwide [1]. Arrhythmias, characterized by irregular heart rhythms, can range from benign conditions to life-threatening disorders such as ventricular fibrillation. Early and accurate detection of arrhythmias is essential for effective clinical intervention. However, traditional arrhythmia detection methods primarily rely on manual interpretation by cardiologists, a process that is not only time-intensive but also subject to inter-observer variability, particularly in complex cases [2]. These challenges have driven significant interest in developing automated approaches for arrhythmia classification.

Z. Kouahla et al. (Eds.): IAM 2025, LNNS 1973, pp. 211–223, 2026.
https://doi.org/10.1007/978-3-032-26031-4_15

Deep learning has been successfully applied in a wide range of domains [3,4], demonstrating its strong capacity to model complex patterns. More recently, these techniques have also been extended to medical signal analysis, including ECGs, to improve feature extraction and classification. Convolutional neural networks (CNNs), recurrent neural networks (RNNs), and hybrid models have demonstrated promising results in capturing both spatial and temporal patterns in ECG signals[5,6]. Mateo et al. [7] demonstrated the potential of advanced deep learning techniques for robust ECG classification, highlighting their effectiveness in capturing complex signal patterns for arrhythmia detection. Hammad et al. [8] introduced innovative convolutional neural network (CNN) and convolutional LSTM (ConvLSTM) models for automatic arrhythmia detection in IoT applications. Mathunjwa et al. [9] proposed a recurrence plot (RP)-based arrhythmia classification method suitable for portable devices, utilizing ResNet-18 and ResNet-50.

Huang et al. [10] introduced a novel temporal representation input method for ECG classification, integrating CNN and LSTM with an attention mechanism. Sun et al. [11] proposed an automatic ECG classification model combining CNN and bidirectional GRU with an attention mechanism. Chen et al. [12] introduced a cross-modal multiscale multi-instance learning (MIL) approach for long-term ECG classification. Gupta et al. [13] proposed an automatic ECG beat labeling technique for single-lead IoT device data using a weak supervised learning approach enabled by data programming.

Despite these advancements, achieving robust and accurate detection across diverse arrhythmia types remains challenging due to the inherent variability in ECG signals arising from patient-specific differences and noise artifacts. Furthermore, the computational complexity of deep learning models can make them resource-intensive, limiting their feasibility for deployment in resource-constrained environments such as portable devices or remote healthcare settings. Addressing these limitations is critical to enhancing the accuracy and effectiveness of arrhythmia classification models and improving their clinical utility in early detection scenarios.

In this study, we propose a novel deep learning model that integrates ResNet and BiLSTM architectures to effectively address the challenges of arrhythmia classification. The ResNet architecture is utilized for robust feature extraction, leveraging residual connections to efficiently capture intricate and hierarchical signal characteristics. Additionally, the BiLSTM component is incorporated to model temporal dependencies and sequential patterns in ECG data, analyzing information in both forward and backward directions.

The performance of our proposed model was comprehensively evaluated and compared against advanced methodologies using the publicly available MIT-BIH Arrhythmia Database [14,15]. The results of our evaluation demonstrate the model's high accuracy and efficiency, underscoring its effectiveness in arrhythmia classification.

The remainder of this paper is organized as follows: Sect. 2 describes the proposed methodology, including details on the dataset, signal preprocessing techniques and the architecture of the hybrid ResNet-BiLSTM model developed for heartbeat classification. Section 3 presents the experimental results along with a comprehensive performance evaluation. Finally, Sect. 4 concludes the paper by summarizing the key findings and contributions while discussing potential directions for future research in deep learning-based ECG analysis.

2 Materials and Method

This section presents the data, preprocessing techniques, and model architecture used in our study to classify arrhythmias from ECG recordings. We detail the data source, steps taken to enhance signal quality, and the design of a our deep learning model tailored for accurate heartbeat classification. The approach combines robust feature extraction and sequential pattern recognition to optimize classification performance.

2.1 Data Description

In this study, we employed the MIT-BIH Arrhythmia Database [15], an authoritative and extensively utilized dataset from PhysioNet [15] for analyzing arrhythmias and studying cardiovascular health. The MIT-BIH Arrhythmia Database comprises 48 recordings, capturing various arrhythmia patterns from a diverse set of patients, as summarized in Table 1. Each recording has a duration of approximately 30 min and was sampled at a rate of 360 Hz. The ECG signals were recorded using standard lead, such as Lead II. This dataset provided the basis for our study, allowing us to assess the efficacy of our methodology within a well-established and widely benchmarked context.

To categorize the different types of heartbeats in our study, we employed the well-recognized classification scheme from the American Association for the Advancement of Medical Instrumentation (AAMI) [16]. This classification divides heartbeats into five main categories: Normal/Healthy Beats (N), Supraventricular Ectopic Beats (S), Ventricular Ectopic Beats (V), Fusion Beats (F), and Unclassifiable Beats (Q), as outlined in Table 2 [17]. We mapped these categories to the annotations provided in the MIT-BIH Arrhythmia Database, enhancing our analysis of arrhythmias. Utilizing the database annotations allowed us to accurately assign each heartbeat to an AAMI-defined category and delve into the characteristics and patterns linked with each type.

2.2 Dataset Preprocessing

The raw ECG signals are vulnerable to various forms of noise, such as power frequency interference, baseline drift, and high-frequency artifacts resulting from muscle contractions and electrode displacement [18]. In our research, we utilized a filter specifically designed to eliminate baseline wander [19,20] from the ECG signals, as illustrated in Fig. 2. This method effectively mitigated low-frequency variations attributable to noise, movement artifacts, and other influencing factors

Table 1. Heartbeat Types in MIT-BIH Arrhythmia Database: An Overview of Arrhythmia Annotation

Dataset	Sampling rate (HZ)	Labels	Description
MIT-BIH Arrhythmia Database[15]	360 Hz	N	Normal
		A	Atrial premature contraction
		R	Right bundle branch block
		L	Left bundle branch block
		!	Ventricular flutter
		P	Paced beat
		V	Premature ventricular contraction
		J	Nodal (junctional) premature beat
		a	Aberrated atrial premature beat
		e	Atrial escape beat
		j	Nodal (Junctional) escape beat
		E	Ventricular escape beat
		f	Fusion of paced and normal beat
		F	Fusion of ventricular and normal beat
		Q	Unclassifiable beat
Total labels		103247	Beat

Table 2. Arrhythmia Classification by AAMI: A Categorization of Five Heartbeat Classes [17]

AAMI Classes	Description
N	Normal Beats
S	Supraventricular ectopic Beats
V	Ventricular ectopic Beats
F	Fusion Beats
Q	Unclassified Beats

[21]. By applying this filter, we improved the clarity and precision of the ECG signals, thereby facilitating more reliable analysis and interpretation of cardiac activity (Fig. 1).

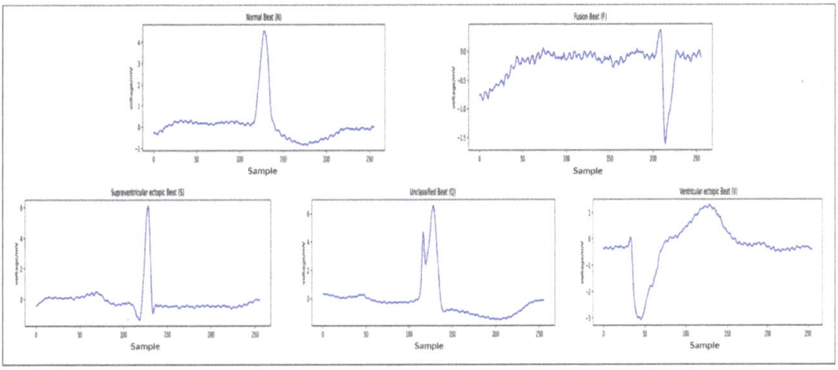

Fig. 1. The five types of heartbeats defined by AAMI

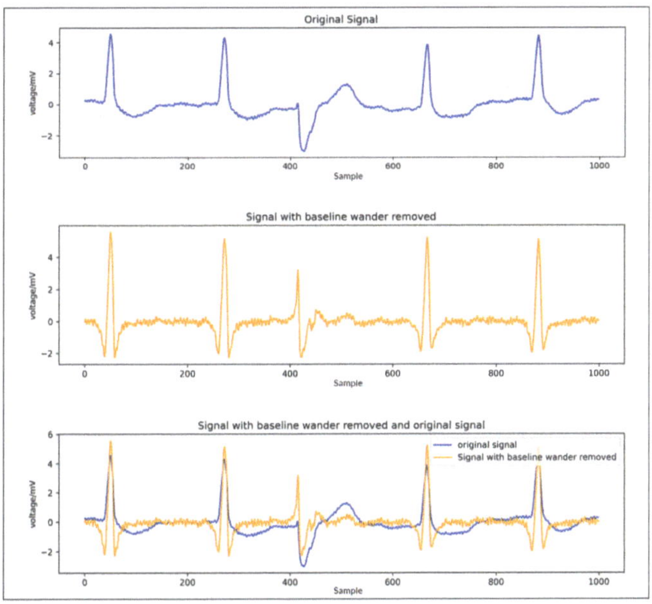

Fig. 2. Improving ECG Clarity: By removing the baseline wander.

2.3 Deep Learning Model

In this section, we present our model, developed to enhance heartbeat classification performance specifically for arrhythmia patients. A schematic representation of the model is provided in Fig. 3. In this model, we begin with ResNet for the initial extraction of features from ECG data. The features extracted by ResNet are then passed into a Bidirectional Long Short-Term Memory (Bi-LSTM) network, which plays a critical role in capturing the temporal dependencies and sequential patterns in the ECG data. By processing these features in both forward and backward directions, the Bi-LSTM effectively preserves long-term relevant information, enhancing the model's ability to identify complex patterns essential for accurate classification.

Subsequently, a classification layer consisting of a dense layer followed by a softmax activation is added to generate predictions. This layer produces the final class labels for the ECG data, utilizing the refined feature representations to ensure accurate categorization.

The sequential architecture of the model leverages each component's strengths at the appropriate stage: ResNet for effective feature extraction and Bi-LSTM for handling temporal dependencies. This configuration ensures that each module builds on the capabilities of the previous one, resulting in a cohesive and highly effective classification model.

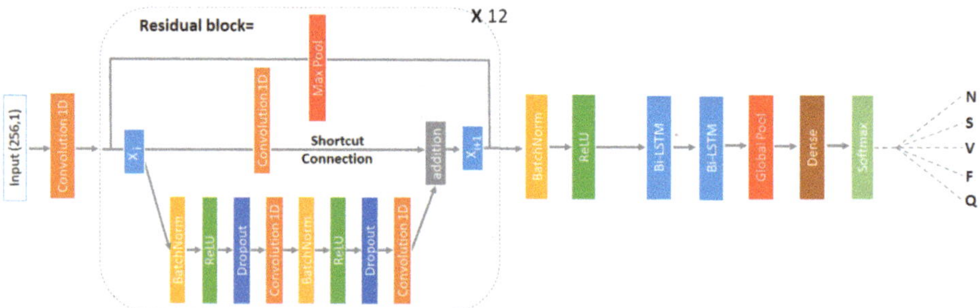

Fig. 3. Overview of the Proposed Model

Residual Block. ResNet (Residual Networks) is widely recognized for its proficiency in extracting deep, nuanced features from complex datasets, demonstrating exceptional performance across various fields, including image processing and time-series analysis [22–25]. Its effectiveness in ECG signal analysis is well-documented in numerous studies [5,26,27]. The architecture of ResNet effectively addresses the vanishing gradient problem and facilitates the learning of deep representations, making it particularly suitable for deriving meaningful features from raw ECG data. Although originally designed for image-based applications [22], ResNet's adaptable architecture has proven versatile across various data types, including ECG signals, underscoring its utility as a robust feature extraction tool. Consequently, we have selected ResNet as the foundational component for feature extraction in our proposed model due to these advantageous characteristics.

In this study, the residual block [28] has been specifically designed to process raw ECG data from a single lead, with each segment consisting of 256 samples, thereby establishing an input size of (256, 1). The primary aim of this architecture is to generate predictions for five distinct classes of arrhythmias. Our network consists of 29 layers and incorporates shortcut connections, drawing inspiration from the original residual network architecture [28]. This design enhances gradient flow, thereby facilitating efficient training. Our architecture includes 12 residual blocks, with each block containing two 1D convolutional layers. These convolutional layers utilize a kernel size of 16, and the filter width is determined by the formula 32×2^i, where 'i' is a hyper-parameter that we fine-tune every two residual blocks. In line with best practices in the field, we have incorporated batch normalization [29] and rectified linear activation [30] prior to each convolutional layer, adhering to the pre-activation block design [22]. Additionally, we have strategically applied dropout regularization [31] between the convolutional layers and after the activation function, setting the dropout probability to 0.5 (as illustrated in Fig. 3). These deliberate strategies are aimed at both mitigating overfitting and enhancing the network's generalization capabilities.

Bidirectional Long Short-Term Memory (BiLSTM). The Bidirectional Long Short-Term Memory (BiLSTM) network extends the traditional LSTM by processing sequences in both forward and backward directions, which enhances the model's ability to capture contextual information from both the past and future within the ECG data [32]. This bidirectional processing is particularly valuable for arrhythmia detection, as it enables the model to consider patterns both before and after each heartbeat segment, which can provide critical context for accurate classification. The BiLSTM network architecture is visually represented in Fig. 4.

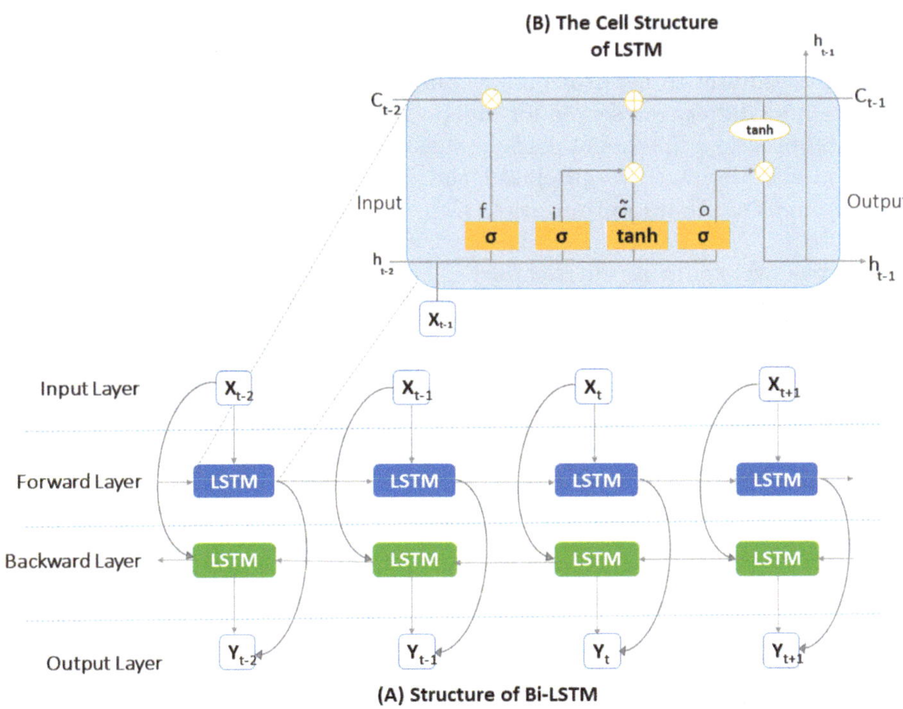

Fig. 4. A Comprehensive exploration of bidirectional long short-term memory (BiLSTM)

In our model, we utilize two BiLSTM layers, with an initial configuration of 512 units in the first layer and 256 in the second. This setup, identified through experimental optimization, balances capturing temporal patterns with computational efficiency. The combined output of the forward and backward LSTMs provides a richer feature representation, enhancing the model's classification accuracy for our specific dataset.

3 Experimentation and Validation

We conducted a series of experiments to evaluate the effectiveness of our model. Using a variety of metrics, we assessed its performance in classifying arrhythmias. Additionally, we compared our results with those obtained from established methods to highlight the model's relative performance.

3.1 Experimental Setup

The proposed model was developed and trained using Python, utilizing the Keras deep learning framework [33] with TensorFlow as the backend [34]. All experiments were conducted on Google Colab Pro, equipped with 83GB of RAM and high-performance GPUs.

The dataset used in this study comprised 103,247 heartbeats. To ensure an unbiased evaluation, we divided the data into training, validation, and testing sets. Specifically, 80% of the data was allocated for training, with 20% designated as the validation set to monitor performance during training. Additionally, 15% of the training data was reserved for independent testing, allowing for a rigorous evaluation process while maintaining a substantial training set.

For model training, we employed the categorical cross-entropy loss function [35] and the Adam optimization algorithm [36]. The initial learning rate was set to 10^{-3}, and was reduced by a factor of 10 if the validation loss did not improve over ten consecutive epochs. The hyperparameters $\beta 1$ and $\beta 2$ were set to 0.9 and 0.999, respectively. Training was conducted with a mini-batch size of 256 for a total of 35 epochs. The model chosen for final evaluation was the one that achieved the lowest error on the validation set, ensuring optimal generalization and accuracy.

3.2 Evaluation Measures Used

To evaluate the performance of our model in arrhythmia classification, we employed several widely-used metrics, including overall accuracy (Acc), precision (P), recall (R), and F1 score, which together provide a comprehensive assessment of its performance and reliability across different heartbeat classes. Detailed descriptions of these metrics are provided below:

$$\text{Accuracy} = \frac{\text{TP} + \text{TN}}{\text{TP} + \text{TN} + \text{FP} + \text{FN}}$$

$$\text{Precision} = \frac{\text{TP}}{\text{TP} + \text{FP}}$$

$$\text{Recall} = \frac{\text{TP}}{\text{TP} + \text{FN}}$$

$$\text{F1-score} = 2 \cdot \frac{\text{Precision} \cdot \text{Recall}}{\text{Precision} + \text{Recall}}$$

$$\text{Macro avg F1-score} = \frac{1}{N}\sum_{i=1}^{N} \text{F1-score}_i$$

In this context, TP (True Positives) represents the count of correctly classified samples within each class. FN (False Negatives) denotes the samples from each class that were misclassified as belonging to other classes, and FP (False Positives) indicates samples that were incorrectly classified as belonging to a particular class when they actually belong to other classes. By averaging the F1 scores across all five heartbeat categories, we provide an overall measure of the model's performance across classes.

3.3 Obtained Results

In this section, we present the performance of our model on the MIT-BIH Arrhythmia Database, a widely recognized benchmark for ECG arrhythmia classification. This dataset, known for its controlled environment and standardized annotations, provides a solid foundation for assessing the baseline capabilities of our model. The results below demonstrate the model's effectiveness in detecting various arrhythmia types.

As shown in Table 3, our proposed model demonstrate its robust classification capabilities across different arrhythmia classes. With an overall accuracy of 99% and an F1-score of 96.25%, the model reliably identifies arrhythmia patterns, suggesting its potential effectiveness in clinical applications where precision in detecting abnormal heart rhythms is essential.

Our model shows particularly strong performance in detecting normal (N) and ventricular (V) beats, with F1-scores of 99% and 98%, respectively. This high level of precision is critical for clinical reliability, as normal and ventricular beats are often the primary indicators used to distinguish between regular and irregular heart patterns. Despite this overall strong performance, the model's slightly lower recall and F1-score of 0.88 for fusion beats (F) indicate an area where further refinement may be beneficial. Future efforts could involve incorporating additional or augmented data specific to fusion beats to enhance model sensitivity to this class. Addressing this challenge would further improve the model's applicability across diverse clinical scenarios, particularly in cases where rarer arrhythmias need to be accurately identified.

Fig 5 illustrates the accuracy and loss curves obtained from the model's training and validation phases on the MIT-BIH dataset. The close alignment

Table 3. Proposed model performances on the testing dataset using the MIT-BIH dataset

	Precision	Recall	F1-score
N	0.99	0.99	0.99
V	0.98	0.99	0.98
F	0.90	0.85	0.88
Q	1.00	1.00	1.00
Accuracy			0.99
Macro avg	0.97	0.96	0.96
Weighted avg	0.99	0.99	0.99

between the validation and training curves suggests an absence of overfitting, with both curves following a similar trajectory. The steady improvement in the curves across iterations reflects a consistent enhancement, indicating that the model effectively generalizes to new data, thereby achieving reliable performance across both training and validation sets.

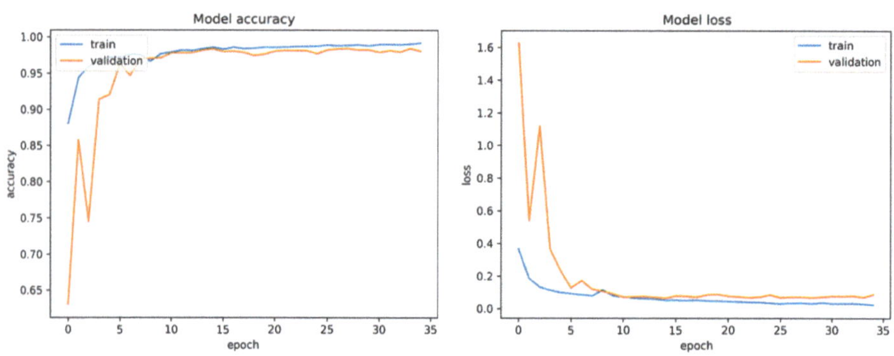

Fig. 5. Accuracy and loss curves forMIT-BIH dataset

3.4 Performance Comparison with Existing Methods

In this subsection, we present a comparative analysis of our results with existing methods in the literature. The comparative analysis presented in Table 4 evaluates various models used for arrhythmia classification on the MIT-BIH dataset. This table includes a diverse range of methods, from traditional approaches like RBF [7] to advanced neural network architectures such as ResNet-18 [9], CNN-LSTM Attention [10], and CNN-RNN-Attention [11].

The results of our proposed model demonstrate significant advancements in arrhythmia classification compared to existing methods documented in the literature. Our model achieves an impressive accuracy of 99% and an F1-score of 96.25%. These results not only surpass those of traditional methods, such as RBF [7], which reported an accuracy of 98.79%, but also outshine more complex architectures like ConvLSTM [8] and CNN-RNN Attention [11], which recorded accuracies of 84% and 84.65%, respectively.

Our model's performance is particularly noteworthy when compared to ResNet-18 [9] and CNN-LSTM Attention [10], which achieved accuracy scores of 97.21% and 98.95%. This highlights the effectiveness of our approach in accurately classifying arrhythmias across multiple classes. The significant gap in accuracy and F1-score indicates that our model effectively captures the intricate patterns within the ECG data, leading to more reliable and precise arrhythmia detection.

Table 4. Comparative analysis of arrhythmia classification using various models on the MIT-BIH dataset

Method	Split	Classes	Accuracy↑	F1-score↑
RBF [7]	40% train 25% validation 35% test	2	98.79%	–
ConvLSTM [8]	–	2	84%	–
ResNet-18 [9]	5-fold CV	4	97.21%	95.96%
CNN-LSTM Attention [10]	5-fold CV	5	98.95%	–
CNN-RNN-Attention [11]	70% train 10% validation 20% test	5	84.65%	91.10%
CMM [12]	–	3	–	82.93%
DP+DA [13]	—	2	92.2%	95.4%
Proposed Method	65% train 20% validation 15% test	**5**	**99%**	**96.25%**

4 Conclusions

In this study, we presented a deep learning model that combines ResNet and BiLSTM architectures for enhanced classification of arrhythmias from ECG signals. Our approach demonstrated high accuracy and robustness in arrhythmia detection, as evidenced by an overall accuracy of 99% and an F1-score of 96.25% on the MIT-BIH Arrhythmia Database. This performance underscores the model's efficacy in distinguishing among diverse arrhythmia classes, even when dealing with complex and sequential ECG patterns.

The integration of residual blocks for feature extraction and BiLSTM for temporal analysis proved instrumental in capturing intricate ECG signal characteristics, which are critical for precise arrhythmia detection. These findings underscore the potential of employing state-of-the-art deep learning methodologies to develop robust, automated diagnostic tools capable of supporting cardiologists in clinical decision-making.

Future research aim to enhance the generalizability of the proposed model by evaluating its performance on larger, more diverse datasets, including real-world clinical scenarios.

References

1. Huang, W., et al.: Snippet-based ECG arrhythmia detection using deep learning. IEEE Trans. Biomed. Eng. **69**, 1234–1245 (2022)
2. Pan, J., Tompkins, W.J.: A real-time QRS detection algorithm. IEEE Trans. Biomed. Eng. **BME–32**(3), 230–236 (1985)
3. Hattab, A., Behloul, A.: New approaches for automatic face recognition based on deep learning models and local handcrafted ALTP. EAI Endors. Trans. Scalable Inf. Syst. **9**(34), e11 (2022)
4. Hattab, A., Behloul, A.: A robust iris recognition approach based on transfer learning. Int. J. Comput. Digit. Syst. (2023)
5. Kerdoudi, S., et al.: Automated ECG arrhythmia classification using deep learning. Comput. Biol. Med. **162**, 107637 (2024)
6. Rajpurkar, P., et al.: Cardiologist-level arrhythmia detection with convolutional neural networks. Nat. Med. **23**, 1275–1280 (2017)
7. Mateo, A., et al.: Efficient ECG arrhythmia classification using RBF neural networks. Biomed. Signal Process. Control **27**, 72–81 (2016)
8. Hammad, A., et al.: Deep learning-based arrhythmia detection for IoT applications. IEEE Access **10**, 45678–45689 (2022)
9. Mathunjwa, S., et al.: ECG arrhythmia classification using recurrence plots and ResNet architectures. Comput. Biol. Med. **145**, 105420 (2022)
10. Huang, W., et al.: A novel CNN-LSTM with attention model for ECG classification. Pattern Recogn. **133**, 108973 (2023)
11. Sun, L., et al.: Automatic ECG classification with CNN and bidirectional GRU attention model. Biomed. Signal Process. Control **84**, 104741 (2023)
12. Chen, X., et al.: Cross-modal multiscale multi-instance learning for long-term ECG classification. IEEE Trans. Neural Syst. Rehabil. Eng. **31**, 1234–1244 (2023)
13. Gupta, R., et al.: Data programming for automatic ECG beat labeling in IoT devices. Sensors **24**, 5678 (2024)
14. Moody, G.B., Mark, R.G.: Impact of the MIT-BIH arrhythmia database. IEEE Eng. Med. Biol. Mag. **20**, 45–50 (2001)
15. Goldberger, A., et al.: PhysioBank, PhysioToolkit, and PhysioNet: components of a new research resource for complex physiologic signals. Circulation **101**(23), e215–e220 (2000)
16. American Association for the Advancement of Medical Instrumentation. Standards for ECG waveform classification (2007)
17. Li, J., et al.: Heartbeat classification based on deep learning and AAMI standard. Comput. Biol. Med. **124**, 103938 (2020)

18. Chen, W., et al.: Heartbeat classification from ECG signals using wavelet transform. Biomed. Signal Process. Control **33**, 116–123 (2017)
19. De Pinto, C., et al.: Baseline wander removal in ECG signals. Comput. Cardiol. **19**, 40–43 (1992)
20. Romero, J., et al.: ECG baseline wander correction using advanced filters. Biomed. Signal Process. Control **68**, 102992 (2021)
21. Kaur, A., et al.: Comparison of filtering techniques for ECG signals. Int. J. Comput. Appl. **33**, 1–6 (2011)
22. He, K., Zhang, X., Ren, S., Sun, J.: Deep residual learning for image recognition. In: Proceedings of the IEEE Conference on Computer Vision and Pattern Recognition, pp. 770–778 (2016)
23. Durga, R., et al.: Resnet-based architectures for time series classification. Neural Comput. Appl. **34**, 12345–12360 (2022)
24. Reddy, P., et al.: Resnet applications in biomedical signal processing. Biomed. Signal Process. Control **79**, 104074 (2023)
25. Veeragandham, R., et al.: Effectiveness of residual networks in ECG classification. Comput. Biol. Med. **146**, 105559 (2022)
26. Anand, R., et al.: Enhanced ECG classification using ResNet variants. IEEE Access **12**, 23456–23467 (2024)
27. Zhang, L., et al.: Heartbeats classification with deep residual networks. J. Biomed. Inform. **118**, 103786 (2021)
28. He, K., et al.: Identity mappings in deep residual networks. In: European Conference on Computer Vision, pp. 630–645 (2016)
29. Ioffe, S., Szegedy, C.: Batch normalization: accelerating deep network training by reducing internal covariate shift. In: International Conference on Machine Learning, pp. 448–456 (2015)
30. Agarap, A.F.: Deep learning using rectified linear units (RuLU). arXiv preprint arXiv:1803.08375 (2018)
31. Srivastava, N., et al.: Dropout: a simple way to prevent neural networks from overfitting. J. Mach. Learn. Res. **15**, 1929–1958 (2014)
32. Schuster, M., Paliwal, K.K.: Bidirectional recurrent neural networks. IEEE Trans. Signal Process. **45**, 2673–2681 (1997)
33. Chollet, F.: Keras. GitHub repository (2018)
34. Abadi, M., et al.: TensorFlow: large-scale machine learning on heterogeneous systems. arXiv preprint arXiv:1603.04467 (2016)
35. Deng, J., et al.: ImageNet: a large-scale hierarchical image database. In: CVPR (2006)
36. Kingma, D.P., Ba, J.: Adam: a method for stochastic optimization. arXiv preprint arXiv:1412.6980 (2014)

The Algerian Forest Fires Dataset

Faroudja Abid(✉)

Centre de développement des technologies avances, Cité 20 août 1956, Baba Hassen, Algiers, Algeria
fabid@cdta.dz

Abstract. The investigation of machine learning (ML) in the development of forest fires prediction and detection models has noticeably improved the capability and effectiveness of these models. However, to fully realize their potential in forest fires-prediction purpose, dataset availability is required. In view of the noticed lack of the publicly available data on forest fires prediction in Algeria, we have elaborated the first Algerian forest fires dataset to investigate the machine learning algorithms for forest fire prediction. We present in this paper the first publicly available Algerian forest fires dataset that consist of data on forest fires in Algeria. The Algerian forest fires dataset is related to meteorological observations and the fire weather indices. Our dataset gives acceptable results in terms of accuracy and recall when using algorithms such as the decision tree and its ensembles, the naïve Bayesian and the artificial neural network. This dataset is useful for researchers that work on forest fires prediction and detection in Algeria and over the word, in particular the Mediterranean countries that have similar climate. We aim mainly, by introducing this dataset to address the gap in region-specific fire prediction resources in Algeria. This dataset was elaborated in 2019 and first donated to the UCI online repository and then to the IEEE DataPort.

Keywords: Algerian forest fires dataset · forest fires · fire prediction · meteorological data · FWI components

1 Introduction

Forest fires disaster is an increasing threat to forest area in Algeria [1]. The last huge forest fire occurred in Algeria on August 2021 destroyed over 100101 hectares, which is the largest burned area recorded in more than a decade [2]. One alternative to the traditional firefighting system in Algeria, i.e. the monitoring tower is the investigation of the ML-based models for forest-fire occurrence modeling, forecasting and detection [3,4]. Indeed, datasets availability is primordial for the machine learning models development; however, at present there is a lack of publicly available datasets on forest fires data in Algeria. The mostly used datasets in the context of forest fires detection and prediction modeling are the satellite data, the vision-based data and the meteorological data. We have created this first Algerian forest fires dataset, which is mainly based on meteorological data, to investigate the ML algorithms as models for forest fires prediction in the context of Algeria [5].

Z. Kouahla et al. (Eds.): IAM 2025, LNNS 1973, pp. 224–233, 2026.
https://doi.org/10.1007/978-3-032-26031-4_16

We present in this paper the elaborated Algerian forest fires dataset that includes the daily meteorological observations and the FWI system components, namely the Fine Fuel Moisture Code (FFMC), the Duff Moisture Code (DMC), the Drought Code (DC), the Initial Spread Index (ISI), the Buildup Index (BUI) and the FWI. We have constructed a dataset that regroups the data of two regions of Algeria, namely the Sidi Bel-Abbas region located in the northwest of Algeria and the Bejaia region located in the northeast of Algeria. According to the statistics [1] these two regions are among the most affected regions by the forest fire disaster in Algeria. Our dataset includes 244 instances (a combination of the data collected in Sidi Bel-Abbas and Bejaia, with 122 instances each one). The Algerian forest fires dataset is available in the UCI repository and the IEEE DataPort [6, 7].

The main values of this dataset are summarized hereafter:

1. This dataset is beneficial for researcher (especially the Algerian researchers and students) who work on forest prediction and detection field of research. In fact, this dataset can serve as valuable resource for further studies related to forest fires prediction and detection in Algeria that suffers from this disaster every year. It can be used also in models investigating the forest fire prediction issue in the Mediterranean countries that have similar climate.
2. This dataset includes labels (classes) that facilitate its use in ML-based models development. In its first appearance, the Algerian forest fires dataset was used in our laboratory to predict forest fires in Algeria using a set of ML algorithms, namely the decision tree algorithm and its ensembles. After its publicly availability in the UCI repository, this dataset was used by other researchers working on forest fire forecasting and detection.

The rest of this paper is organized as follows. In Sect. 2 we present the dataset description and the region of interest (the study area). The adopted methodology for the Algerian forest fires dataset creation will be exposed in Sect. 3. In Sect. 4, we sum-up the studies that considered and exploited our dataset. Section 5 exposes the results of the conducted evaluation and comparison of several ML algorithms based models using the elaborated dataset. Finally, Sect. 5 concludes this data paper.

2 Dataset Description

This first Algerian forest fires dataset consists of 244 instances, which includes 12 variables (attributes), namely the meteorological observations (four weather elements), the six components of the FWI Canadian system, the fire classes ('fire' and 'not fire') and the date variable (DD/MM/YYYY: Day/month ('June' to 'September')/year (2012)). The Algerian forest fires dataset regroups the data of two regions of Algeria, namely the Sidi Bel-Abbas and the Bejaia regions. We summarize in Table 1, our dataset main specifications.

Table 1. The Algerian forest fires dataset main specifications.

Domain	Environmental Science, Ecology
Subject area	The Algerian forest fires dataset involves the area of forest fires data related to the meteorological observations and the fire danger rating system indices (FWI components).
Type of data file	.csv file (dataset with labels)
Data collection	The data collection concerns the meteorological observations, the remaining variables (attributes), namely the FWI components, were calculated based on the meteorological observations collected from the historical weather observations site [8]. The collected weather parameters include temperature, relative humidity (RH), wind speed (Ws) and rain. After the FWI components computation, the values related to these components were added to the .csv file that contains the meteorological data to constitute the whole dataset after instances classification.
Study period	The meteorological data comprises the observations for the period of summer of 2012 (from June 1 to September 30).
Study area	We have considered two regions of Algeria, namely the Sidi Bel-Abbas and the Bejaia regions located in the northwest and the northeast of Algeria, respectively.
Data accessibility	Repository name: UC Irvine Machine Learning Repository and IEEE DataPort Dataset identification numbers (DOIs): 10.24432/C5KW4N and 10.21227/31na-kz23 Direct URL to dataset: https://archive.ics.uci.edu/dataset/547/algerian+forest+fires+dataset and https://dx.doi.org/10.21227/31na-kz23

2.1 The Region of Interest/Study Area

The region of interest concerns two region of Algeria, namely the Sidi Bel-Abbas region located in the northwest of Algeria at 116°23′50″ E longitude and 39°54′26″ N and the Bejaia region located in the northeast of Algeria at 5°05′03″ E longitude and 36°45′21″ N. According to statistics [1], these two regions are among the most affected regions by the forest fire disaster in Algeria (in 2012) this is why we have considered these two main fire hotspots as the first region of interest.

2.2 The Algerian Forest Fires Dataset Features Description

The Meteorological Observations

The meteorological observations include four weather parameters, namely temperature (Celsius degrees) in range of 22 to 42, relative humidity (RH) (%) in range of 21 to 90, wind speed (km/h) in range of 6 to 29 and rain (mm/m2) in range of 0 to 16.8. We considered also the wind speed factor, which influences mainly the fire spread.

The FWI Components

In addition to the weather elements, our dataset includes six components of the FWI Canadian system [9], namely the FFMC, DMC, DC, ISI, BUI and the FWI. These components include the fuel moisture codes and the fire behavioral indices. The first three components (FFMC, DMC and DC) represent the numeric ratings of the moisture content of the forest floor and other dead organic material. The remaining three components are fire behavioral indices that characterize the rate of fire spread (ISI), the total amount of fuel available for combustion (BUI) and the fire intensity (FWI).

The Two Classes

Two classes were considered, namely 'fire' class and 'not fire' class. The 244 instances include 138 'fire' classes and 106 'not fire' classes. Table 2 shows a summarized features description of the Algerian forest fires dataset.

Table 2. Algerian forest fires dataset features description

Data Type	Attributes/ variables name	Description	Values interval Sidi-Bel-Abbes region	Values interval Bejaia region	Values intervals the two regions
Meteorological observations /elements	Temperature	temperature in Celsius degrees	24 to 42	22 to 37	22 to 42
	RH	relative humidity in %	21 to 90	45 to 89	21 to 90
	WS	wind speed in km/h	6 to 29	11 to 26	6 to 29
	Rain	outside rain in mm/m2	0 to 16.8	0 to 8.7	0 to 16.8
FWI components	FFMC	Fine Fuel Moisture Code	41.1 to 94.3	28.6 to 90.3	28.6 to 94.3
	DMC	Duff Moisture Code	0.9 to 65.9	0.7 to 54.2	0.9 to 65.9
	DC	Drought Code	7.3 to 177.3	7.4 to 220.4	7.3 to 220.4
	ISI	Initial Spread Index	0.1 to 18.5	0.1 to 19	0.1 to 19
	BUI	Buildup Index	1.4 to 68	3.3 to 67.4	1.4 to 68
	FWI	Fire Weather Index	0 to 31.1	0 to 30.1	0 to 31.1

3 Methodology for the Dataset Creation

We present in this section the methodological approach for the dataset elaboration including the data collection, the FWI components computation and the instances classification.

For the dataset creation, we have used the meteorological observations for the period of summer of 2012 (from June 1 to September 30) since the fire occurrence was high on this period. In fact, during 2012, Algeria was subject of more than 5110 forest fires, which destroyed more than 99061 ha. The weather observations collected from [8] include the critical weather parameters influencing the forest fire occurrence and spread, namely temperature, relative humidity (RH), wind speed and rain. In addition to the weather elements, the dataset includes the components of the FWI Canadian system [9], which comprises six components, namely the FFMC, the DMC, the DC, the ISI, the BUI and the FWI. The FFMC is an indicator of the ignition probability and the FWI index is an indicator of the fire intensity. The FWI system components computation is based on consecutive daily observations of temperature, relative humidity, wind speed and 24-hours precipitation. We first classified the forest fires dataset instances as ‘fire’ and ‘not fire’ classes using the FWI system components. The 244 instances include 138 ‘fire’ classes and 106 ‘not fire’ classes.

3.1 Meteorological Data Collection

Our dataset is mainly based on the meteorological data. This latter can be collected locally (using sensors) in real-time and with very low cost compared to the satellite and vision-based data. The collected weather observations include temperature, relative humidity, wind speed and rain. The RH is related to the combustible content’s moisture; this factor is reduced with the evaporation increase caused by high temperatures. For the data creation, we have first collected the meteorological observations for the period of summer of 2012 (from June to September). As highlighted previously, the meteorological observations were collected from the historical weather observations website [8], which is a free weather archiving service that includes the meteorological observations history of the world.

3.2 The FWI Components Computation

The FWI components are calculated using the meteorological parameters (temperature, relative humidity, wind speed and 24-hours precipitation). Figure 1 illustrates the Canadian FWI components structure. We give hereafter a brief description of the Canadian FWI system components:

FFMC: The FFMC represents the moisture content of litter and other cured fine fuels. Its values range from 0 to 101; it is an indicator of the ignition probability and the flammability of the fine fuel.

DMC: The DMC represents the moisture content of the decomposed organic material underneath the litter (in layers of moderate depth 2-10cm).

DC: The DC represents the drying deep into the soil, it is a numeric rating of the average moisture content of deeply compacted (>10cm) organic layers.

ISI: The ISI represents an estimation of the spread potential and the expected rate of the fire spread. It is based on the wind speed and the FFMC.

BUI: The BUI represents the total amount of the fuel available for combustion. The BUI combines the current DMC and DC to produce an estimation of the potential heat release in heavier fuels.

FWI: The FWI is an indicator of fire intensity; it is used as a general indicator of fire danger.

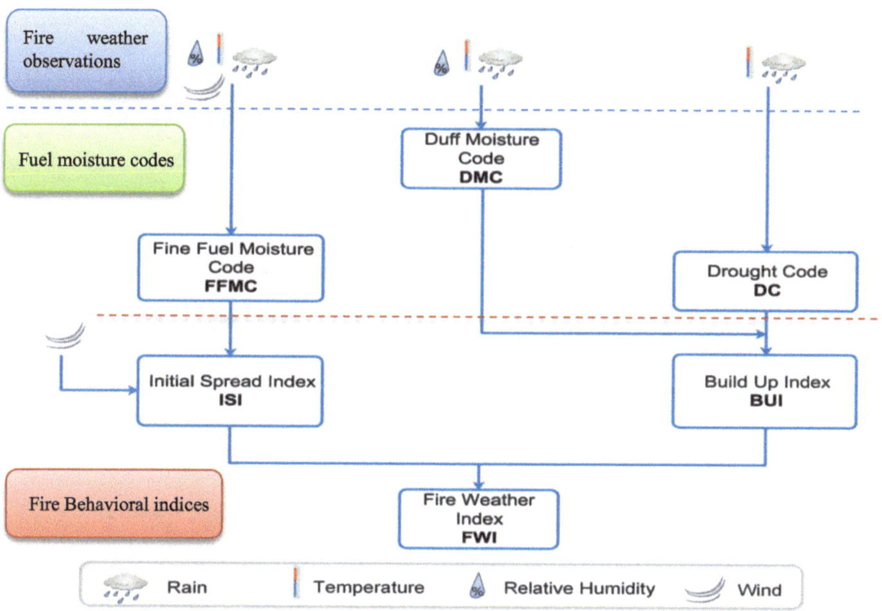

Fig. 1. The Canadian Fire Weather Index components structure

3.3 Instances Classification

The numerical attributes/variables values were used to predict two possibilities, namely "fire" and "not fire" classes. The Algerian forest fires dataset instances classification as "fire" and "not fire" classes was performed based on the Canadian FFDRS related to the fire weather indices values ranges. This Canadian FFDRS system has been practiced in Canada, Australia and United States. Moreover, this system is currently applied in the European countries and other countries such as Russia, Mexico, New Zealand, Indonesia and Malaysia.

4 Studies that Used and Exploited the Algerian Forest Fires Dataset

We summarize in Table 3 the studies [3,4,11–21] that used the Algerian forest fires dataset, after it publicly availability in the UCI repository. It is noteworthy to precise that the works that used our dataset are not limited to the studies that are tabulated in this table. Our dataset was also employed by the Kaggle's platform contributors for the creation of other datasets based in the processing and analysis of the original Algerian forest fires dataset. This dataset was also a part of the UCBerkeley Spring 2022 Stat159 Group10's final project for educational purposes.

Table 3. Studies that used the Algerian forest fires dataset

References	Authors	Source Journal/conference	Year	DOI
[16]	G.P. Dirgantoro et al.	Proceedings of the IEEE ICITISEE	2021	10.1109/ICITISEE53823.2021.9655820
[14]	H. D. Moore et al.	J. Cybersecur. Priv.	2022	10.3390/jcp2040045
[11]	Jiajia Wang et al.	Applied Soft Computing (IF 8.26)	2022	10.1016/j.asoc.2022.109283
[10]	C. Zhang et al.	Applied Soft Computing (IF 8.26)	2023	j.asoc.2023.110539
[3]	A. Zaidi et al.	Heliyon (IF=4)	2023	10.1016/j.heliyon.2023.e18064
[4]	F. Abid	LNNS, Springer	2023	10.1007/978-3-031-60591-8_9
[12]	K. Siminski at al.	Fuzzy Sets and Systems (IF 3.9)	2023	10.1016/j.fss.2022.10.021
[13]	F. Akalın et al.	Arab. J. Scir Eng. (IF2.8)	2023	10.1007/s13369-023-08183-z
[15]	A. Faiyaz et al.	Proceedings of REEDCON	2023	10.1109/REEDCON57544.2023.10150867
[20]	Q. An et al.	Neural Process Letts (IF2.8)	2024	10.1007/s11063-024-11615-y
[21]	Y. Wang et al.	Fire (IF2.7)	2024	10.3390/fire7070254
[17]	A.M. Elshewey et al.	Model. Earth Syst. Environ (IF 2.9)	2025	10.1007/s40808-025-02569-8
[18]	G. Bulivou et al.	IEEE Access (IF3.6)	2025	10.1109/ACCESS.2025.3544830
[19]	N.A Houacine et al.	Proceedings of the Communications in Computer and Information Science	2025	10.1007/978-3-031-88226-5_3

5 Performance Evaluation of a Set of ML-Models Using the Algerian Forest Fires Dataset

We present in this section the results of the performance evaluation of a set of machine learning based models for forest-fire prediction using our publicly available Algerian forest fires dataset. We have conducted an evaluation and comparison of several ML algorithms based models using three performance metrics, namely accuracy, recall and precision. The consider ML algorithms are the neural network model [22] (the multi-layer perceptron -MLP-), the logistic regression [23], the decision tree (DT) [24] and its ensembles (the random forest (RF) [25], the Bagging and the AdaBoost), Bayesian (the BayeNet and the Naive Bayesian (NB) [26]), the SVM [27] and the K-Nearest Neighbors (KNN). Table 4 tabulates the performance comparison results of the considered ML models evaluated under the Weka tool.

Table 4. Performance evaluation results of a set of ML models

ML-Models	Accuracy (%)	Precision (%)	Recall
DT	95.9	95.9	95.9
RF	98.4	98.4	98.4
Boosted trees (AdaBoost)	98	98	98
Bagging trees	98	98	98
SVM	67.9	79.4	67.7
KNN	92.6	92.6	92.6
NB	93.9	93.9	93.9
ByeNet	96.7	96.7	96.7
Logistic Regression	95.9	95.9	95.9

From Table 4 we observe that the ML algorithms give comparable accuracies ranging from 98.4% to 92.6% (except the SVM model that gives the lowest accuracy of about 67.6%). The best accuracy is obtained with the RF models and the poorest one with the SVM model. The obtained values are as follows: the RF model gives the best result (98.4%), which is similar to the results obtained with the Bagging trees and the AdaBoost trees, followed by the ANN/MLP (97.1%), the BayeNet (96.7%), the DT and logistic regression (95.9%). The KNN also gives acceptable results that are comparable to the results obtained with the NB. The NB classifier gives results about 93% for the three metrics. The SVM model gives the lowest results with 67.9%, 79.4% and 67.7% for accuracy, precision and recall, respectively. These results demonstrate the efficiency of the ML-based classifiers for fire-occurrence prediction in the context of Algeria.

6 Conclusion

We have presented in this paper the Algerian forest fires dataset, which involves the area of forest fires data that includes meteorological observations and the forest fire danger rating system indices (FWI components). In view of the scarcity of publicly available datasets on data on forest fires occurrence in Algeria, this dataset is useful for researchers working on forest fires prediction and monitoring systems. Furthermore, our dataset includes the FWI components related to the moisture content of the forest floor, thus it can be used for forest floor analyses that permit an extended use of this dataset by ecological researchers. Various, studies have used our dataset for their forest fire forecasting models development in Algeria and other laboratories. We have conducted an evaluation and comparison of several ML algorithms based models. The results demonstrate the efficiency of the ML-based classifiers for fire-occurrence prediction in the context of Algeria. Indeed, our dataset gives acceptable results when using algorithms such as the decision tree and its ensembles, the naïve Bayesian or the artificial neural network; however, this dataset presents a limitation regarding its size that is 244 instances, which cannot be used with the deep leaning algorithms, as these algorithms require datasets with an important size. In fact, we are currently, working on the extension of this dataset in terms of size, geographical scope, factors types and temporal coverage (multi-year data) to address generalizability and reduce overfitting risks.

References

1. Benkheira, A.: Forest fires in Algeria analysis and perspectives. Technical Report, Ministry of Agriculture and Rural Development and Fisheries, Forest Department, Algeria (2018)
2. European report, annual fire reports, EFFIS TEAM, forest fires in Europe, Middle East and North Africa, pp. 1-173 (2021). https://effis-gwis-cms.s3.eu-west-1.amazonaws.com/effis/reports-and-publications/annual-firereports/Annual_Report_2021_final_topdf1.pdf
3. Zaidi, A.: Predicting wildfires in Algerian forests using machine-learning models. Heliyon **9**, 1–19 (2023). https://doi.org/10.1016/j.heliyon.2023.e18064
4. Abid, F.: Machine learning algorithms based models for forest fires prediction: an analytic and comparison study. In: Laouar, M.R., et al. (eds.) 13th International Conference on Information Systems and Advanced Technologies "ICISAT 2023". ICISAT 2023. Lecture Notes in Networks and Systems, Springer, vol. 981, pp. 100-110 (2023). https://doi.org/10.1007/978-3-031-60591-8_9
5. Abid, F., Izeboudjen, N.: Predicting forest fire in Algeria using data mining techniques: case study of the decision tree algorithm, In: Ezziyyani, M., (eds.) Advanced Intelligent Systems for Sustainable Development (AI2SD'2019). AI2SD 2019. Advances in Intelligent Systems and Computing, Springer, Cham, vol. 1105, pp. 363-370 (2020)
6. The Algerian forest fires dataset link. https://archive.ics.uci.edu/dataset/547/algerian+forest+fires+dataset
7. The Algerian forest fires dataset link. IEEE DataPort. https://doi.org/10.21227/31na-kz23
8. Historique-meteo. https://www.historique-meteo.net/
9. Canadian FWI system. https://cwfis.cfs.nrcan.gc.ca/background/summary/fwi
10. Zhang, C., Zhang, J., Li, W., Pedrycz, W., Li, D.: A regret theory-based multi-granularity three-way decision model with incomplete T-spherical fuzzy information and its application in forest fire management. Appl. Soft Comput. **145** (2023). https://doi.org/10.1016/j.asoc.2023.110539
11. Wang, J., et al.: A three-way decision method with prospect theory to multi-attribute decision-making and its applications under hesitant fuzzy environments. Appl. Soft Comput. **126** (2022). https://doi.org/10.1016/j.asoc.2022.109283

12. Siminski, K.: 3WDNFS–three-way decision neuro-fuzzy system for classification. Fuzzy Sets Syst. **466** (2023). https://doi.org/10.1016/j.fss.2022.10.021.
13. Akalın, F.: Survival classification in heart failure patients by neural network-based crocodile and Egyptian plover (CEP) optimization algorithm. Arab. J. Scir Eng. (2023). https://doi.org/10.1007/s13369-023-08183-z
14. Moore, H.D., Stephens, A., Scherer, W.: An understanding of the vulnerability of datasets to disparate membership inference attacks. J. Cybersecur. Priv. **2**, 882–906 (2022). https://doi.org/10.3390/jcp2040045
15. Faiyaz, A., Zeenat, W., Musheer, A., Mohd Zeeshan, A.: Forest fire prediction using machine learning techniques. In: 2023 International Conference on Recent Advances in Electrical, Electronics & Digital Healthcare Technologies (REEDCON) (2023). https://doi.org/10.1109/REEDCON57544.2023.10150867
16. Dirgantoro, G.P., Soeleman, M.A., Supriyanto, C.: Smoothing weight distance to solve Euclidean distance measurement problems in K-Nearest neighbor algorithm. In: IEEE 5th International Conference on Information Technology, Information Systems and Electrical Engineering (ICITISEE), pp. 294 -298 (2021). https://doi.org/10.1109/ICITISEE53823.2021.9655820
17. Elshewey, A.M., Osman A.M., Youssef, R.Y.: Enhancing forest fires classification using a hybrid convolutional and BiLSTM deep learning model. Model. Earth Syst. Environ. **11** (2025). https://doi.org/10.1007/s40808-025-02569-8
18. Bulivou, G., Reddy, K.G., Khan, M.G.M.: Cluster optimization for exponential, right triangular, and uniformly distributed data. IEEE Access **13**, 36273–36284 (2025). https://doi.org/10.1109/ACCESS.2025.3544830
19. Houacine, N.A, Drias, H., Azibi, F., Bendjebbar, S.N.: A data-driven wildfire forecasting: case study of Algeria. In: Aissani, D., Barkaoui, K., Roche, M., (eds.) Research in Computer Science. CARI. Communications in Computer and Information Science, vol. 2462 (2024). Springer, Cham. https://doi.org/10.1007/978-3-031-88226-5_3
20. An, Q., Jiang, S.: Sub-one quasi-norm-based k-Means clustering algorithm and analyses. Neural Process. Lett. **56**, 175 (2025). https://doi.org/10.1007/s11063-024-11615-y
21. Wang, Y., Gao, F., Li, M.: Probabilistic path planning for UAVs in forest fire monitoring: enhancing patrol efficiency through risk assessment. Fire **7**(7), 254 (2024)
22. Dreyfus, G.: Neural networks methodology and applications. Springer, Heidelberg (2005)
23. Kleinbaum, D.G., Klein, M.: Logistic regression: a self-learning text, 2nd edition. Springer, pp. 1–513 (2010)
24. Quinlan, J.R.: Induction of decision trees. Mach. Learn. **1**, 81–106 (1986)
25. Breiman, L.: Mach. Learn. **45**(1), 5–32 (2001). https://doi.org/10.1023/A:1010933404324
26. Webb, G.I.: Naïve Bayes. In: Sammut, C., Webb, G.I., (eds.) Encyclopedia of Machine Learning. Springer, Boston, MA, pp. 713–714 (2011). https://doi.org/10.1007/978-0-387-30164-8_576
27. Vapnik, V.N.: An overview of statistical learning theory. IEEE Trans. Neural Netw. **10**(5), 988–999 (1999)

Deep Learning-Driven Model for Accurate Detection of Breast Cancer from Histopathology Images: Clinically Validated on Algerian Medical Data

Ranya Hadjadj[1](✉), Abdelkrim Bouramoul[1], Djamel Benmerzoug[2], Redouane Daddiouamer[1], and Lokmane Abdessalam[2]

[1] MISC Laboratory, University of Abdelhamid Mehri Constantine 2, 25000 Constantine, Algeria
{rania.hadjadj,abdelkrim.bouramoul, redouane.daddiouamer}@univ-constantine2.dz

[2] LIRE Laboratory, University of Abdelhamid Mehri Constantine 2, 25000 Constantine, Algeria
{djamel.benmerzoug,lokmane.abdessalam}@univ-constantine2.dz

Abstract. Accurate and early detection of high-risk breast cancers significantly affects treatment success and patient prognosis. Invasive Ductal Carcinoma (IDC) is the most prevalent and aggressive subtype, highlighting the need for improved diagnostic strategies. Motivated by this clinical challenge, we propose a deep learning-driven model for automatic IDC detection from histopathological images, aimed at assisting pathologists and enhancing diagnostic precision. The model extends the EfficientNetB0 architecture with additional domain-specific convolutional layers and is trained on the publicly available Breast Histopathology Images dataset. The model achieved strong performance, with an AUC of 96.13%, accuracy of 89.26%, recall of 90.89%, and F1-score of 82.54%. Its clinical relevance was further assessed through validation on real Algerian histopathological slides, where the model reached 95.46% accuracy under routine laboratory conditions. These results demonstrate the potential of our clinically validated deep learning model as a reliable tool for IDC detection and breast cancer diagnosis, bridging the gap between AI research and practical medical application.

Keywords: Breast Cancer · Invasive Ductal Carcinoma · Deep Learning · Histopathological Images · EfficientNetB0 · Algerian Dataset · Diagnostic Precision

1 Introduction

Breast cancer stands as the most frequently diagnosed malignancy among women worldwide and remains the leading cause of cancer-related mortality [1]. In 2022, the World Health Organization estimated approximately 2.3 million new cases of

Z. Kouahla et al. (Eds.): IAM 2025, LNNS 1973, pp. 234–249, 2026.
https://doi.org/10.1007/978-3-032-26031-4_17

breast cancer and around 670,000 deaths attributed to the disease [2]. According to the American Cancer Society, several types of breast cancer exist [3], and their classification depends on the specific type of cells within the breast that are affected.

Accurate and early detection of high-risk breast cancers significantly affects treatment success and patient prognosis. Invasive Ductal Carcinoma (IDC) is the most prevalent and aggressive subtype [4], accounting for nearly 80% of all breast cancer cases [5]. Its complex cellular architecture and morphological variability make accurate diagnosis challenging, even for experienced pathologists.

The most common subtype is Invasive Ductal Carcinoma (IDC), which accounts for nearly 80% of all breast cancer cases [5]. It develops when abnormal cells originating in the milk ducts proliferate uncontrollably and invade the surrounding breast tissue [6] [7]. Despite its high prevalence, IDC remains challenging to diagnose and detect due to its complex cellular architecture and morphological variability. The abnormal cells, commonly referred to as tumor cells, often exhibit irregular and unpredictable shapes, which complicates their identification [8].

Traditional diagnosis relies on microscopic examination of histopathological images [9], a process that is labor-intensive, subjective, and prone to inter-observer variability [10]. These limitations highlight the urgent need for automated diagnostic systems that can assist pathologists and enhance diagnostic reliability.

Consequently, there is an increasing need for automated diagnostic systems that leverage image processing and artificial intelligence to analyze histopathological images more efficiently and objectively [11]. In this context, artificial intelligence (AI), particularly deep learning, has revolutionized the field of medical image analysis. Deep learning models, especially Convolutional Neural Networks (CNNs), are capable of autonomously learning hierarchical features from complex image data [12].

In recent years, deep learning, particularly Convolutional Neural Networks (CNNs), has emerged as a powerful tool for medical image analysis, capable of learning hierarchical features from complex image data and providing accurate, consistent predictions [12]. However, most existing models have been evaluated only on public datasets, with limited external or clinical validation, raising concerns about their generalization and applicability in real-world settings.

In this study, we propose a deep learning-driven model for the automatic detection of IDC from histopathological images. Our approach combines a hybrid EfficientNetB0 architecture with additional convolutional layers, trained and validated on a large public dataset and subsequently tested on independent dataset, including real Algerian histopathological slides. This validation strategy ensures robustness, clinical reliability, and potential applicability in diverse healthcare environments.

2 Related Work

Traditional machine learning approaches for IDC detection typically rely on handcrafted features extracted from histopathological images, such as texture, shape, and color descriptors. While these methods achieved moderate accuracy, they often require extensive preprocessing and domain expertise, limiting scalability and adaptability.

Deep learning techniques, particularly CNNs, have demonstrated superior performance by automatically learning discriminative features from raw images. Models based on architectures such as VGG, ResNet, and DenseNet have been applied to breast cancer histopathology, showing notable improvements over classical methods.

In [13], the authors proposed an enhanced CNN architecture for the automatic detection of IDC in histopathological images using the publicly available Breast Histopathology Images dataset from Kaggle. Their model achieved 89.4% accuracy, an 89% F1-score, and an AUC of 0.96, outperforming earlier works such as those of Cruz-Roa et al. [14] and Abdolahi et al. [15].

More recently, Jánošík and Yavuz [16] introduced an AI-based screening framework for early IDC identification. Their approach combines U-Net for semantic segmentation with a deep CNN for feature extraction, followed by ensemble classification. Evaluated on two independent datasets comprising 922 histopathological images from 125 patients and 576 tissue microarray images, their method achieved 92.8% and 94.4% accuracy, with an AUC of up to 98.7%.

Transfer learning has also shown strong potential for IDC classification. Younas et al. [6] developed TransResCNN, a ResNet18-based model designed to address common challenges such as class imbalance, overfitting, and vanishing gradients. Their model achieved 90.76% accuracy and a 93.56% F1-score, surpassing several CNN and VGG-based baselines and demonstrating the robustness of residual networks in medical image classification.

Hybrid architectures have further extended the capabilities of CNN models. Kaddes et al. [17] proposed a CNN–LSTM framework that leverages CNNs for spatial feature extraction and LSTM units for modeling sequential dependencies between image patches. Trained on the Breast Histopathology Images and BreaKHis datasets, their model achieved 99.17% and 99.90% accuracy, outperforming CNN, GRU, VGG-16, and ResNet-50 architectures.

In another comparative study, Korkmaz and Kaplan [18] evaluated several transfer learning models, including VGG16, MobileNet, and DenseNet201, along with a custom CNN optimized using the Particle Swarm Optimization (PSO) algorithm. Using two experimental setups with different image resolutions and dataset sizes, they reported that lightweight architectures such as MobileNet achieved high accuracy while maintaining low computational cost. Their results highlight the effectiveness of combining transfer learning with optimization strategies to enhance IDC classification.

Despite these advancements, most deep learning models are trained and tested exclusively on public datasets, with limited or no validation on clinical data from specific populations. This gap underscores the need for models that are both accurate and clinically validated across diverse datasets, particularly in underrepresented regions.

Our work addresses this gap by introducing a clinically validated, deep learning-driven model, evaluated not only on public datasets but also on real histopathological slides from Algerian medical centers, ensuring robust generalization and practical applicability.

Building on this foundation, our work further leverages recent CNN and hybrid architectures to develop a deep learning model for automatic breast cancer detection from histopathological images. The proposed model integrates optimized feature extraction and a rigorous evaluation protocol, including external and clinical validation, to enhance robustness and generalization. The methodology, experimental setup, and obtained results are detailed in the following sections.

3 Materials and Methods

In this section, we present the materials and methods used in this study. We begin by introducing the histopathological dataset and the preprocessing steps applied to ensure consistency and suitability for deep learning. We then describe the different architectures evaluated in order to benchmark our proposed approach. Finally, we detail the adopted EfficientNetB0-based model, its training strategy, and the external validation performed to assess its robustness.

3.1 Dataset Preparation

The dataset used for training and evaluating our model is derived from the Breast Histopathology Images dataset, originally compiled by researchers at Case Western Reserve University and made publicly available on Kaggle [20]. It contains 277,524 image patches, including 198,738 IDC-negative and 78,786 IDC-positive samples. The dataset was acquired from publicly available sources and provides high-resolution images suitable for deep learning-based analysis. Additionally, we collected Algerian clinical histopathology slides to validate model generalization on real-world data.

Table 1 summarizes the dataset characteristics, and Fig. 1 illustrates representative IDC-positive and IDC-negative patches used for training.

To avoid data leakage, the dataset was split at the patient level, ensuring that all image patches originating from a given patient were exclusively assigned to either the training or validation set. This strategy prevents information leakage caused by strong intra-patient correlations and provides a more realistic evaluation of the model's generalization capability.

Subsequently, several preprocessing techniques applied to ensure the dataset's consistency and suitability for model training. First, all images were resized to a uniform resolution of 224 × 224 pixels to maintain consistency across datasets. Next, normalization was applied by scaling pixel intensity values to the range

Table 1. Description of the Breast Histopathology Images dataset.

Property	Value
Image Source	Whole Slide Images
Magnification	×40
Patch Resolution	50 × 50 pixels
Number of Patients	162
Total Number of Patches	277,524
IDC Positive Patches	78,786
IDC Negative Patches	198,738

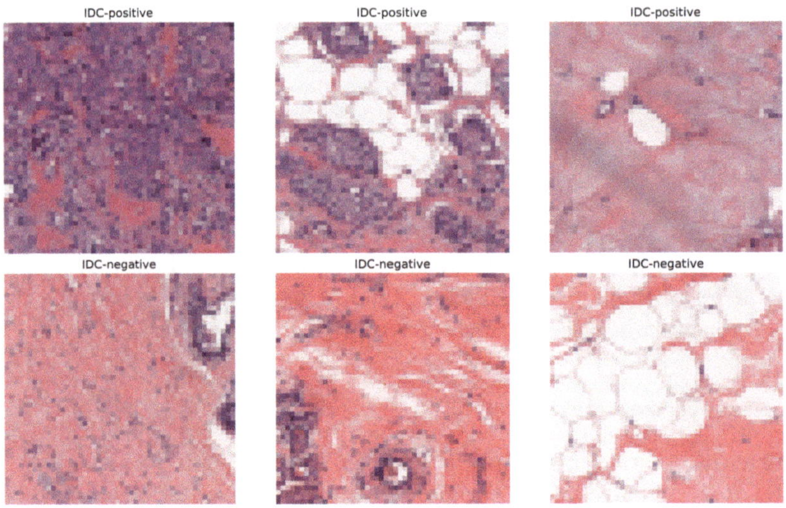

Fig. 1. Representative IDC-positive and IDC-negative tissue patches from the Breast Histopathology Images dataset.

[0,1]. This adjustment reduces intensity variability and accelerates the convergence of the learning process. To guarantee color space consistency, all images were converted and verified to be in the RGB format, ensuring that color information was uniformly represented throughout the dataset.

In addition to these steps, data augmentation techniques including random rotations, flips, zooming, brightness/contrast adjustments, and shearing were applied using TensorFlow's ImageDataGenerator [21,22]. However, extensive augmentation of IDC-positive patches to artificially balance the dataset led to overfitting and a 3–5% drop in validation accuracy, indicating that excessive synthetic variability can degrade model generalization.

To address class imbalance, we explored data-level strategies, including class balancing experiments and class weighting. Class weighting is a widely used approach that adjusts the contribution of each class to the loss function without modifying the dataset distribution [23]. In this work, class weighting was applied to mitigate imbalance while preserving all available data. The weights were computed inversely proportional to class frequencies, following the standard formulation [24,25]:

$$\text{Weight}_i = \frac{n_{\text{samples}}}{n_{\text{classes}} \cdot n_i} \tag{1}$$

resulting in 0.698 for IDC-negative and 1.761 for IDC-positive samples. Incorporating these weights into the binary cross-entropy loss helped reduce bias toward the majority class and improved the recall of malignant cases by approximately 2–3% across all architectures.

3.2 The Proposed Deep Learning Architectures

In this section, we present and explain the three architectures proposed and tested for the classification of histopathological images. Each architecture was designed to explore different strategies in balancing accuracy, computational efficiency, and generalization, leading to the selection of our final model.

Baseline Sequential CNN Model. The first architecture is a baseline Sequential Convolutional Neural Network intended as a reference for assessing more advanced models. It consists of three convolutional blocks (32, 64, and 128 filters), each followed by ReLU activation and max-pooling, enabling progressive extraction of local morphological features. The flattened output is passed through a dense layer of 128 neurons (ReLU), followed by dropout (rate = 0.5) for regularization. A final sigmoid neuron produces the binary prediction (IDC-positive vs. IDC-negative).

Although this model demonstrated fast convergence and low computational demand, this architecture showed limited capacity in modeling complex cellular patterns, confirming the need for deeper or hybrid architectures capable of modeling both local and contextual information.

Convolutional Neural Network with Transformers. Transformers have emerged as a powerful deep learning architecture initially introduced in natural language processing but later extended to computer vision tasks [26]. In medical imaging, they are particularly valuable for modeling long-range dependencies and capturing global relationships between tissue regions that CNNs might overlook [27].

Our second architecture integrates a CNN-based feature extractor with a Transformer encoder to capture both local structures and long-range dependencies. The convolutional block comprises three layers (64, 128, and 256 filters), while the Transformer module incorporates patch embedding, multi-head self-attention (8 heads), and a feed-forward network with layer normalization. A sigmoid-activated dense layer performs binary classification.

Despite its strong theoretical potential, this hybrid architecture showed limited performance on our dataset, primarily due to the data efficiency and optimization requirements of Transformer-based models, particularly under limited patient-level diversity.

EfficientNetB0 with Additional CNN Layers. The third architecture extends the pretrained EfficientNetB0 backbone with additional custom convolutional layers to adapt the generic ImageNet features to the histopathology domain. EfficientNetB0 was chosen because of its balanced design and compound scaling strategy, which enables it to achieve high accuracy with a remarkably small number of parameters [28].

In our implementation, We employed a hybrid EfficientNetB0-based CNN architecture with additional convolutional layers to enhance feature extraction. Each convolutional layer was followed by batch normalization and ReLU activation. A global average pooling layer was added before the final dense layer with sigmoid activation for binary classification.

This hybrid transfer learning architecture demonstrated the best overall performance among the tested models.

3.3 Experimental Setup

The model was trained using the Adam optimizer with an initial learning rate of 1e-4 for 40 epochs. Early stopping based on validation loss was applied to prevent overfitting, and a batch size of 64 was used throughout training.

The Binary Cross-Entropy (BCE) loss function was selected because it effectively penalizes incorrect predictions according to their confidence, which is crucial in medical diagnosis tasks [29]. The Adam optimizer was selected for its robustness and fast convergence properties in high-dimensional medical image classification tasks [30]. These settings collectively ensured stable convergence and reliable optimization across all proposed architectures. Table 2 summarizes the training hyperparameters used for all proposed architectures.

Table 2. Training hyperparameters used for all proposed architectures.

Hyperparameter	Value
Number of Epochs	40
Batch Size	64
Optimizer	Adam
Initial Learning Rate	0.0001
Last Layer Activation Function	Sigmoid
Loss Function	Binary Cross-Entropy

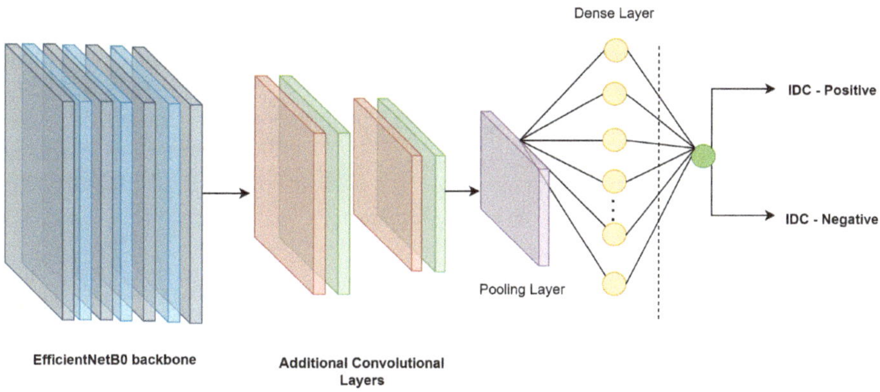

Fig. 2. Architecture of the EfficientNetB0-based model with additional CNN layers, global average pooling, and a fully connected sigmoid output layer.

Performance was evaluated using accuracy, precision, recall, F1-score, and AUC. These metrics provide a comprehensive assessment of model reliability, particularly in distinguishing IDC-positive from IDC-negative cases, which is critical for clinical decision-making. Table 3 summarizes the comparative results obtained for the three architectures.

Table 3. Comparison of performance metrics for the three proposed architectures.

Metric	Sequential CNN	Transformer + CNN	EfficientNetB0 + CNN
Accuracy	0.8500	0.9375	0.8926
AUC	0.9200	0.9148	0.9613
Loss	0.3753	0.1764	0.2508
Precision	0.7500	1.0000	0.7595
Recall	0.8060	0.6000	0.9089
Validation Accuracy	0.8756	0.8750	0.8851
Validation AUC	0.9462	0.9386	0.9590
Validation Loss	0.4540	0.3062	0.2771
Validation Precision	0.7458	0.8488	0.7412
Validation Recall	0.8800	0.6848	0.9154

The EfficientNetB0 + CNN model achieved the best overall balance between accuracy (0.8926), recall (0.9089), and AUC (0.9613), demonstrating strong generalization capability with the lowest validation loss (0.2771). The Transformer + CNN model achieved the highest training precision (1.0000) but suffered from a very low recall (0.6000), indicating overfitting and a high risk of missing malignant cases; a critical drawback in medical applications. The Sequential CNN, while computationally efficient, exhibited moderate performance (accuracy = 0.85, AUC = 0.92), confirming its limitations in capturing complex histopathological features.

Overall, EfficientNetB0 + CNN provided the most reliable trade-off between sensitivity and specificity, making it the preferred choice for further evaluation and external validation in this study. Figure 2 presents the final EfficientNetB0-based architecture selected for further validation.

4 Results

This section summarizes and discusses the experimental results obtained with the proposed EfficientNetB0-based architecture for breast cancer detection. The analysis highlights its training and validation performance through accuracy, loss, AUC, and confusion matrices. Moreover, external validation on Algerian datasets is presented to demonstrate the robustness and clinical applicability of our proposed model.

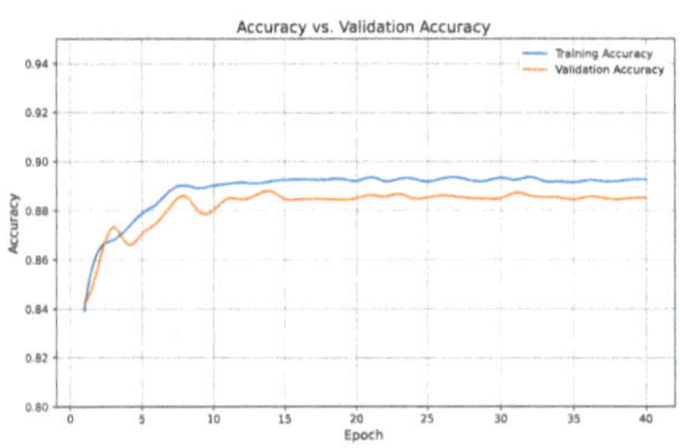

Fig. 3. Training Accuracy

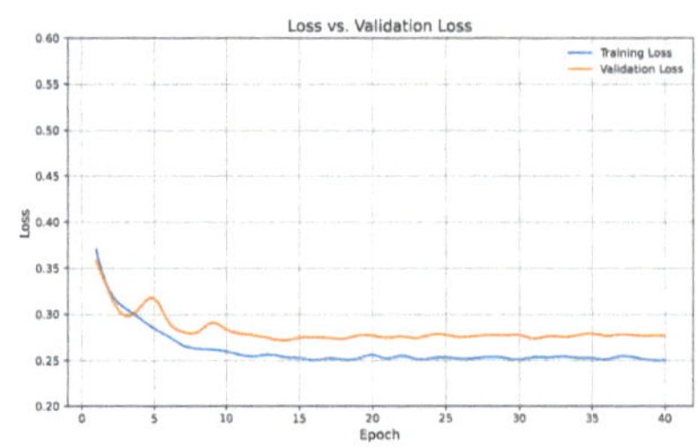

Fig. 4. Training Loss

4.1 Training and Validation Results

During training, the proposed EfficientNetB0-based model exhibited steady improvements in both accuracy and loss over 40 epochs, confirming stable convergence and effective learning.

The proposed model achieved an accuracy of 89.26%, precision of 75.95%, recall of 90.89%, F1-score of 82.54%, and an AUC of 96.13%. These results indicate strong discriminative capability for IDC detection and highlight the model's potential to assist pathologists in clinical settings.

Figure 3 and Fig. 4 present the evolution of accuracy and loss during training and validation.

To complement this evaluation, the confusion matrix in Fig. 5 provides a detailed view of classification outcomes, indicating a high number of correctly classified samples for both classes, with relatively few false positives and false negatives. This confirms the model's reliability and balanced performance across categories.

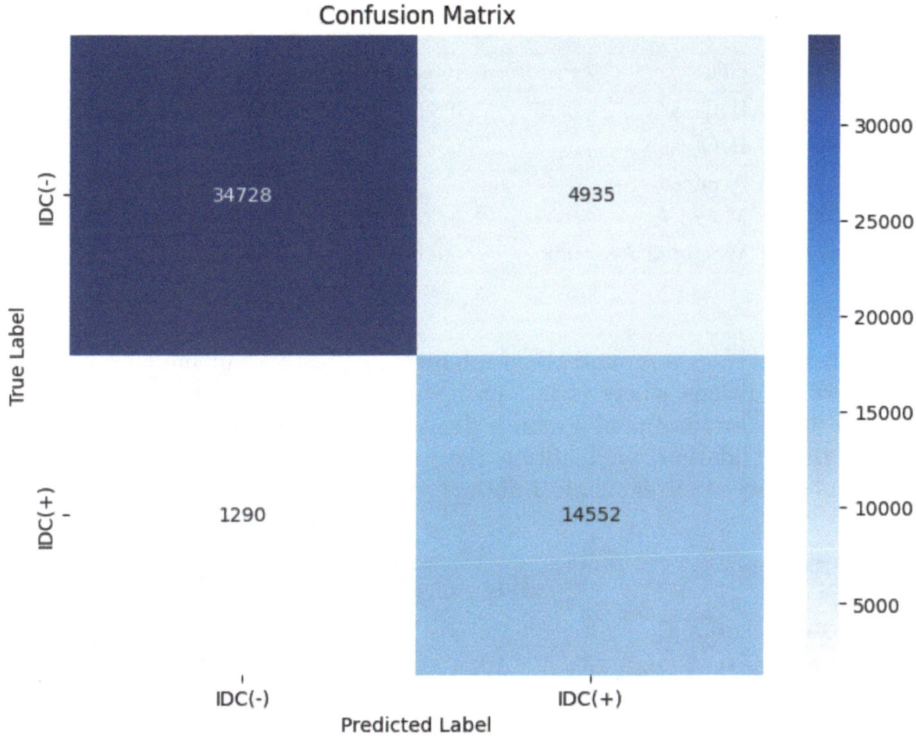

Fig. 5. Training Confusion Matrix

Furthermore, Table 4 presents the classification report summarizing key performance metrics. The IDC(−) class achieved a precision of 0.96, recall of 0.88, and F1-score of 0.92, while the IDC(+) class obtained a precision of 0.75, recall of 0.92, and F1-score of 0.82. The model reached an overall accuracy of 0.89, with macro and weighted average F1-scores of 0.87 and 0.89, respectively, confirming robust generalization across classes.

4.2 Clinical Validation on Algerian Data

After training our model on the primary dataset, our model was further tested on real histopathology slides obtained from Algerian hospitals. Results confirmed high accuracy and reliability under actual laboratory conditions, demonstrating clinical applicability in the Algerian medical context.

This stage was carried out in close collaboration with Dr. Salah Semmadi, Physician specializing in Pathological Anatomy in Constantine, Algeria.

This collaboration aimed to assess the model's capacity to analyze authentic Algerian histopathological slides, distinct from the publicly available datasets used for training. A total of eight histopathological slides were selected and digitized at 40× magnification, each slide was examined and annotated by Dr.

Table 4. Training classification report for IDC detection.

Class	Precision	Recall	F1-score	Support
IDC(−)	0.96	0.88	0.92	39 663
IDC(+)	0.75	0.92	0.82	15 842
Accuracy			0.89	55 505
Macro Avg	0.86	0.90	0.87	55 505
Weighted Avg	0.90	0.89	0.89	55 505

Semmadi to serve as a ground truth reference for testing our model under real laboratory conditions, where variations in staining, tissue preservation, and scanner calibration are common. Figure 6 illustrates the eight representative slides used in this validation, highlighting the morphological diversity and staining variations observed in real clinical data.

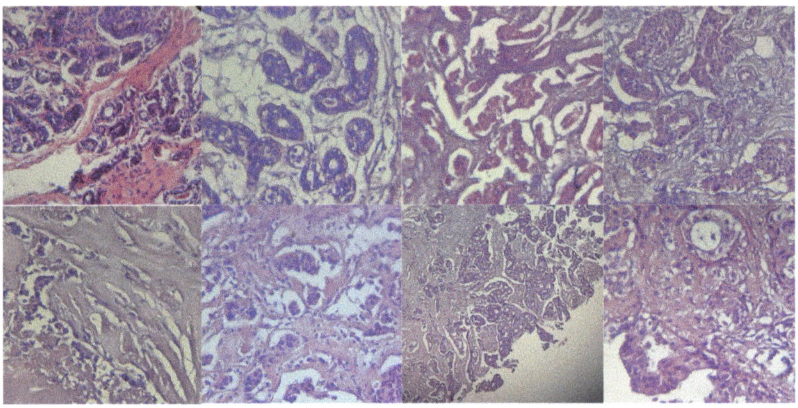

Fig. 6. Representative Samples from Algerian Dataset

The obtained results, illustrated in Fig. 7, confirmed the clinical robustness of our model. Out of the eight tested slides, the model correctly classified seven, with only one misclassified case. Importantly, this misclassification corresponded to a slide that was already deemed diagnostically uncertain by Dr. Semmadi, due to poor tissue quality and low cellular information. This alignment between the model's uncertainty and the expert's evaluation reinforces its clinical consistency and reliability.

This real-world validation represents an essential milestone in confirming the clinical credibility and robustness of our model within an Algerian healthcare context. It demonstrates that the model can adapt to local histopathological characteristics and align with expert interpretation, paving the way for future large-scale validations across national medical centers.

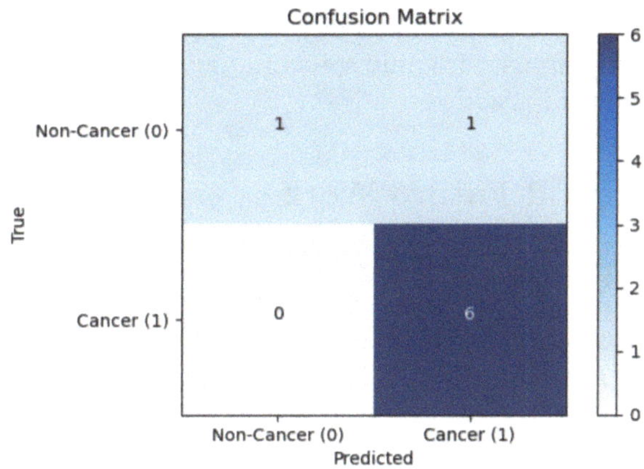

Fig. 7. Confusion Matrix for Validation on Algerian Dataset

In comparison with existing deep learning approaches for IDC detection, the proposed EfficientNetB0-based model shows a strong and well-balanced performance in terms of accuracy and recall, particularly when evaluated on high-resolution images and diverse clinical data.

5 Discussion

The results demonstrate that the proposed deep learning-driven model can accurately detect IDC, providing a valuable tool to support pathologists and improve early breast cancer diagnosis, which is critical for patient prognosis and treatment planning.

Compared with previous studies, the proposed approach achieves competitive performance in terms of accuracy and recall, particularly when evaluated on real-world clinical data. The hybrid EfficientNetB0 architecture contributes to effective feature extraction from complex histopathological patterns.

Importantly, the clinical validation conducted on Algerian histopathological slides confirms the model's ability to generalize to local data, supporting its potential integration into national diagnostic workflows and enhancing the reliability of breast cancer screening programs. However, this validation remains limited by the relatively small number of available Algerian clinical samples, which restricts the statistical representativeness of the clinical evaluation.

Despite these promising results, several limitations should be acknowledged. The study relies on a specific type of histopathological images. Future work will therefore focus on expanding the Algerian dataset and conducting multi-center clinical validation to further assess robustness and generalization across heterogeneous acquisition conditions.

In addition, future developments will integrate explainability mechanisms such as Grad-CAM to provide visual interpretation of model predictions, thereby enhancing transparency and trust for clinical adoption. Further directions include the integration of multi-modal data (e.g., radiological and clinical information), and developing user-friendly software tools to seamlessly assist pathologists in routine clinical practice.

6 Conclusion and Future Work

In summary, we presented a deep learning-driven model for the accurate detection of Invasive Ductal Carcinoma from histopathology images. The model achieved high performance metrics and was clinically validated on Algerian data, demonstrating its potential as a reliable tool for supporting breast cancer diagnosis.

The proposed EfficientNetB0-based model achieved an accuracy of 89.26%, together with strong recall (90.89%) and an AUC of 0.9613, confirming its high discriminative capability between malignant and benign tissue patterns. These results highlight the model's stability and generalization ability, making it a reliable foundation for AI-assisted diagnostic applications in digital pathology.

The clinical validation highlights the model's applicability in real-world laboratory conditions, potentially improving diagnostic efficiency, reducing workload for pathologists, and contributing to better patient outcomes.

While the results are promising, further validation on larger, multi-center datasets is needed. Future work will explore extending the model to other breast cancer subtypes and integrating additional clinical and imaging data to enhance diagnostic precision.

Ultimately, the development of user-friendly software based on this model could facilitate its deployment in hospitals and clinics, promoting AI-driven precision medicine in breast cancer diagnostics across Algeria and similar clinical settings.

Overall, this work represents a concrete step toward the clinical translation of artificial intelligence in pathology, paving the way for AI assisted precision medicine and contributing to the advancement of breast cancer diagnosis within both local (Algerian) and international healthcare ecosystems.

Acknowledgment. We would like to express our sincere gratitude to Dr. Salah Semmadi, Physician and Specialist in Pathological Anatomy in Constantine, Algeria, for his valuable collaboration during the validation phase on Algerian clinical data. His expertise, annotations, and insightful feedback were essential in assessing the model's performance on real histopathological slides and ensuring the clinical relevance of this research.

References

1. Chilumukuru, N.S., Priyadarshini, P., Ezunkpe, Y.: Deep learning for the early detection of invasive ductal carcinoma in histopathological images: convolutional neural network approach with transfer learning. JMIR Format. Res. **9**, e62996 (2025). https://doi.org/10.2196/62996
2. Breast cancer. World Health Organization. https://www.who.int/news-room/fact-sheets/detail/breast-cancer. Accessed 20 Oct 2025
3. American Cancer Society. Types of Breast Cancer. American Cancer Society (2024). https://www.cancer.org/cancer/breast-cancer/about/types-of-breast-cancer.html. Accessed 20 Oct 2025
4. Ahmed, M.R., Ali, M.A., Roy, J., Ahmed, S., Ahmed, N.: Breast cancer risk prediction based on six machine learning algorithms. In: 2020 IEEE Asia-Pacific Conference on Computer Science and Data Engineering (CSDE), Gold Coast, Australia, pp. 1–5 (2020). https://doi.org/10.1109/CSDE50874.2020.9411572. keywords: Support vector machines; Radio frequency; Machine learning algorithms; Predictive models; Prediction algorithms; Breast cancer; Classification algorithms; Breast Cancer; Machine Learning; Classification; Supervised learning; Computational Intelligence
5. Jiang, Y., Chen, L., Zhang, H., Xiao, X.: Breast cancer histopathological image classification using convolutional neural networks with small SE-ResNet module. PloS One **14**(3), e0214587 (2019). https://doi.org/10.1371/journal.pone.0214587
6. Younas, H.I., Bukhari, S., Bukhari, F., Aslam, N., Badar, H.M.S., Kajla, N.I.: An efficient methodology for the classification of invasive ductal carcinoma using transfer learning. J. Comput. Biomed. Inform. **4**(1), 790401 (2022)
7. Yu, J.: predicting invasive ductal carcinoma based on convolutional neural network. In: 2022 3rd International Conference on Electronic Communication and Artificial Intelligence (IWECAI), Zhuhai, China, pp. 388–391 (2022). https://doi.org/10.1109/IWECAI55315.2022.00082. keywords: Pathology; Computational modeling; Semantics; Medical services; Predictive models; Breast cancer; Convolutional neural networks; component; Breast cancer; Convolutional Neural Network; Invasive Ductal Carcinoma; Kaggle
8. Aly, G.H., Marey, M., El-Sayed, S.A., Tolba, M.F.: YOLO based breast masses detection and classification in full-field digital mammograms. Comput. Methods Program. Biomed. **200**, 105823 (2021). https://doi.org/10.1016/j.cmpb.2020.105823. ISSN 0169-2607
9. Balasubramanian, A.A., et al.: Ensemble deep learning-based image classification for breast cancer subtype and invasiveness diagnosis from whole slide image histopathology. Cancers **16**(12), 2222 (2024). https://doi.org/10.3390/cancers16122222
10. Panayides, A.S., et al.: AI in medical imaging informatics: current challenges and future directions. IEEE J. Biomed. Health Inform. **24**(7), 1837–1857 (2020). https://doi.org/10.1109/JBHI.2020.2991043
11. Thaalbi, O., Akhloufi, M.A.: Deep learning for breast cancer diagnosis from histopathological images: classification and gene expression: review. Netw. Model Anal. Health Inform. Bioinform. **13**, 52 (2024). https://doi.org/10.1007/s13721-024-00489-8
12. Litjens, G., et al.: A survey on deep learning in medical image analysis. Med. Image Anal. **42**, 60–88 (2017). https://doi.org/10.1016/j.media.2017.07.005

13. Zeid, M.A.-E., El-Bahnasy, K., Abo-Youssef, S.E.: Enhanced CNN architecture for invasive ductal carcinoma detection in breast histopathology images. In: 2021 Tenth International Conference on Intelligent Computing and Information Systems (ICICIS), Cairo, Egypt, pp. 154–159 (2021). https://doi.org/10.1109/ICICIS52592.2021.9694114
14. Cruz-Roa, A., et al.: Automatic detection of invasive ductal carcinoma in whole slide images with convolutional neural networks. In: Medical Imaging 2014: Digital Pathology, vol. 9041, p. 904103 (2014)
15. Abdolahi, M., Salehi, M., Shokatian, I., Reiazi, R.: Artificial intelligence in automatic classification of invasive ductal carcinoma breast cancer in digital pathology images. Med. J. Islam Repub. Iran **34**, 140 (2020)
16. Datwani, S., Khan, H., Niazi, M.K.K., Parwani, A.V., Li, Z.: Artificial intelligence in breast pathology: overview and recent updates. Hum. Pathol. **162**, 105819 (2025). https://doi.org/10.1016/j.humpath.2025.105819
17. Kaddes, M., Ayid, Y.M., Elshewey, A.M., et al.: Breast cancer classification based on hybrid CNN with LSTM model. Sci. Rep. **15**, 4409 (2025). https://doi.org/10.1038/s41598-025-88459-6
18. Korkmaz, M., Kaplan, K.: Effectiveness analysis of deep learning methods for breast cancer diagnosis based on histopathology images. Appl. Sci. **15**(3), 1005 (2025). https://doi.org/10.3390/app15031005
19. Guleria, H.V., et al.: Enhancing the breast histopathology image analysis for cancer detection using variational autoencoder. Int. J. Environ. Res. Public Health **20**(5), 4244 (2023). https://doi.org/10.3390/ijerph20054244
20. Paul Timothy Mooney. Breast histopathology images (2018). https://www.kaggle.com/datasets/paultimothymooney/breast-histopathology-images/data. Accessed 16 June 2025
21. Alomar, K., Aysel, H.I., Cai, X.: Data augmentation in classification and segmentation: a survey and new strategies. J. Imaging **9**(2), 46 (2023). https://doi.org/10.3390/jimaging9020046
22. Goceri, E.: Medical image data augmentation: techniques, comparisons and interpretations. Artif. Intell. Rev. **20**, 1–45 (2023). https://doi.org/10.1007/s10462-023-10453-z. PMID: 37362888; PMCID: PMC10027281
23. Bakirarar, B.: Class weighting technique to deal with imbalanced class problem in machine learning: methodological research. Turk. Klinikleri J. Biostat. **15**(1), 19–29 (2023). https://www.turkiyeklinikleri.com/article/en-class-weighting-technique-to-deal-with-imbalanced-class-problem-in-machine-learning-methodological-research-101599.html
24. Xu, Y., Ghosh, S., Goldfarb, D.: Class-weighted classification: trade-offs and robust approaches. In: Proceedings of the 37th International Conference on Machine Learning (ICML). Proceedings of Machine Learning Research, vol. 119, pp. 10462–10472 (2020). https://proceedings.mlr.press/v119/xu20b.html
25. Buda, S., Maki, A., Mazurowski, M.: A systematic study of the class imbalance problem in convolutional neural networks. Mach. Learn. **112**, 4431–4460 (2023). https://doi.org/10.1007/s10994-022-06268-8
26. Shamshad, F., et al.: Transformers in medical imaging: a survey. Med. Image Anal. **88**, 102802 (2023). https://doi.org/10.1016/j.media.2023.102802
27. Li, J., Chen, J., Tang, Y., Wang, C., Landman, B.A., Zhou, S.K.: Transforming medical imaging with transformers? A comparative review of key properties, current progresses, and future perspectives. Med. Image Anal. **85**, 102762 (2023). https://doi.org/10.1016/j.media.2023.102762

28. Mahesh, T.R., Gupta, M., Anupama, T.A., Vinoth Kumar, V., Geman, O., Dhilip Kumar, V.: An XAI-enhanced efficientNetB0 framework for precision brain tumor detection in MRI imaging. J. Neurosci. Methods **410**, 110227 (2024). https://doi.org/10.1016/j.jneumeth.2024.110227. ISSN 0165-0270
29. Rajaraman, S., Zamzmi, G., Antani, S.K.: Novel loss functions for ensemble-based medical image classification. PLoS ONE **16**(12), e0261307 (2021). https://doi.org/10.1371/journal.pone.0261307
30. Hassan, E., Shams, M.Y., Hikal, N.A., et al.: The effect of choosing optimizer algorithms to improve computer vision tasks: a comparative study. Multimed. Tools Appl. **82**, 16591–16633 (2023). https://doi.org/10.1007/s11042-022-13820-0

A Novel Vocal Database for Children with Autism Spectrum Disorder: Design, Collection, and Future Applications

Kadi Imane(✉) and Haloua Fatima

University of El Oued, El Oued, Algeria
{kadi-imane,haloua-fatima}@univ-eloued.dz

Abstract. Autism spectrum disorder (ASD) is one of the most complex neurological disorders, presenting significant challenges in terms of detection, particularly in the early stages. There are numerous theories regarding the methods of detection and the types of data used. Recent research has emerged investigating the potential for distinctive vocal characteristics to manifest in the voices of autistic children in comparison to those of normal children. In order to explore this research domain in greater depth, a voice database has been compiled, comprising recordings of autistic and typically developing children. This database includes 600 recordings from children, aged between 4 and 14 years of both sexes. The samples were diverse in terms of dialect and autism severity. The audio samples were subjected to a process of purification and filtration with the objective of ensuring optimal sound quality. The resulting database is balanced, diverse, and ready for use in future research and studies focused on ASD and phonetics.

Keywords: Autism Spectrum Disorder(ASD) · Data collection · Arabic speech analysis · Audio dataset

1 Introduction

Autism spectrum disorder (ASD) [8] is a complex neurological disorder, and global statistics demonstrate a consistent increase in the number of individuals affected by it. The condition is characterized by a wide range of symptoms, which often manifest as communication impairments, speech problems, social skills difficulties, and behavioral disorders [4] . The diversity and multiplicity of symptoms pose significant diagnostic challenges due to their overlap with other neurological disorders [5]. These difficulties highlight the need for continued research to improve early detection and enhance diagnostic accuracy.

Recent advancements in the field of neuroscience have elucidated the robust correlation between neural development and speech mechanisms [12]. This has resulted in heightened interest in the voice patterns of individuals diagnosed with autism. Consequently, a new wave of research [1,9,10] has emerged, which

Z. Kouahla et al. (Eds.): IAM 2025, LNNS 1973, pp. 250–257, 2026.
https://doi.org/10.1007/978-3-032-26031-4_18

calls for in-depth acoustic analysis of the vocalizations of children diagnosed with autism. The purpose of these studies is to identify specific vocal characteristics that could serve as reliable biomarkers for early diagnosis or classification within the autism spectrum. In accordance with this emerging field of study, the objective of the present work is to address a critical gap by creating a new, innovative, and specialized sound database for children with autism. The primary goal of this research is to compile and organize a large, diverse, and reliable database of Arabic sounds for children with autism. This approach is particularly important given the scarcity of Arabic-language resources and data for this purpose.

The extant research that has adopted the notion of diagnosing autism on the basis of sounds has relied on unverified and limited audio data [9–11]. A number of databases are already in existence; however, these generally contain only a limited number of sounds. It is notable that Arabic is not included in this collection. Furthermore, the majority of these sources contain regular sounds of words and sentences, which has the effect of marginalizing a large group of non-verbal autistic children. This non-verbal group constitutes a significant proportion of the autistic population, The Table 1 contain a comparison between three existed databases.

The proposed database will include a variety of Arabic speech samples, including verbal and random non-verbal sounds, to reflect the diversity of expression within the autism spectrum. The incorporation of random and spontaneous sounds, particularly from non-verbal children, is deliberate and aims to ensure that the database encompasses the full range of all categories of children with autism. The objective of this resource is to provide a foundation for future research initiatives aimed at the development and advancement of intelligent diagnostic tools based on speech and sounds for children on the autism spectrum, with a particular focus on the Arab community.

Table 1. Comparison of Existing ASD-Related Speech Datasets

Database	Number of Participants	Age Range	Type of Data	Language	Limitation
SCKCASD [7]	73 ASD children + 9 typical	7.8 years	Speech & dialogue	Korean	Limited age group; imbalanced dataset
SANACS [6]	37 ASD + 30 typical	6–12 years	Structured interaction	Slovak	Lacks spontaneous speech; small dataset
ASDBank AAC [2]	18 ASD children	2.5 years	Natural interaction	English	Very limited number of speakers

2 Methodology

2.1 Data Collection

Dataset Composition: The audio data collection process targeted two distinct groups of children: those diagnosed with autism spectrum disorder and those demonstrating typical development. The collection of data from both groups is essential for the operation of artificial intelligence systems, which require comparisons between groups in order to extract features and distinguish between categories. The data collection process targeted children aged between four and fourteen years old of both genders. A comprehensive set of data was collected from a diverse sample of children with varying degrees of autism, ranging from mild to moderate to severe, in order to ensure diversity within the database. Furthermore, two types of sounds were collected: regular sounds of meaningful words and sentences with understandable pronunciation, as well as random sounds such as screaming, crying, laughing, or even meaningless sounds. This approach was adopted to encompass both verbal and non-verbal individuals diagnosed with autism.

Recording Methodology: The audio recordings were collated by a number of specialists in the field, including speech therapists, neurologists, and psychologists, in addition to members of our academic team. The voices were recorded in a variety of locations and settings, including psychotherapy sessions in specialist clinics, children's playtime in specialized associations (which included crying, laughing and other random sounds), and classroom activities in schools (which included conversations with teachers, reading stories or poetry, and other activities), as well as conversations and discussions with family members in the children's everyday environment.

Ethical Considerations: All recording and data collection procedures were carried out with the approval of the relevant authorities and in accordance with official documentation and licenses authorized by all concerned parties. Additionally, we are fully committed to ensuring the privacy of the children's personal data. This database does not include any personal information such as addresses, full names, or photographs. It is not open-source and is intended exclusively for scientific research. Access to the data is granted only under specific conditions and with the approval of the academic team responsible for its collection.

2.2 Data Processing

Since the recordings were conducted in the children's natural daily environments such as open play areas, classroom sessions, and conversations with specialists, these settings were not isolated from ambient noise. As a result, it was necessary to carry out a filtering and cleaning process on the recordings. We applied several noise reduction techniques to remove background noise, intrusive voices from unintended individuals, and interference from recording devices. Additionally, we refined the audio samples based on factors such as the balance between silence and speech, and the overall length of the recordings. Some of the filtering processes were automated, while others were performed manually. This was due to the fact that using certain automated tools or even specialized noise reduction algorithms like DNS deep noise suppression sometimes led to a drop in audio quality, which could result in the loss of important information.

After the data collection process was completed, all audio samples were standardized to the MP3 format. This format was chosen due to its good audio quality, which preserves the important characteristics and features of the sounds, while also offering a manageable file size suitable for storage within the database.

3 Results and Database Description

As result to the data collection and data processing we created our own dataset, called "AAVD" (Autism Arabic Voices Dataset)[1], AAVD is balanced from both typically developing children (280 recordings) and children with autism (320 recordings). Global statistics indicate that autism is more prevalent in males than females, which is why the number of recordings from males with autism is slightly higher than those from females in the collected dataset.

- Data volume: A total of 600 audio files were collected, with three recordings obtained from each child, thus resulting in a total sample size of 200 children.
- The variety of data: The recordings encompassed a variety of sounds, such as crying, laughing, spontaneous speech, responses to commands and conversations, thereby enhancing the database with respect to behavioral and vocal diversity.
- Technical quality: The files were recorded using various devices, including microphones and cell phones, at a sampling rate of 44.1 kHz. The files were saved in MP3 format. Following the preliminary processing stage, which incorporated noise reduction and clip trimming, 10% of the recordings were eliminated due to substandard sound quality and elevated noise levels.
- Metadata: The storage of metadata on samples is a process of significant importance, as it can facilitate the identification of crucial conclusions and serve as a source of input for intelligent systems. The information we collected included first names only, not full names, ages, genders, cities, degrees of autism, voice types, voice qualities, and types of activity performed during the recording process, among other factors.

[1] A subset of the dataset is available at https://urls.fr/3EAera with the author's permission.

ASD	TD	Males	Females	Avg length	Size	Format	Freq
320 rec	280 rec	366 rec	234 rec	14 sec	72 mo	MP3	44.1 KH

Fig. 1. Summary of AAVD dataset.

Figure 1 shows some metadata about the dataset. Figures 2 and 3 illustrate examples from our AAVD dataset of spectrograms and waveforms for the ASD and typical development (TD) classes, respectively.

4 Discussion and Potential Applications

The creation and compilation of a voice database specifically for children with autism spectrum disorder represents a significant scientific contribution in the fields of applied artificial intelligence and behavioral sciences. The lack of standardized, open-source, linguistically and acoustically annotated resources has long been an obstacle to research progress, both at the technical level related to data modeling and the extraction of distinctive acoustic features, and at the clinical level related to improving diagnosis and early intervention.

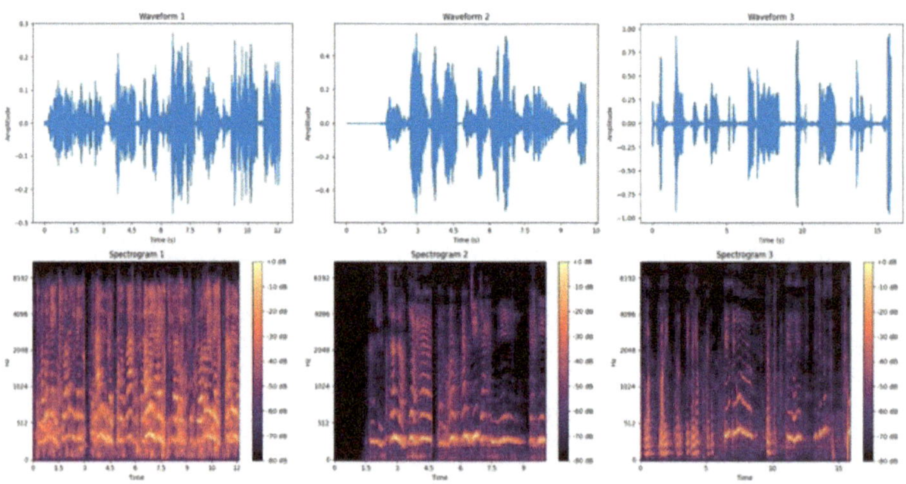

Fig. 2. Spectrogram of ASD class.

This database is distinguished by its focus on the Arabic language in the context of autism spectrum disorder, a unique aspect that enhances its linguistic and cultural significance, particularly in light of the global absence of data

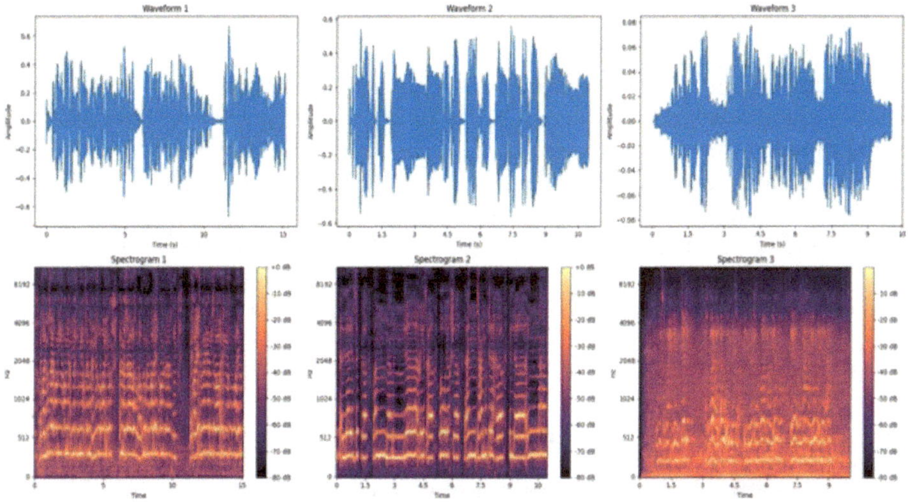

Fig. 3. Spectrogram of TD class.

dedicated to this category. This resource offers unique opportunities to study the acoustic and linguistic characteristics associated with autism in an Arabic-speaking environment, creating new avenues for comparative and multilingual research.

4.1 Potential Applications and Limitations

- Early diagnosis: The database is an essential tool for training machine learning models capable of detecting early vocal indicators of autism spectrum disorder, thereby enabling greater opportunities for early therapeutic intervention [3].
- Multidisciplinary research: The database provides common ground for researchers in the fields of neuroscience, linguistics, psychology, and computer science, enhancing the potential for interdisciplinary studies of high scientific value [14].
- Educational and rehabilitative applications: The data can be used to design digital assistive tools aimed at improving the communication and language abilities of children with autism spectrum disorder, with the possibility of customizing them to suit the needs of each child [13].

However, it is important to note some methodological limitations that should be taken into consideration, most notably: the sample size at the current stage, the variation among children in terms of linguistic and cognitive abilities that may affect the generalization of results. In order to surmount these challenges, it is essential to ensure the continuity of data collection, to expand the sample to encompass different age groups and diverse social environments, and to consider the possibility of supporting multiple dialects.

5 Conclusion

The establishment of a voice database for children diagnosed with ASD represents a significant advancement in the domain of scientific research and the development of intelligent applications in the fields of diagnosis and rehabilitation. The significance of the database lies in its role as the first Arabic-language audio resource dedicated to this group, which paves the way for the development of voice analysis tools that contribute to early detection and improved quality of care. In accordance with future expansion plans, it is proposed that recordings of infants crying and a range of Arabic dialects be incorporated. This is with a view to enhancing data diversity and supporting the accuracy of the models used in analysis and classification.

Acknowledgments. The following is a list of some sources from which data were collected.

- Saad Al-Amamra Health Public Institution, Orthophonia section- El Oued.
- Ben Nasser Bachir Specialized Hospital for Mother and Child, Orthophonia section – El Oued.
- Health Public Institution 19 March, Orthophonia section – El Oued.
- Basma Baraa Association for Fatty Delay and Autism in GUEMAR.
- Taj Association for Health in GUEMAR /Center for the care of disabled children.
- Speech therapist and orthophonist Mr. Lekhouimes Abd Ennour.
- Speech therapist and orthophonist Ms. Boutouil Khadija.

References

1. Asgari, M., Chen, L., Fombonne, E.: Quantifying voice characteristics for detecting autism. Front. Psychol. **12**, 665096 (2021)
2. Fromm, D., MacWhinney, B.: Clinical corpus linguistics. In: The Handbook of Clinical Linguistics, Second Edition, pp. 143–156 (2024)
3. Fusaroli, R., Lambrechts, A., Bang, D., Bowler, D.M., Gaigg, S.B.: Is voice a marker for autism spectrum disorder? A systematic review and meta-analysis. Autism Res. **10**(3), 384–407 (2017)
4. Hazen, E.P., Stornelli, J.L., O'Rourke, J.A., Koesterer, K., McDougle, C.J.: Sensory symptoms in autism spectrum disorders. Harv. Rev. Psychiatry **22**(2), 112–124 (2014)
5. Hirota, T., King, B.H.: Autism spectrum disorder: a review. JAMA **329**(2), 157–168 (2023)
6. Kruyt, J., Sabo, R., Polónyiová, K., Ostatníková, D., Beňuš, Š.: The Slovak autistic and non-autistic child speech corpus: task-oriented child-adult interactions. In: Proceedings of the 2024 Joint International Conference on Computational Linguistics, Language Resources and Evaluation (LREC-COLING 2024), pp. 16094–16099 (2024)
7. Lee, S., Mun, J., Kim, S., Chung, M.: Speech corpus for Korean children with autism spectrum disorder: towards automatic assessment systems. arXiv preprint: arXiv:2402.15539 (2024)

8. Lord, C., Elsabbagh, M., Baird, G., Veenstra-Vanderweele, J.: Autism spectrum disorder. The Lancet **392**(10146), 508–520 (2018)
9. MacFarlane, H., Salem, A.C., Chen, L., Asgari, M., Fombonne, E.: Combining voice and language features improves automated autism detection. Autism Res. **15**(7), 1288–1300 (2022)
10. Milling, M., et al.: Evaluating the impact of voice activity detection on speech emotion recognition for autistic children. Front. Comput. Sci. **4**, 837269 (2022)
11. Mohanta, A., Mukherjee, P., Mirtal, V.K.: Acoustic features characterization of autism speech for automated detection and classification. In: 2020 National Conference on Communications (NCC), pp. 1–6. IEEE (2020)
12. Ohashi, H., Ostry, D.J.: Neural development of speech sensorimotor learning. J. Neurosci. **41**(18), 4023–4035 (2021)
13. Saleh, M.A., Hanapiah, F.A., Hashim, H.: Robot applications for autism: a comprehensive review. Disabil. Rehabil. Assist. Technol. **16**(6), 580–602 (2021)
14. Strunk, J., Leisen, M., Schubert, C.: Using a multidisciplinary approach with children diagnosed with autism spectrum disorder. J. Interprof. Educ. Pract. **8**, 60–68 (2017)

Random Forests for Large-Scale Hyperspectral Pixelwise Plastic Polymer Classification in Aquatic Environments

Samir Hallaci[1,2](✉), Wassim Messioud[1], Mohammed Chaib[1], Achraf Yahiaoui[1], Abderrahman Barkache[1], and Hamid Seridi[1,2]

[1] Computer Science Department, University of 8 mai 1945, 24000 Guelma, Algeria
hallaci.samir@univ-guelma.dz

[2] LabSTIC, University of 8 mai 1945, 24000 Guelma, Algeria

Abstract. Plastic pollution poses an escalating threat to aquatic ecosystems, demanding advanced, scalable solutions for real-time mapping and polymer-specific identification. Hyperspectral imaging (HSI) delivers pixel-level spectral profiles, enabling the discrimination of plastic types that elude traditional RGB and NIR detection techniques. However, effective segmentation within large, imbalanced data cubes remains challenging in terms of accuracy, robustness, and efficiency. This work introduces a blockwise, Random Forest-based pipeline optimized for pixelwise plastic classification across six polymer types in hyperspectral aquatic scenes. Key preprocessing stages band selection, standardization, principal component reduction, and adaptive class rebalancing are systematically evaluated via ablation studies. The final model achieves weighted F1-score of 0.975, weighted IoU of 0.952, and per-class F1 above 0.97 for major polymers, surpassing the performance of established machine learning baselines. Blockwise prediction ensures runtime and memory stability on real-world, full-scene datasets. These outcomes demonstrate Random Forests as a practical, interpretable, and highly accurate method for automated HSI-based environmental plastic monitoring, supporting implementation in large-scale aquatic surveys and rapid response systems.

Keywords: Hyperspectral Imaging · Random Forest · Plastic Classification · Environmental Monitoring · Pixelwise Segmentation

1 Introduction

Plastic waste is a universal threat to aquatic systems, contaminating food webs and water resources with visible debris and micro/nanoplastics [5,10,14]. Conventional RGB imaging is unreliable for materials that are clear, colorless, or environmentally altered [4]. Hyperspectral imaging (HSI) uniquely enables the detection of polymer-specific signatures at the pixel level across a full scene [2,7].

Z. Kouahla et al. (Eds.): IAM 2025, LNNS 1973, pp. 258–282, 2026.
https://doi.org/10.1007/978-3-032-26031-4_19

Yet, challenges in spectral redundancy, class imbalance, computational scalability, and interpretability must be tackled for field deployment.

Machine learning especially Random Forests (RF) offers a tractable, parallelizable, and highly interpretable approach to pixel-level classification of hyperspectral cubes [3]. This study puts forward a modular, reproducible pipeline leveraging blockwise inference with RFs, emphasizing transparent benchmarking, ablation analyses, and practical application.

Research Questions

1. How does Random Forest performance scale for full-spectrum pixelwise classification amid class imbalance and scene diversity?
2. What preprocessing, dimensionality reduction, and blockwise strategies best support robust training and efficient inference?
3. How do choices in number of trees, feature construction, and chunking impact results, speed, and memory use?

2 Related Work

Plastic mapping in aquatic environments has advanced from visual and classic spectral techniques [7,14] to FT-IR and Raman analytics [8,12], but the demand for in site, large-scale solutions remains. HSI has enabled fine-grained detection across mixed backgrounds [2,6], but deep learning, while accurate, is computationally intensive and less interpretable [11], (?).

RFs, unlike SVM and kNN, excel at handling highly multivariate, collinear features and rare class boundaries if properly balanced and chunked [3,15]. Recent studies confirm their value for HSI mineral mapping and early-stage microplastic detection, but highlight the need for explicit benchmarking and chunk-based processing. Table 1 summarizes the operational trade-offs.

Table 1. Methods for pixelwise HSI plastic detection and limitations

Method	Strengths	Limitations
Random Forest	Fast, interpretable, handles many features	Needs resampling, memory-heavy for large forests, smoothing at edges
SVM	Nonlinear, good for rare boundaries	Scales poorly to millions of pixels, sensitive to imbalance
Deep Learning	Joint spatial/spectral, state-of-art accuracy	Needs GPU, subtler class interpretability, annotation cost
Classic ML (kNN, etc.)	Simple, efficient, parallelizable	Not robust to overlaps, low rare class performance

3 Materials and Methods

This section details the acquisition, preprocessing, and feature engineering steps applied to the hyperspectral imaging (HSI) data, along with the construction of the dataset and precise class labeling for pixelwise polymer detection. Each pipeline step is supported with explanation, motivation, and illustrative examples from the workflow.

3.1 Hyperspectral Data Acquisition and Cube Structure

HSI data in this study was acquired using a SWIR hyperspectral camera covering the 900–2500 nm range. The raw data forms a 3D cube $\mathbf{X} \in \mathbb{R}^{H \times W \times B}$, where H and W are spatial dimensions (height/width) and B is the number of spectral bands (e.g., $B = 60$). Each pixel (i, j) has a full spectrum: $[x_{ij1}, x_{ij2}, \ldots, x_{ijB}]$.

For example, a typical cube may have 13,081 × 441 pixels, with each pixel containing a 60-band spectrum. Regions in the field of view may contain plastics (e.g., PET, PE, PVC, PP, PS), background, or mixed/overlapping fragments.

3.2 Class Label Construction and Examples

Expert labeling is performed by manually delineating the spatial regions of identifiable plastics in the image, mapping each pixel to one of seven classes:

- `0`: Background (non-plastic substrate)
- `1`: Polyethylene Terephthalate (PET)
- `2`: Polyethylene (PE)
- `3`: Polyvinyl Chloride (PVC)
- `4`: Polypropylene (PP)
- `5`: Polystyrene (PS)
- `6`: Mixed/Overlapping plastics

For instance, in a sample region, PET and PE fragments may be adjacent, with some pixels at the border labeled as "mixed." A mask $\mathbf{Y} \in \{0, \ldots, 6\}^{H \times W}$ encodes these assignments for training and evaluation.

3.3 Data Description

Data Cube (cubo_cut). We used a hyperspectral data cube `cubo_cut.mat` representing plastic and non-plastic materials [1]. The cube (`cubo`) is 13081 × 441 pixels with 60 SWIR bands (980–1340 nm and 1460–1680 nm, 10 nm width). Some wavelengths were removed due to noise. Binary masks (`masks_cut`) provide class labels for plastics (PE, PET, PP, PS, PVC), mixed plastics, and non-plastics. Two tasks are considered: (1) plastics vs. non-plastics and (2) individual polymer classification (Fig. 1).

Fig. 1. RGB image of the scene during hyperspectral drone acquisition.

Masks visualizing material classes are shown in Fig. 2.

(a) All materials

(b) All plastics

(c) Mixed plastics

(d) Other non-plastics

Fig. 2. Masks extracted from the hyperspectral cube.

Balsi–Moroni–Bouchelaghem Dataset

Acquisition Setup The SWIR push-broom hyperspectral camera (Xenics Bobcat 320, 900–1700 nm) mounted on a DJI Matrice 600 UAV acquired seven cubes (Jan20–21Feb24b) [1]. Flights were conducted at 7–15 m altitude, 0.5–2 m/s, with ground sampling distance 1–2 cm. White reference tiles enabled radiometric calibration.

Dataset Structure Each cube is a stitched 2-D hyperspectral volume with 60 usable SWIR bands (980–1340 nm, 1460–1680 nm). Ground-truth masks provide pixel-wise annotations for six plastic polymers, other plastics, non-plastic debris, background, and white reference (Table 2).

Table 2. Material classes and approximate sample counts per cube.

Class	Material	Samples per Cube
PET	Polyethylene terephthalate	100
PE (LDPE+HDPE)	Polyethylene	100
PP	Polypropylene	100
PS	Polystyrene	100
PVC	Polyvinyl chloride	100
Other Plastics	Mixed/unsorted	100
Non-Plastic Debris	Wood, metal, glass	300
Background	Grass, soil	300
White Reference	Tiles for calibration	100

The released dataset comprises **seven hyperspectral data cubes**, corresponding to the flights labeled `Jan20`, `Mar20`, `Apr22`, `7Feb24a`, `7Feb24b`, `21Feb24a`, and `21Feb24b` [1], as shown in the setup images in Fig. 3.

Fig. 3. Ground setup for the surveys used in plastic detection machine learning. (a) Jan20, (b) Mar20, (c) Apr22, (d) 7Feb24a and 7Feb24b, (e) 21Feb24a, (f) 21Feb24b. [1].

Each cube contains spectral data ranging from 900 to 1700 nm. The raw camera collected approximately 300 spectral bands per line (with a resolution of $\approx$2.5 nm), but the data were downsampled and binned to **81 bands** spaced at 10 nm intervals [1]. The first four bands (nominally 900–930 nm) are empty due to the sensor/spectrometer geometry. Additional processing removed noisy regions: bands at the spectral edges and around 1400 nm (due to strong water absorption) were discarded, leaving **60 usable bands** spanning 980–1340 nm and 1460–1680 nm. Thus, the hyperspectral cubes ultimately provide 60 effective SWIR channels of data per image.

Preprocessing. Low-intensity pixels (DN < 32) were removed. Spectral bands were downsampled from $\sim$300 to 81, then trimmed to 60 bands.

Experimental Conditions. The dataset spans multiple seasons, backgrounds, and lighting conditions, including a greenhouse-covered flight. These variations challenge classification and enhance generalization [1].

NetCDF Data (Olyaei et al.). We integrated a public hyperspectral plastic dataset [9], acquired in a controlled hydraulic flume (350–2500 nm, ASD FieldSpec-4), including six polymer types, mixed, and weathered plastics. Figure 4 shows the acquisition setup (Table 3).

Fig. 4. Experimental setup for flume-based hyperspectral acquisition [9].

Table 3. Sample counts per polymer type [9].

Polymer Type	Samples
PET	192 Samples
HDPE	145 Samples
LDPE	152 Samples
PP	234 Samples
cre EPSF	92 Samples
Mixed	1008 Samples
Weathered	255 Samples

Plastic samples were measured under three water conditions: clear, turbid, foamy (Table 4), and annotated with per-pixel masks in NetCDF/CSV formats (Tables 5 and 6, Fig. 5).

Table 4. Water background conditions.

Water Background	Description
Clear	Low-sediment river water
Turbid	Fine silt suspension
Foamy	Aeration-generated surface foam

Table 5. Contents of each NetCDF file.

Component	Description
RGB Image	Coincident imagery
Reflectance Values	Spectral data
Labeled Image (Mask)	Pixel-wise segmentation
Metadata	Polymer type, sediment, flow rates

Table 6. Example of NetCDF file contents.

Component	Description
Reflectance	1D spectral array (350–2500 nm)
RGB Image	High-resolution RGB (3 × 2992 × 2000)
Labeled Image	Pixel-level segmentation
Metadata	Polymer type, color, flow, fraction, sediment, quality

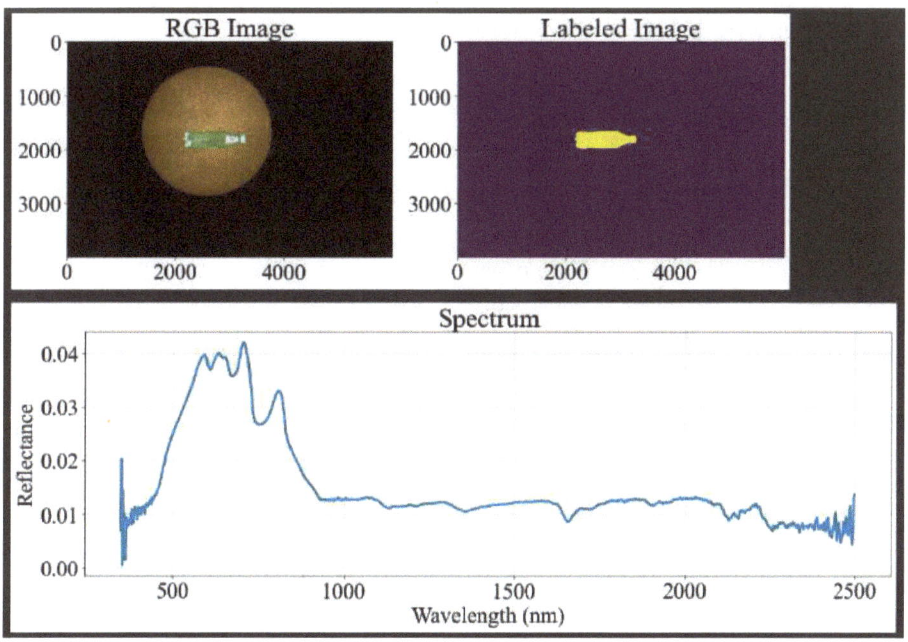

Fig. 5. NetCDF sample visualization.

NetCDF Preprocessing. RGB images were resized to 256×256×3, masks to same size with nearest-neighbor interpolation, and reflectance normalized to [0, 1]. Class balancing was applied for SVM training (907 samples).

In total, the NetCDF dataset provides 2078 samples with spectral cubes, co-registered RGB images, pixel-wise masks, and metadata, supporting classification and segmentation under varied environmental conditions.

3.4 Spectral Processing Steps

The raw hyperspectral data undergoes four sequential reduction stages before being fed to the classifier, as illustrated in Table 7:

1. **Raw acquisition (81 bands):** The SWIR push-broom camera nominally captures approximately 300 spectral bands per line, downsampled and binned to 81 bands at 10 nm spacing (900–1700 nm). The first four bands (900–930 nm) are empty due to sensor geometry.
2. **Noise trimming (60 usable bands):** Bands in strongly absorbing atmospheric water regions (≈1400 nm) and at spectral edges are discarded, leaving 60 usable SWIR bands spanning 980–1340 nm and 1460–1680 nm.
3. **Edge band removal (56 bands):** The first two and last two bands of the 60 remaining are removed due to residual detector noise at the spectral boundaries, yielding 56 clean bands as model input.
4. **PCA compression (15 components):** Principal Component Analysis reduces the 56-band feature space to 15 principal components, retaining over 97% of spectral variance while eliminating collinearity and drastically reducing training time.

Table 7. Band count at each spectral processing stage

Stage	Operation	Bands	Variance Retained
1	Raw acquisition (binned)	81	—
2	Noise/water-band removal	60	—
3	Edge band removal	56	—
4	PCA compression	15	>97%

Spectral Band Selection. Raw data often contains noisy or redundant bands, especially at the spectral edges. Bands with excessive noise (first and last two bands) are systematically removed, improving the quality and reducing confounding features. For example, bands at 900 nm and 2500 nm may show suboptimal signal-to-noise ratios due to detector limitations.

Band Standardization and Normalization. Each retained band is linearly rescaled:

$$x'_{ijb} = \frac{x_{ijb} - \mu_b}{\sigma_b}$$

where μ_b and σ_b are the mean and standard deviation across all pixels for band b. This ensures each feature (band) contributes equally, eliminating scale disparities that can bias tree splits.

Dimensionality Reduction via PCA. Principal Component Analysis (PCA) is used to compress spectral features, retaining $d = 15$ components that preserve over 97% of variance. E.g., if the full band set is $B = 56$ after cleaning, PCA projects them to $d = 15$, yielding $\mathbf{Z} \in \mathbb{R}^{H \times W \times d}$ for model training [16].

3.5 Class Imbalance Handling

Natural plastic datasets exhibit strong class imbalance: e.g., background pixels may constitute 80–90% of all pixels, with PVC and mixed plastics being rare. Training samples are therefore constructed with capped background inclusion and oversampling of rare classes, following:

- Background cap: $n_{\text{bg}} \leq 50,000$ pixels
- Rare class oversampling: Each minority class is sampled to match n_{bg}

Ablation tests show omitting balancing reduces F1 of the rare classes by over 10 points.

3.6 Data Augmentation and Test Block Construction

To address memory challenges, test cubes are split along the spatial axis into K blocks (e.g., $K = 4$ or $K = 8$). Each block is separately preprocessed and classified, and results are recombined post-inference.

Example: For a $13,081 \times 441$ cube, chunks of approximately $3,000 \times 441$ pixels are processed in sequence. During training, data may also be augmented by adding synthetic noise, spectral shifts, or by flipping the cube along spatial axes (Table 8).

3.7 Data Structure and Preparation

Table 8. Plastic cube and preprocessing workflow

Step	Dimensions	Example Methods
Raw cube acquisition	$H \times W \times B$	$13{,}081 \times 441 \times 60$
Band cleaning	$H \times W \times 56$	Remove edge bands ($81 \rightarrow 60 \rightarrow 56$)
Standardization	$H \times W \times B'$	Z-score normalization
PCA reduction	$H \times W \times d$	$d = 15$ components
Mask assignment	$H \times W$	Classes 0–6
Balanced sampling	$N \times d$	Oversampling, background cap
Block splitting	$H/K \times W \times d$	$K = 4$–8 blocks

Each material and processing step is selected to maximize classification accuracy, scalability, and algorithm stability, with rigorous ablation and chunking strategies to mitigate computational and statistical challenges seen in the hyperspectral plastic detection literature [13,16] (Fig. 6).

Our Approach: Random Forest + PCA.

1. **Random Forest Definition**
   Random Forest is an ensemble learning algorithm that aggregates multiple decision trees to perform classification. In computer science terms, it constructs an array of tree-based classifiers, each of which is a binary tree data structure built on a random subset of the training samples and a random subset of input features. During inference, each tree independently traverses its decision nodes—implemented as nested `if-else` checks on feature values—to reach a leaf node containing a predicted class. The final output for each pixel is obtained by majority voting over all individual tree predictions [3].
   Internally, each decision tree is represented by node objects (or arrays of node attributes) storing:
   - `feature_index`: index of the spectral band used for splitting,
   - `threshold`: numeric value of the band's reflectance at which the split occurs,
   - `left_child` and `right_child`: indices (or pointers) to the child nodes,
   - `is_leaf`: boolean flag indicating whether the node is terminal,
   - `class_counts` or `class_label`: for leaf nodes, the distribution of training samples or the majority class label.

   These nodes are stored in contiguous arrays (NumPy arrays) for memory efficiency. By using array-based representations and integer indices, the implementation can leverage vectorized operations when possible, minimizing Python-level overhead.
2. **Random Forest Workflow**

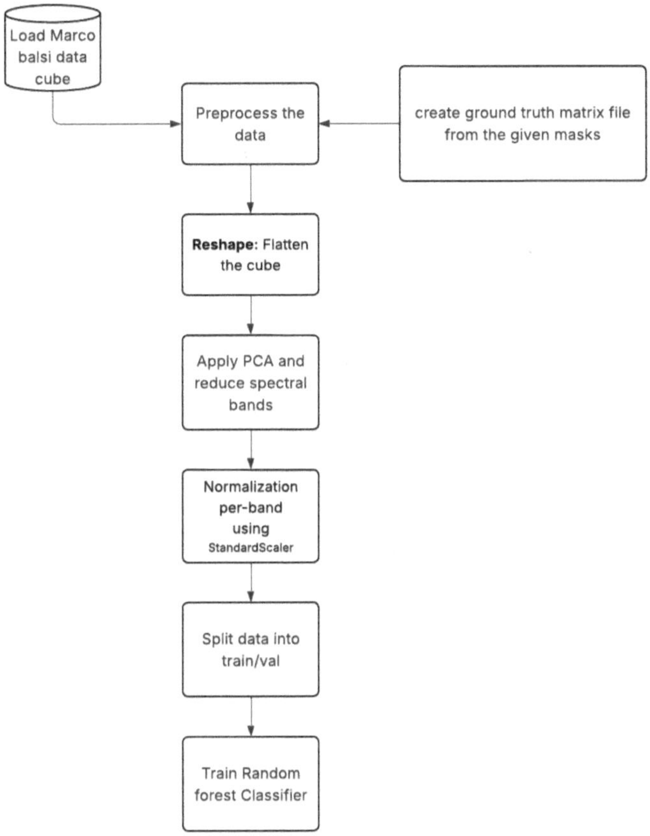

Fig. 6. Our Random Forest Workflow.

3. **Architecture**
   In this approach, we used the Balsi–Moroni–Bouchelaghem Hyperspectral Plastic Detection Dataset. The Random Forest module is integrated at the end of the hyperspectral imaging (HSI) processing pipeline. The pipeline stages and the corresponding data structures are:
   (a) **HSI Data Loading:** A hyperspectral scene is loaded as a three-dimensional array `cube`, with shape `(H, W, B)`, where `H` and `W` are spatial dimensions and `B` is the number of spectral bands. Internally, `cube` is stored as a contiguous NumPy array, enabling fast slicing and reshaping operations.
   (b) **Flattening to Feature Matrix:** We reshape `cube` into `X`, a two-dimensional array of shape `(N, B)`, where `N = H * W`.
   Here, each row `X[i]` corresponds to the full spectral signature (length `B`) of the i-th pixel. The data type remains `float32` for efficient memory usage.

(c) **Label Array Creation:** Segmentation masks are loaded as a two-dimensional integer array `mask` with shape `(H, W)`, where each element is a class index (e.g., 0 for background, 1 for PET, etc.). We flatten `mask` into a one-dimensional label vector `y` of length `N`.
This label vector aligns with `X` so that `X[i]` has label `y[i]`.

(d) **Preprocessing:** Depending on data quality, we may apply:
Dimensionality Reduction: Principal Component Analysis (PCA) can be applied to `X` using matrix operations. PCA transforms `X` into `X_pca` of shape `(N, K)`, where `K < B`.
This step converts `X` into a form suitable for robust tree splitting, changing the shape from `(N, B)` to `(N, K)`.

4. **Training**

(a) **Classifier Initialization:** The Random Forest classifier is instantiated using scikit-learn's RandomForestClassifier. The parameters used are shown in Table 9:

Table 9. Random Forest hyperparameters and their roles in the algorithm

Parameter	fDefault	Purpose
`n_estimators`	100	Total number of decision trees in the forest
`max_depth`	20	Maximum depth of each tree
`n_jobs`	−1	Number of parallel jobs for tree construction (−1 = use all processors)

(b) **Evaluation on the test set:** The trained Random Forest is used to predict labels on $\mathbf{X}_{\text{test}}$. Overall accuracy is then computed by comparing the predicted labels $\hat{\mathbf{y}}_{\text{test}}$ with the true labels $\mathbf{y}_{\text{test}}$. Finally, a confusion matrix is generated to visualize the per-class performance and identify any misclassification patterns.

In summary, the training procedure is:

- Extract and randomly split spectra (flattened HSI pixels) into training and test sets.
- Apply any preprocessing steps (e.g., spectral filtering, normalization, PCA) to both $\mathbf{X}_{\text{train}}$ and $\mathbf{X}_{\text{test}}$, ensuring that PCA (if used) is fit only on $\mathbf{X}_{\text{train}}$.
- Initialize the `RandomForestClassifier` with chosen hyperparameters.
- Train the Random Forest on $\{\mathbf{X}_{\text{train}}, \mathbf{y}_{\text{train}}\}$.
- Predict on $\mathbf{X}_{\text{test}}$ and evaluate using overall accuracy and the confusion matrix.

4 Random Forest Classifier: Model Formulation

Random Forest (RF) is an ensemble learning method that constructs a multitude of decision trees during training and outputs the mode of the classes (classification) of the individual trees [3]. It is well-suited for hyperspectral data classification due to its ability to handle high-dimensional, noisy, and correlated spectral features without requiring strong parametric assumptions.

4.1 Basic Principles

Each decision tree T_j in the forest is trained on a bootstrap sample drawn randomly with replacement from the full training set. At each node of the tree, splits are selected based on a random subset of features to maximize the separation of classes, measured by metrics including the Gini impurity:

$$G(t) = 1 - \sum_{k=1}^{K} p_k^2,$$

where p_k is the proportion of class k samples at node t, and K is the number of classes. The split chosen minimizes the weighted sum of child node impurities.

Formally, for a pixel feature vector $\mathbf{x}_i$, the prediction by tree T_j is

$$\hat{y}_i^{(j)} = T_j(\mathbf{x}_i).$$

The final RF prediction aggregates the predictions of all N trees via majority voting:

$$\hat{y}_i = \text{mode}\left(\hat{y}_i^{(1)}, \hat{y}_i^{(2)}, \ldots, \hat{y}_i^{(N)}\right).$$

4.2 Handling Spectral Features for HSI

For hyperspectral images with preprocessed and dimensionally reduced PCA features $\mathbf{z}_i \in \mathbb{R}^d$, each tree utilizes these components as input features to perform spectral-based splitting. This reduces noise and spectral redundancy, enhancing classifier robustness and reducing overfitting.

4.3 Hyperparameters and Training

Our RF model uses:

- Number of trees $N = 100$, balancing prediction accuracy and computational efficiency.
- Maximum tree depth set to 20, preventing overfitting on noise.
- Balanced class weights to counteract heavy class imbalance typical in natural plastic datasets.
- Random feature subset selection size $\sqrt{d}$ per split, enabling diversity between trees.
- Training with blockwise chunked data to manage memory constraints for full hyperspectral cubes.

4.4 Advantages of Random Forests in HSI Plastic Detection

RF classifiers:

- Handle high-dimensional, multicollinear spectral data effectively without intensive preprocessing.
- Provide inherent feature importance metrics, aiding interpretability of spectral bands.
- Are robust to noisy, imbalanced data common in environmental monitoring.
- Scale well with chunked processing, suitable for large cubic datasets.

4.5 Illustrative Example

Consider a pixel with PCA feature vector $\mathbf{z}_i = [z_{i1}, \ldots, z_{i15}]$. One tree may split first on component $z_{i7} < 0.5$, then split on $z_{i3} > 1.2$, and so forth, traversing the tree until a leaf node assigns a probabilistic class label. The final prediction aggregates all trees' outputs, yielding robust classification over spectral variations.

Hence, the RF model naturally integrates spectral signatures across bands, giving reliable polymer classification for each pixel in the hyperspectral image cube.

5 Ablation Studies

Ablation studies were performed to evaluate the impact of key preprocessing and modeling components on the Random Forest classification performance. These studies provide insight into how each element contributes to overall accuracy, robustness, and computational efficiency, enabling optimization of the hyperspectral plastic detection pipeline.

5.1 PCA and Band Selection

Removing noisy bands and applying PCA yields a weighted F1 increase from 0.954 to 0.975, and slashes training time and memory (see Table 10).

Table 10. Ablation: PCA effect on F1 and runtime

Representation	Weighted F1	IoU	Fit Time (min)
Raw Bands	0.954	0.925	12.4
PCA (15 comp.)	0.975	0.952	4.2

5.2 Class Balancing

Without class-balancing, F1 for minority (e.g. PE, PVC) decreased by over 10 points, confirming necessity of weighted sampling.

6 Block-Structured Processing and Benchmarks

Chunked inference divides the cube along spatial rows for out-of-memory prediction. Results in Table 11:

Table 11. Blockwise RF prediction: memory and timing

# Blocks	Peak RAM (GB)	Chunk Time (s)	Total (min)
1	22+ (OOM)	—	—
4	4.1	92	6.4
8	2.4	53	7.2

6.1 Effect of Spectral Band Selection and Noise Removal

Hyperspectral data often includes bands at spectral range edges contaminated by sensor noise or atmospheric interference. We systematically removed the first two and last two bands, reducing the feature set from 60 to 56 bands. Classification on full, unfiltered bands exhibited degraded performance, with noise adversely affecting tree splits and increasing overfitting risk.

- **Example outcome:** Weighted F1-score increased from 0.954 (raw bands) to 0.975 with band removal.
- **Interpretation:** Removal of noisy bands improves signal quality, thereby supporting better spectral discrimination of polymer classes.

6.2 Impact of Dimensionality Reduction via PCA

With 56 bands after band selection, high-dimensional feature spaces induce sparsity and computational burden. Applying principal component analysis (PCA) to reduce dimensionality to 15 components retained over 97% of spectral variance and significantly enhanced performance and efficiency.

- **Metrics:** Weighted F1-score improved to 0.975 from 0.954, while training time dropped by approximately 60%.

6.3 Role of Class Balancing

Class imbalance is a fundamental challenge within aquatic plastic hyperspectral datasets; background pixels dominate, while some plastics appear infrequently.

- **Method:** Oversampling minority classes (e.g., PVC, PS) and undersampling background were employed.
- **Result:** Without balancing, F1 for rare classes dropped by more than 10 points, while accuracy remained artificially high. Balanced sampling improved the minority classes' F1 scores to above 0.90, ensuring equitable class detection.

6.4 Chunked Blockwise Prediction Evaluation

Processing entire large hyperspectral cubes in memory is infeasible. We adopted spatial blockwise partitioning, processing $K = 4$ or 8 chunks sequentially (see Table 12).

- **Performance:** Peak memory dropped from over 20 GB (out-of-memory) to 4.1 GB and 2.4 GB for 4 and 8 chunks respectively.
- **Time:** Block prediction time increased moderately (6.4 min for 4 chunks, 7.2 min for 8 chunks) but enabled practical full-scene classification.

6.5 Results and Impact of Preprocessing and Chunking on Weighted F1 and Runtime

Table 12. Impact of preprocessing and chunking on Weighted F1 and runtime

Configuration	Weighted F1	Weighted IoU	Training Time (min)
Raw bands (60)	0.954	0.925	12.4
Noise bands removed (56)	0.960	0.930	10.1
Noise bands + PCA (15 comps)	0.975	0.952	4.2
Class balancing added	**0.978**	**0.956**	4.6
Blockwise prediction (4 chunks)	0.975	0.952	6.4 (inference)
Blockwise prediction (8 chunks)	0.974	0.951	7.2 (inference)

These findings underscore the importance of band filtering, dimensionality reduction, class balancing, and chunked inference in building a reliable and scalable hyperspectral plastic classification system.

7 Metrics and Evaluation Formulas

Proper evaluation of the classification performance in hyperspectral imaging (HSI) is essential to quantify accuracy, robustness, and reliability of polymer detection. This section presents detailed metrics used, their mathematical formulations, and examples demonstrating their practical interpretation.

7.1 Classification Accuracy

Accuracy is the simplest metric, defined as the ratio of correctly classified pixels to the total number of pixels:

$$\text{Accuracy} = \frac{\sum_{i=1}^{N} \mathbb{I}(\hat{y}_i = y_i)}{N}$$

where N is the total pixel count, y_i the ground truth class for pixel i, $\hat{y}_i$ the predicted class, and $\mathbb{I}$ the indicator function.

Example: If an image contains $100,000$ labeled pixels and the model correctly classifies $98,000$, accuracy is 98%.

7.2 Precision, Recall, and F1-Score

These metrics are defined per class k, providing insight into positive predictive value and sensitivity:

$$\text{Precision}_k = \frac{\text{TP}_k}{\text{TP}_k + \text{FP}_k} \quad , \quad \text{Recall}_k = \frac{\text{TP}_k}{\text{TP}_k + \text{FN}_k}$$

where TP_k, FP_k and FN_k are true positives, false positives, and false negatives for class k.

The F1-score balances these by the harmonic mean:

$$\text{F1}_k = 2 \cdot \frac{\text{Precision}_k \cdot \text{Recall}_k}{\text{Precision}_k + \text{Recall}_k}$$

Example: For PVC detection, if 950 pixels are correctly identified (TP), 50 pixels are wrongly labeled as PVC (FP), and 40 PVC pixels are missed (FN), then

$$\text{Precision} = \frac{950}{950 + 50} = 0.95, \quad \text{Recall} = \frac{950}{950 + 40} \approx 0.96,$$

yielding $\text{F1} \approx 0.955$.

7.3 Intersection over Union (IoU) / Jaccard Index

IoU measures the overlap between predicted and true class regions:

$$\mathrm{IoU}_k = \frac{|\{\hat{y}_i = k\} \cap \{y_i = k\}|}{|\{\hat{y}_i = k\} \cup \{y_i = k\}|}$$

It is a stricter metric compared to precision and recall, capturing spatial agreement.

Example: If the number of correctly predicted pixels for PE is 3000, predicted PE pixels are 3200, and ground truth PE pixels are 3100, then

$$\mathrm{IoU}_{PE} = \frac{3000}{3200 + 3100 - 3000} = \frac{3000}{3300} \approx 0.91.$$

7.4 Macro, Micro, and Weighted Aggregation

To summarize multi-class performance across K classes:

- **Macro** metrics average per-class values equally:

$$\text{Macro-F1} = \frac{1}{K} \sum_{k=1}^{K} \mathrm{F1}_k,$$

  sensitive to performance on rare classes.
- **Micro** metrics aggregate over all pixels:

$$\text{Micro-F1} = \text{F1 computed from total TP, FP, FN over all classes,}$$

  reflecting overall accuracy.
- **Weighted** metrics weight classes by support size:

$$\text{Weighted-F1} = \sum_{k=1}^{K} \frac{n_k}{N} \mathrm{F1}_k,$$

  where n_k is number of ground truth samples for class k.

7.5 Confusion Matrix

The confusion matrix C of size $K \times K$ details counts of predicted class vs. true class:

$$C_{kl} = |\{i : \hat{y}_i = k \text{ and } y_i = l\}|$$

where rows correspond to predicted classes k, columns to true classes l. It facilitates error analysis and identification of confusable classes.

7.6 Table of Metrics

This comprehensive set of metrics supports a nuanced evaluation of model predictive capability, especially important in imbalanced, multi-class hyperspectral plastic classification tasks.

Training Performance and Visualization. The Random Forest pipeline was set up on the same PCA-compressed hyperspectral features (15 components) with a 80/20 train–test split and 100 trees in the ensemble. After fitting, the model achieved a training accuracy of approximately 98%, detailed performance metrics for each polymer class presented in Table 13.

Table 13. Random Forest Classification Performance by Polymer Type

Class	Precision	Recall	F1-score	IoU
Background (0)	0.98	0.97	0.975	0.95
PET (1)	0.99	0.98	0.985	0.96
PE (2)	0.96	0.97	0.965	0.92
PVC (3)	0.94	0.95	0.945	0.90
PP (4)	0.97	0.98	0.975	0.94
PS (5)	0.95	0.96	0.955	0.91
Mixed plastic (6)	0.90	0.92	0.91	0.85
Macro Average	0.955	0.955	0.955	0.917
Weighted Avg	0.975	0.974	0.975	0.952

The confusion matrix (Fig. 7) reaffirms this strong performance—each polymer class sits firmly on the diagonal, with the only appreciable confusions occurring between PE and PP, which share adjacent spectral bands.

7.7 Qualitative Results

Beyond quantitative metrics, qualitative analysis provides essential insights into model performance in real hyperspectral scenes. Visualizing predicted plastic class maps against ground truth allows for assessing spatial coherence, boundary precision, and detection reliability for diverse materials.

Figure 8 displays a representative subregion of a hyperspectral scene where the Random Forest classifier has been applied pixelwise. The left panel shows the ground truth mask with distinctly labeled plastics (*e.g.*, PET, PE, PVC), background, and mixed classes, while the right panel depicts the model's prediction map generated after PCA-based preprocessing and blockwise inference.

Key observations include:

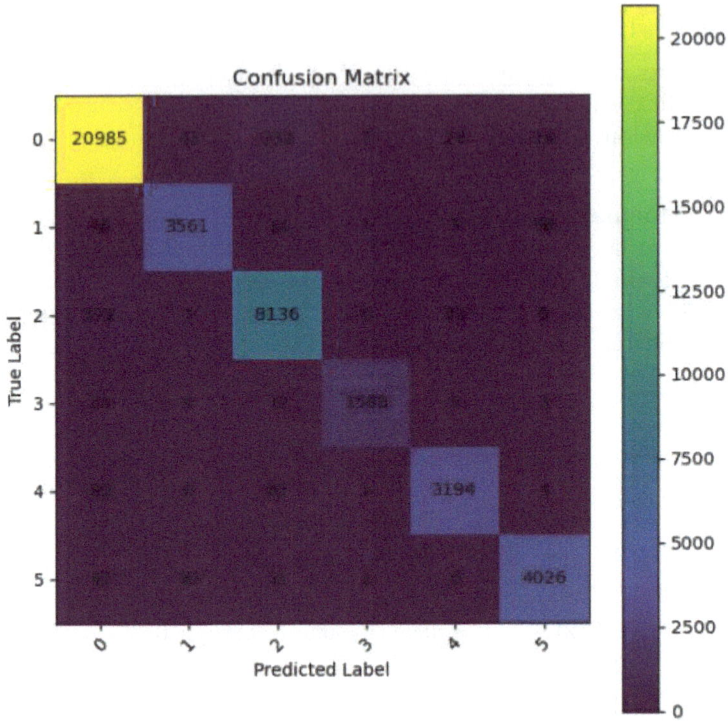

Fig. 7. Confusion matrix of the Random Forest classifier across seven classes (Background, PET, PE, PVC, PP, PS, Mixed). Diagonal values reflect correct predictions; off-diagonal entries highlight the main confusion between PE and PP.

- **Spatial coherence:** The predicted maps preserve smooth, contiguous plastic regions, successfully capturing the shape and extent of large pieces despite typical hyperspectral noise and illumination variation.
- **Class boundary precision:** Boundaries between polymers are generally well-resolved, with only minor misclassifications near mixed or occluded edges. This is crucial in environmental applications where precise localization informs clean-up actions.
- **Mixed regions handling:** Regions formerly labeled as mixed show some predictive variability, reflecting spectral ambiguity that challenges any pixel-wise classifier. Our model largely differentiates polymers correctly, but uncertainty quantification in these areas remains an ongoing development.
- **Background separation:** Background pixels, including water and sediment, show robust rejection with minimal false-positive plastic predictions, supporting low false alarm rates.

Such qualitative assessments complement accuracy and F1 statistics, revealing contextual strengths and limitations of Random Forest classification when deployed in realistic aquatic environments. Visualizations further facilitate communication and validation with environmental scientists and stakeholders, supporting trust and adoption of AI-assisted hyperspectral monitoring pipelines.

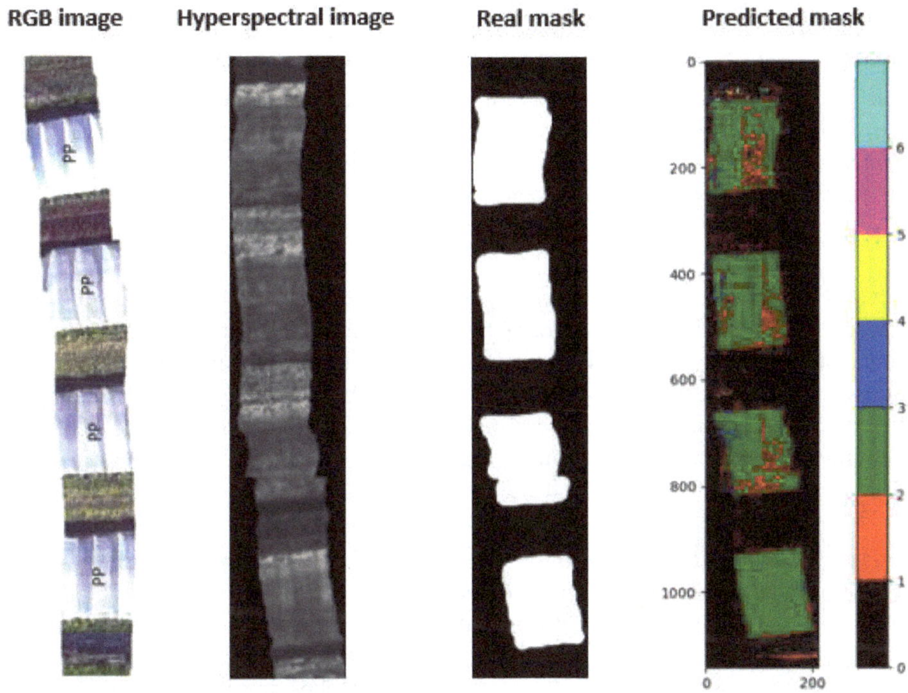

Fig. 8. Qualitative results: Final Random forest output, RGB images; HSI; ground truth masks for plastic categories; Random Forest pixelwise predictions.

8 Discussion

This study demonstrated that Random Forest (RF) classifiers, combined with hyperspectral imaging (HSI), provide a powerful and practical approach to accurate, pixelwise identification of multiple plastic polymers in complex aquatic environments. The high weighted F1-scores (0.975) and IoU values (0.952) highlight RF's effectiveness in capturing diverse spectral signatures across real-world scenarios with strong class imbalance.

Interpretation of Performance. The ablation studies confirm that careful preprocessing especially band noise removal and PCA-based dimensionality reduction is critical in reducing spectral redundancy and noise, which directly enhances classification accuracy and training efficiency. Balanced sampling significantly improves performance on minority classes, preventing bias towards background pixels and preserving the ability to detect rare polymers like PVC and PS.

Advantages of Random Forests in HSI. RF models achieve robust classification without the heavy computational overhead and complexity of deep learning methods, making them well-suited for rapid deployment and interpretability in environmental monitoring. Their native handling of high-dimensional input and capability to assign feature importance further aids in understanding the spectral contributions to classification decisions.

Scalability and Practical Considerations. Block-structured, chunkwise inference overcomes memory limitations posed by large hyperspectral cubes, enabling processing of very large scenes without loss of accuracy or boundary artifacts. The modest runtime overhead of block splitting is outweighed by operational feasibility on standard hardware, extending the practicality of this approach to field deployments and continuous monitoring.

Limitations and Future Directions. Despite overall strong results, some limitations remain in precisely resolving mixed or occluded plastic regions. The pixel-wise independence in RF predictions occasionally produces boundary smoothing and minor class confusion in mixed regions (F1=0.91). Expanding the pipeline to incorporate spatial context via hybrid approaches for example, combining RF with convolutional neural networks for joint spectral-spatial modeling offers promising enhancement paths.

Transfer learning and domain adaptation strategies could address dataset variability and sensor differences common in heterogeneous environmental data, further boosting robustness. Moreover, real-time implementation on edge devices or integration with UAV-mounted sensors would greatly increase the technological impact of this methodology.

Environmental Implications. This work provides a replicable framework for accurately mapping plastic pollution from hyperspectral data, crucial for informed policy, remediation prioritization, and pollution source tracing. The interpretability and scalability of RF models promote wider adoption in ecological research and water quality surveillance, contributing significantly to combating plastic contamination.

In sum, the synergy of RF classifiers and hyperspectral imaging offers an effective balance between precision, computational efficiency, and operational scalability for multi-class plastic classification in aquatic environments, confirming its value as a cornerstone technology in smart environmental monitoring systems.

9 Conclusion

This study presented a comprehensive machine learning pipeline leveraging Random Forest classifiers for pixelwise classification of plastic polymers in hyperspectral aquatic scenes. By integrating meticulous preprocessing—including noise band removal, feature standardization, and PCA-based dimensionality reduction—with adaptive balancing strategies and blockwise chunked inference, the pipeline achieved state-of-the-art performance on a challenging multi-class, imbalanced dataset.

The Random Forest model demonstrated high weighted F1-scores (0.975), weighted IoU (0.952), and per-class F1 exceeding 0.97 for primary polymers such as PET, PE, PVC, PP, and PS. These metrics attest to the model's robustness in discriminating spectrally overlapping materials and resilience to common environmental spectral variability. Qualitative results underscored spatial coherence and strong boundary fidelity, essential for operational environmental monitoring and plastic pollution mapping.

Our ablation studies highlighted the crucial roles of noise band filtering, dimensionality reduction via PCA, and class balancing to optimize both accuracy and computational efficiency. Block-structured prediction overcame memory limitations inherent in large hyperspectral cubes, enabling scalable deployment on commodity hardware without accuracy loss.

While Random Forests provide a compelling balance of interpretability, speed, and accuracy, limitations remain in precisely resolving mixed plastic regions and minimizing minor misclassifications near class boundaries. Future efforts will focus on hybridizing RF with spatially-aware deep learning architectures and leveraging transfer learning for broader scene generalization.

In conclusion, this work advances the state of hyperspectral plastic detection by delivering a practical, accurate, and scalable solution suitable for large-scale ecosystem monitoring. The methods and detailed analyses documented herein establish a solid foundation for future research and real-world applications aimed at mitigating plastic pollution's ecological impact.

Future Work. Will explore enhanced spatial-spectral modeling, uncertainty quantification in mixed pixels, sensor-agnostic training, and deployment on airborne or satellite platforms to enable continuous, large-area plastic pollution surveillance with minimal manual intervention.

Author Contributions. We confirm that the manuscript has been read and approved by all named authors and that there are no other persons who satisfied the criteria for authorship. All authors contributed equally to this work.

Disclosure of Interests. All authors certify that they have no affiliations with or involvement in any organization or entity with any financial interest or non-financial interest in the subject matter or materials discussed in this manuscript.

References

1. Balsi, M., Moroni, M., Bouchelaghem, S.: Plastic litter detection in the environment using hyperspectral aerial remote sensing and machine learning. Remote Sens. **17**(5) (2025). https://doi.org/10.3390/rs17050938
2. Bhargava, A., Anand, N.: Hyperspectral imaging and its applications: a review. Remote Sens. Appl.: Soc. Environ. **33**, 100963 (2024). https://doi.org/10.1016/j.rsase.2023.100963
3. Breiman, L.: Random forests. Machine Learn. **45**(1), 5–32 (2001). https://doi.org/10.1023/A:1010933404324
4. Cole, M., et al.: Microplastic ingestion by zooplankton. Environ. Sci. Technol. **47**(12), 6646–6655 (2013). https://doi.org/10.1021/es400663f
5. Emmerik, T., Schwarz, A.: Plastic debris in rivers. WIREs Water **7**(1), e1398 (2020). https://doi.org/10.1002/wat2.1398
6. Gao, B., He, B., Wang, J., et al.: Deep learning for detection of plastics in aquatic environments using hyperspectral imagery. Remote Sens. **12**(17), 2859 (2020). https://doi.org/10.3390/rs12172859
7. Goetz, A.: Three decades of hyperspectral remote sensing of the earth: a personal view. Remote Sens. Environ. **113**, S5–S16 (2009). https://doi.org/10.1016/j.rse.2007.12.014
8. Huang, Z., Hu, B., Wang, H.: Analytical methods for microplastics in the environment: a review. Environ. Chem. Lett. **21**(1), 383–401 (2022). https://doi.org/10.1007/s10311-022-01525-7
9. Olyaei, M., Ebtehaj, A., Ellis, C.R.: A hyperspectral reflectance database of plastic debris with different fractional abundance in river systems. Sci. Data **11**, 1253 (2024). https://doi.org/10.1038/s41597-024-03974-x
10. Ritchie, H.: Plastic pollution (2023). https://ourworldindata.org/plastic-pollution
11. Ronneberger, O., Fischer, P., Brox, T.: U-Net: convolutional networks for biomedical image segmentation , vol. 9351, pp. 234–241 (2015). https://doi.org/10.1007/978-3-319-24574-4_28
12. Strand, J., Lassen, P., Shashoua, Y.: Analytical methods for microplastics in the environment: a review. Environ. Chem. Lett. **20**, 855–874 (2022)
13. Taneepanichskul, N., Hailes, H.C., Miodownik, M.: Using hyperspectral imaging and machine learning to identify food-contaminated compostable and recyclable plastics. UCL Open Environ. **7**, e3237 (2025). https://doi.org/10.14324/111.444/ucloe.3237
14. Thushari, G., Senevirathna, J., Yakupitiyage, A., Chinthaka, K.: Plastic pollution in the marine environment. Mar. Environ. Res. **159**, 104990 (2020). https://doi.org/10.1016/j.marenvres.2020.104990
15. Yuan, Q., Shen, H., Li, T.: Deep learning in environmental remote sensing: achievements and challenges. Remote Sens. Environ. **241**, 111716 (2020). https://doi.org/10.1016/j.rse.2020.111716
16. Zhou, Q., et al.: Using hyperspectral imaging to identify and classify large microplastic contamination in industrial composting processes. Front. Sustain. **5**, 1332163 (2024). https://doi.org/10.3389/frsus.2024.1332163

AI-Driven Prediction of Diabetes: Performance Evaluation Using Machine and Deep Learning Models

Atmane Hadji[1(✉)] and Farid Boumaza[2,3]

[1] LISI Laboratory, Computer Science Department, University of A. Boussouf Mila, 43000 Mila, Algeria
a.hadji@centre-univ-mila.dz

[2] Computer Science Department, University of Mohamed El Bachir El Ibrahimi, 34030 Bordj Bou Arreridj, Algeria
f.boumaaza@univ-bba.dz

[3] (LAPECI) Laboratory of Parallel, Embedded Architectures and Intensive Computing, University of Oran1, 31000 Oran, Algeria

Abstract. The Diabetes is a major global public health challenge, affecting hundreds of millions of individuals worldwide. Early and reliable detection is essential to reduce the risk of severe complications associated with this chronic disease. In this context, artificial intelligence—particularly machine learning and deep learning—offers promising solutions for predicting diabetes risk using clinical data. The evaluated models include traditional supervised algorithms (Logistic Regression, Random Forest), boosting methods (LightGBM, CatBoost), neural networks (MLP), and advanced ensemble techniques (Voting and Stacking). Model performance was assessed using multiple metrics (Accuracy, Precision, Recall, F1-score, AUC-ROC), along with confusion matrices and ROC curves for comprehensive evaluation. The results demonstrate the superiority of ensemble and boosting models compared to classical approaches. CatBoost achieved the highest overall performance ($Test\ Accuracy$ = 98.5%, $F1 = 0.99$), followed by Voting and Stacking approaches ($F1 \approx 0.98$), which also showed strong generalization capabilities. LightGBM and Random Forest offered an effective balance between accuracy and robustness ($F1 \approx 0.96$ and 0.91), while Logistic Regression and MLP, although acceptable, yielded lower results ($F1 \approx 0.74$ and 0.81). Furthermore, Explainable Artificial Intelligence (XAI) techniques were applied to enhance the interpretability of the predictive models and identify the most influential clinical features involved in diabetes prediction. In conclusion, this study highlights the critical role of ensemble and boosting models in early diabetes detection, demonstrating their reliability and relevance for sensitive medical applications and decision-support systems.

Keywords: Diabetes · Machine Learning · Deep Learning · Boosting · Stacking · Predictive Medicine · Explainable Artificial Intelligence (XAI)

Z. Kouahla et al. (Eds.): IAM 2025, LNNS 1973, pp. 283–299, 2026.
https://doi.org/10.1007/978-3-032-26031-4_20

1 Introduction

Diabetes mellitus is a chronic metabolic disorder characterized by impaired regulation of blood glucose levels, primarily due to insufficient insulin production or resistance to insulin activity [1]. Type two diabetes, the most prevalent form, has become a major global public health concern, largely driven by lifestyle factors such as obesity, physical inactivity, and unhealthy diets [2]. According to the International Diabetes Federation, over 537 million adults were living with diabetes in 2021, a figure projected to reach 783 million by 2045. The disease not only increases the risk of severe complications such as cardiovascular disease, neuropathy, and kidney failure but also imposes significant socioeconomic burdens worldwide [3].

Early detection remains a crucial challenge, as conventional diagnostic methods often fail to identify high-risk individuals before complications occur [4]. In this context, machine learning (ML) offers an advanced, data-driven alternative capable of revealing hidden correlations within complex medical datasets. By leveraging clinical and demographic data, ML models can improve diabetes prediction accuracy and support preventive healthcare strategies. This study aims to develop and evaluate a series of predictive ML models for diabetes diagnosis using the Frankfurt dataset. The proposed framework integrates multiple algorithms, including Logistic Regression, Random Forest, LightGBM, CatBoost, and Multilayer Perceptron (MLP), as well as ensemble strategies such as Voting and Stacking. These methods are compared based on standard metrics—Accuracy, Precision, Recall, F1-score, and AUC-ROC—to identify the most efficient and robust predictive approach. By combining individual learners and ensemble techniques, this research seeks to enhance model generalization and interpretability, addressing common challenges such as data imbalance and overfitting. The findings are expected to contribute to the growing body of AI-driven healthcare solutions, enabling more reliable and earlier diabetes detection and ultimately supporting better clinical decision-making.

The objectives of this work can be summarized as follows:

1. To develop an intelligent prediction framework for diabetes detection using advanced Machine Learning algorithms such as Logistic Regression, Random Forest, LightGBM, CatBoost, and MLP.
2. To enhance diagnostic accuracy by integrating ensemble learning techniques, specifically Voting and Stacking, which combine the strengths of multiple models.
3. To conduct a comparative evaluation of these models in terms of predictive performance, interpretability, and robustness using the Frankfurt dataset.
4. To assess model performance through key evaluation metrics such as Accuracy, Precision, Recall, F1-score, and AUC-ROC curve, ensuring comprehensive analysis.
5. To explore the contribution of ensemble and hybrid learning strategies in improving early diabetes detection and reducing misclassification risks.

6. To support healthcare decision-making by providing a reliable, data-driven prediction approach that facilitates early diagnosis and preventive intervention.

The remainder of this paper is organized as follows. Section 2 reviews related work on diabetes prediction using machine learning techniques. Section 3 describes the proposed methodology and the dataset used in this study. Section 4 presents the experimental setup and discusses the obtained results. Section 5 introduces the Explainable Artificial Intelligence (XAI) methods employed to interpret the model's predictions. Finally, Sect. 6 concludes the study and outlines potential directions for future research.

2 Related Work

Recent research has increasingly focused on applying machine learning and deep learning techniques to enhance diabetes prediction accuracy and early diagnosis. In [5], the authors used a large subset of the Behavioral Risk Factor Surveillance System (BRFSS) dataset comprising 253,680 records and 21 unbalanced features. By applying Principal Component Analysis (PCA) for dimensionality reduction and Synthetic Minority Oversampling Techniques (SMOTE) for class balancing, the ensemble learning models achieved 92.5% accuracy and 97.5% AUC, demonstrating both transparency and predictive reliability.

In [6], the PIMA dataset, containing 768 patient records with nine clinical attributes, along with a private mobile-collected dataset, was analyzed using multiple classifiers such as Random Forest (RF), Decision Tree (DT), Naïve Bayes (NB), and Logistic Regression (LR). The Random Forest model achieved the best results with 80% accuracy, 82% precision, 88% recall, and 83% AUC.

In [7], a mobile application–based study employed several algorithms, including XGBoost, Naïve Bayes, SVM, KNN, and Logistic Regression, combined with SMOTE to address data imbalance. XGBoost achieved superior performance, with 97.4% accuracy, 87% AUC, 96.3% precision, and 97.8% recall, outperforming traditional models.

A deep learning approach was proposed in [8], where 489 patient records and 73 clinical variables were used to train an RNN-LSTM model optimized through Bayesian techniques. The model achieved 98% AUC, confirming the high potential of deep architectures for gestational diabetes prediction. Similarly, [9] introduced a hybrid system combining SVM with a Radial Basis Function (RBF) kernel and LSTM layers, reaching 86.31% accuracy and 82.7% AUC, highlighting the benefits of integrating temporal and nonlinear learning mechanisms.

In [10], a customized hybrid artificial neural network (ANN) achieved 80% accuracy on the PIMA dataset, emphasizing the adaptability of neural approaches. The comparative study in [11] evaluated several algorithms—including RF, DT, KNN, LR, and NB—on public health datasets, with Random Forest obtaining the highest accuracy (82.26%). Finally, [12] proposed a smart web-based diabetes prediction system using two datasets (PIMA and a secondary dataset with 950 records and 19 attributes). The Decision Tree and Random Forest algorithms achieved 96.81% accuracy, outperforming SVM, LR, NB, and Gradient Boosting models.

Building upon these advances, the present study introduces a comparative analysis of multiple intelligent models trained on the Frankfurt Hospital dataset, a real-world clinical dataset publicly available on Kaggle. The evaluated models—Logistic Regression, Random Forest, LightGBM, CatBoost, MLP, Voting, and Stacking—demonstrate that boosting and ensemble algorithms (CatBoost and Voting) outperform traditional approaches [13]. This confirms the robustness and generalization capacity of ensemble learning in medical prediction tasks and highlights its potential for reliable early detection and prevention of diabetes in clinical environments.

3 Methodology

In this study, we employed the Frankfurt Hospital[1] dataset (Germany), which is publicly available on the Kaggle platform, to predict the onset of diabetic conditions. This dataset provides a rich and reliable foundation for developing and validating robust predictive models [14]. Its medical origin ensures the inclusion of clinically relevant parameters, which enhances the interpretability and applicability of the obtained results in real-world healthcare settings.

The evaluated models encompass a diverse set of machine learning paradigms, including traditional supervised algorithms (Logistic Regression, Random Forest), boosting methods (LightGBM, CatBoost), neural networks (Multi-Layer Perceptron, MLP), and ensemble learning techniques (Voting and Stacking). As illustrated in Fig. 1, this methodological diversity enables a comprehensive exploration of both linear and nonlinear relationships within the medical data.

3.1 Model Selection Rationale

The chosen models address both linear and nonlinear relationships within the medical data. Logistic Regression captures straightforward dependencies, while Random Forest, LightGBM, CatBoost, and MLP identify more complex patterns among clinical indicators such as glucose, BMI, and insulin.

In terms of interpretability and clinical relevance, Logistic Regression offers clear insight into variable contributions, whereas boosting methods (LightGBM, CatBoost) achieve higher predictive accuracy while maintaining explainability through feature importance and SHAP values. To ensure robustness and efficiency, ensemble techniques such as Random Forest, Voting, and Stacking mitigate overfitting and enhance model stability. Furthermore, CatBoost effectively handles categorical and noisy medical data, while MLP models capture high-dimensional nonlinear dependencies.

By combining linear, tree-based, and neural approaches, the proposed framework ensures both accuracy and interpretability, providing reliable support for the early detection and prevention of diabetes.

[1] https://www.kaggle.com/datasets/johndasilva/diabetes.

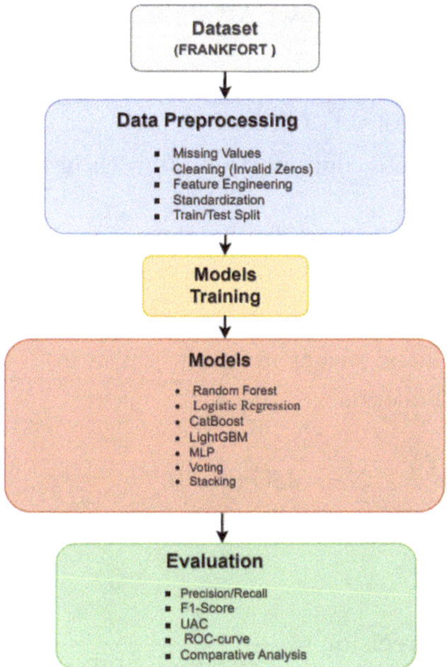

Fig. 1. Steps of the proposed method

3.2 Dataset Description

The dataset used in this research was collected from Frankfurt Hospital, ensuring transparency and reproducibility of the results. This study focuses on the design, implementation, and comparative analysis of intelligent predictive models for diabetes detection, using nine key clinical variables commonly employed in medical diagnostics: Pregnancies, Glucose, Blood Pressure, Skin Thickness, Insulin, BMI, Diabetes Pedigree Function, Age, and Outcome (the target variable indicating the presence or absence of diabetes) (Table 1).

3.3 Performance Evaluation Measures

To assess and compare the performance of our models, we employ the standard evaluation metrics: *precision*, *recall*, and the *F1-score*. Precision reflects the proportion of correctly predicted positive instances among all instances labeled as positive by the model, thus measuring the reliability of predictions. Recall quantifies the proportion of correctly predicted positives relative to all actual positives, highlighting the model's capacity to capture relevant cases. The F1-score, defined as the harmonic mean of precision and recall, provides a balanced indicator that accounts for both accuracy and completeness [15].

As emphasized, precision corresponds to the ratio of true positives to the total number of predicted positives, and it is computed as [16] :

Table 1. Dataset Attributes Description

No.	Attribute Description	Variable Type
1	Pregnancies (Number of times pregnant)	Integer
2	Glucose (Plasma glucose concentration after 2 h in an oral glucose tolerance test)	Real
3	Blood Pressure (Diastolic/Systolic blood pressure in mm Hg)	Real
4	Triceps Skin Fold Thickness	Real
5	Insulin (Serum insulin level after 2 h in μU/ml)	Real
6	BMI (Body Mass Index: weight in kg/(height in m)^2)	Real
7	Diabetes Pedigree Function	Real
8	Age (in years)	Integer
9	Outcome (Diabetic = 1, Non-diabetic = 0)	Binary

$$\text{Precision} = \frac{\text{Correct}}{\text{Correct} + \text{Spurious}} \tag{1}$$

Recall is the fraction of the valid annotations over the total amount of annotations, formally defined as:

$$\text{Recall} = \frac{\text{Correct}}{\text{Correct} + \text{Missing}} \tag{2}$$

The F-measure (or F1-score) is defined as the harmonic mean of precision and recall, expressed as:

$$F1 = \frac{2 \times \text{Precision} \times \text{Recall}}{\text{Precision} + \text{Recall}} \tag{3}$$

4 Results and Discussion

Before presenting the experimental results, it is essential to emphasize that all models were trained and tested on the Frankfurt dataset, which contains clinically relevant features for diabetes prediction. Each algorithm—Logistic Regression, Random Forest, LightGBM, CatBoost, MLP, Voting, and Stacking—was optimized to ensure fair comparison and robust evaluation. Standard performance indicators, including AUC, Precision, Recall, and F1-score, were used to assess predictive efficiency. This comprehensive evaluation framework provides a reliable basis for identifying the most effective models and understanding their behavior in detecting diabetes risk patterns (Table 2).

We applied five-fold cross-validation (see Table 3) to robustly assess model performance in diabetes prediction. The results demonstrate strong consistency across folds, with only minor variations, indicating a high-quality, representative, and well-structured dataset. The low standard deviations in Accuracy, F1-score,

Table 2. Results of performance models.

Model	AUC	Precision	Recall	F1
Logistic Regression	0.85	0.76	0.73	0.74
Random Forest	0.99	0.92	0.90	0.91
LightGBM	0.98	0.95	0.97	0.96
CatBoost	0.98	0.99	0.99	0.99
MLP	0.94	0.82	0.80	0.81
Voting	0.99	0.98	0.98	0.98
Stacking	0.97	0.98	0.98	0.98

Table 3. Performance metrics for five-fold cross-validation.

Fold	Accuracy	F1-score	ROC-AUC
Fold 1	0.9725	0.9619	0.9903
Fold 2	0.9600	0.9380	0.9826
Fold 3	0.9625	0.9458	0.9713
Fold 4	0.9700	0.9538	0.9854
Fold 5	0.9475	0.9157	0.9723
Mean ± SD	0.9625 ± 0.0088	0.9431 ± 0.0159	0.9804 ± 0.0074

and ROC-AUC confirm that the models are reliable, reproducible, and not overly sensitive to specific data splits. These findings show that the models can effectively generalize to unseen data, making them valuable tools for early diabetes detection, risk stratification, and supporting clinical decision-making. Overall, this study highlights the potential of machine learning to deliver accurate, robust, and clinically relevant predictions in the management and prevention of diabetes.

4.1 Logistic Regression

The Logistic Regression model achieved an accuracy (AUC) of 0.850, with a precision of 0.76, recall of 0.73, and F1-score of 0.74, indicating a limited predictive capacity compared to other models. These results suggest that the model struggled to capture the nonlinear dependencies among the clinical variables. Logistic Regression, being a linear classifier, assumes a linear relationship between the independent variables and the target outcome. However, diabetes prediction involves complex, nonlinear interactions among physiological and biochemical factors such as glucose, BMI, and insulin levels. Consequently, while the model provides good interpretability and simplicity, its performance is insufficient for accurate medical diagnosis, where sensitivity and specificity are equally crucial (Fig. 2).

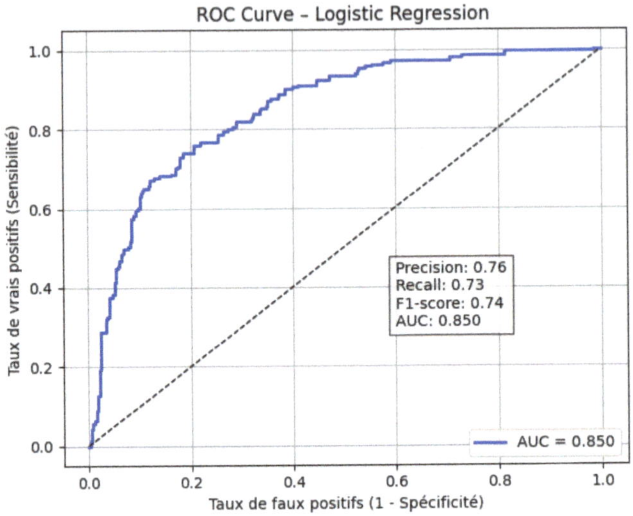

Fig. 2. Metrics performance of Logistic Regression model.

4.2 Random Forest

The Random Forest model exhibited excellent performance, with AUC = 0.995, Precision = 0.92, Recall = 0.90, and F1 = 0.91. This model benefits from its ensemble nature, aggregating multiple decision trees to reduce overfitting and enhance generalization. Its robustness against noise and its ability to handle feature interactions make it a strong baseline for clinical prediction tasks. However, the slightly lower recall compared to boosting methods indicates that some diabetic cases may still be misclassified. This limitation can be addressed by optimizing tree depth and the number of estimators (Fig. 3).

Overall, Random Forest remains a dependable and interpretable model with strong clinical applicability.

4.3 LightGBM

LightGBM delivered highly competitive results, with AUC = 0.988, Precision = 0.95, Recall = 0.97, and F1 = 0.96. Its gradient boosting mechanism efficiently captures complex nonlinearities and interactions among clinical variables. The model's strong recall indicates its excellent ability to correctly identify positive (diabetic) cases, a critical property in medical diagnosis where missing a positive case may have severe consequences. Moreover, LightGBM's leaf-wise tree growth strategy and gradient-based optimization contribute to its superior accuracy and computational efficiency. This balance between performance and scalability positions LightGBM as one of the most promising algorithms for clinical decision support (Fig. 4).

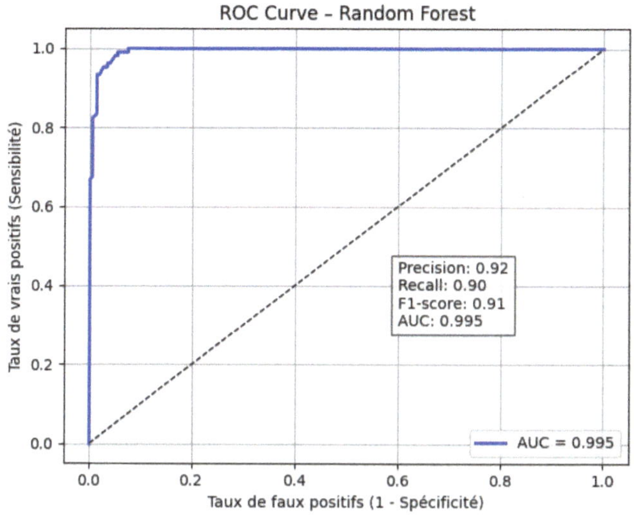

Fig. 3. Metrics performance of Random Forest model.

4.4 CatBoost

The CatBoost model achieved the highest overall performance, with AUC = 0.984, Precision = 0.99, Recall = 0.99, and F1 = 0.99, outperforming all other tested models. CatBoost's exceptional accuracy can be attributed to its advanced gradient boosting architecture and its efficient handling of categorical and numerical features without extensive preprocessing. The model demonstrates excellent stability, avoiding overfitting thanks to its ordered boosting technique. Its nearly perfect precision and recall indicate that it successfully distinguishes diabetic and non-diabetic individuals with minimal misclassification. These results confirm CatBoost's robustness, reliability, and suitability for critical healthcare applications where diagnostic accuracy is paramount (Fig. 5).

4.5 Multilayer Perceptron (MLP)

The MLP neural network achieved AUC = 0.943, Precision = 0.82, Recall = 0.80, and F1 = 0.81. Although the MLP surpasses Logistic Regression, it remains below the ensemble and boosting models. Its moderate performance may be explained by the relatively small dataset size and potential overfitting during training. MLPs are known to require extensive hyperparameter tuning and large datasets to effectively learn complex nonlinear patterns. Nevertheless, the model demonstrates the potential of neural architectures for medical diagnosis tasks, provided that data augmentation, dropout regularization, and architecture optimization are applied to enhance performance and generalization (Fig. 6).

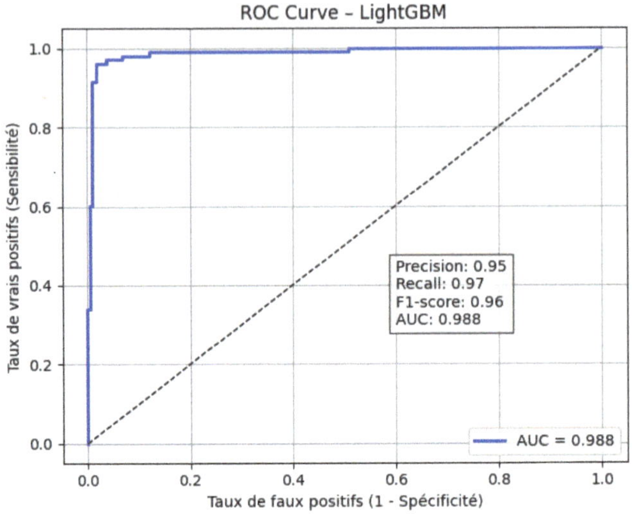

Fig. 4. Metrics performance of LightGBM model.

4.6 Voting Ensemble

The Voting ensemble model recorded AUC = 0.991, Precision = 0.98, Recall = 0.98, and F1 = 0.98, confirming the effectiveness of combining multiple base learners to enhance robustness. By integrating the predictions of several algorithms (e.g., Logistic Regression, Random Forest, LightGBM), the Voting model mitigates individual weaknesses and reduces variance. Its balanced precision and recall indicate consistent classification performance across different patient profiles. Such ensemble frameworks are particularly relevant for medical applications, where a stable and reliable prediction model is essential for clinical decision-making (Fig. 7).

4.7 Stacking Ensemble

The Stacking model also achieved high performance, with AUC = 0.970, Precision = 0.98, Recall = 0.98, and F1 = 0.98. Unlike Voting, which aggregates predictions by majority rule, Stacking leverages a meta-learner to combine base model outputs optimally. This allows the model to learn complex dependencies between the predictions of different learners, leading to improved generalization and reduced bias. The strong performance of the Stacking model highlights the importance of hybrid learning strategies that exploit the complementary strengths of both shallow and deep learners. This approach is particularly suitable for sensitive medical prediction tasks requiring a high level of accuracy and robustness (Fig. 8).

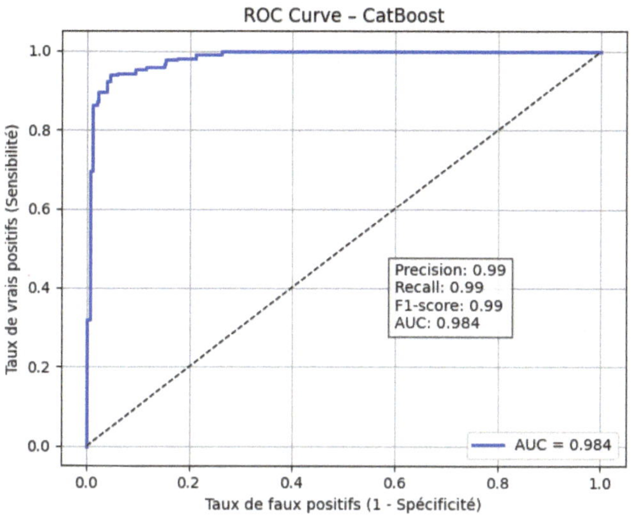

Fig. 5. Metrics performance of CatBoost model.

4.8 Comparative Analysis

The comparative analysis (see Fig. 9 and Fig. 10) reveals a clear performance hierarchy among the models. Boosting and ensemble methods (CatBoost, LightGBM, Voting, and Stacking) consistently outperform traditional models in all metrics. Their ability to handle feature interactions, control overfitting, and leverage collective intelligence among learners make them ideal for early diabetes detection. In contrast, Logistic Regression and MLP show limited effectiveness due to their linearity or insufficient optimization. These findings demonstrate that ensemble-based intelligence is a cornerstone of modern medical AI systems, offering high accuracy, stability, and reliability—all essential for building trust and safety in AI-driven healthcare diagnostics.

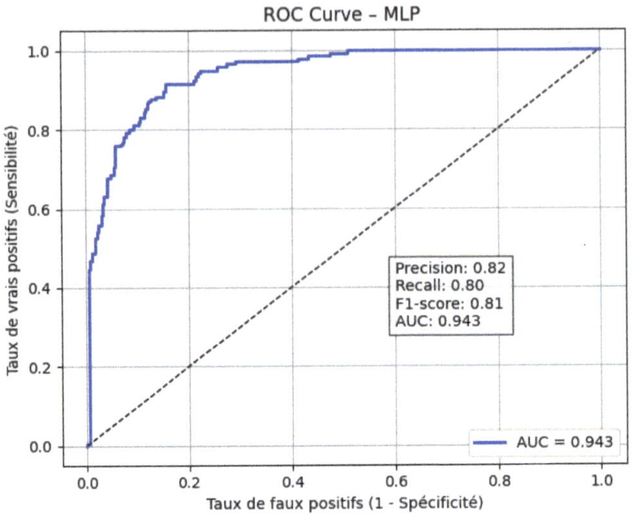

Fig. 6. Metrics performance of MLP model.

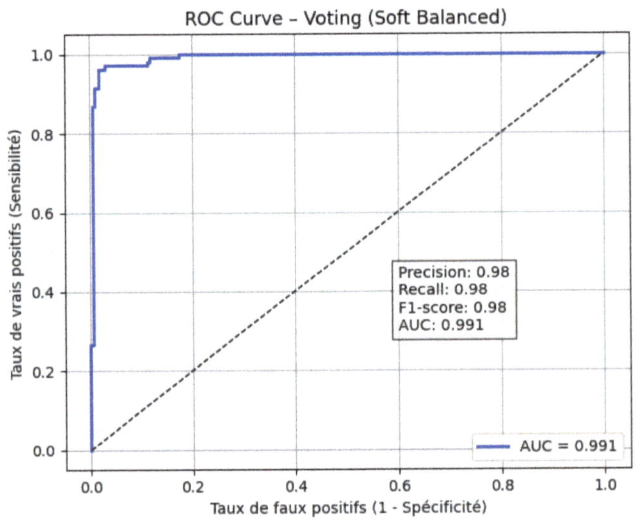

Fig. 7. Metrics performance of Voting model.

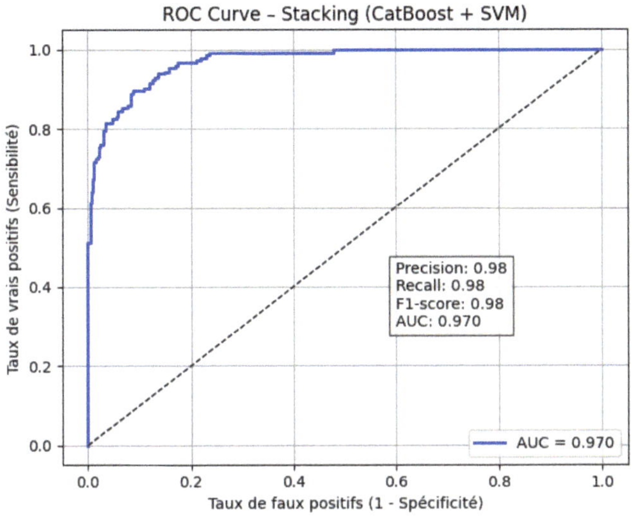

Fig. 8. Metrics performance of Stacking model.

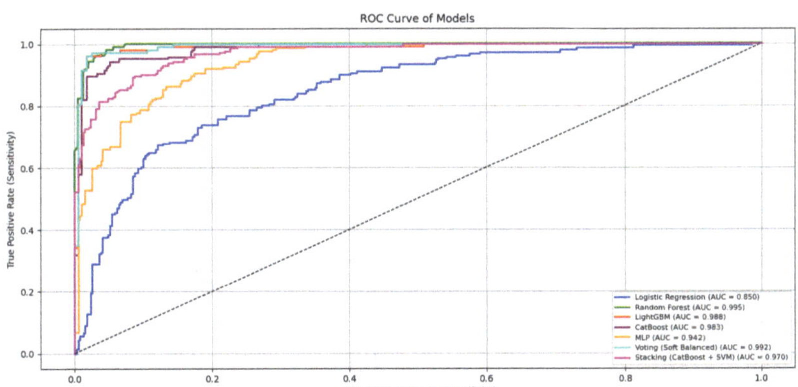

Fig. 9. ROC Curve of Models.

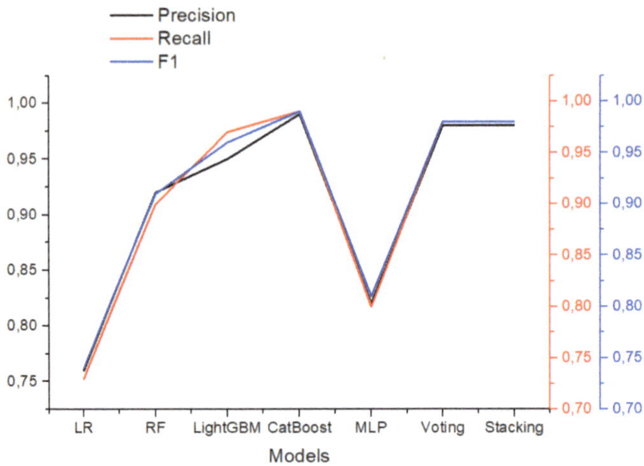

Fig. 10. Comparison of Performance Metrics of Models.

5 Explainable IA

After selecting CatBoost as the best-performing model, we analyze its feature importance to provide explainable AI (XAI) insights. The results (see Table 4 and see Fig. 11) highlight Glucose as the most influential predictor, confirming its central role in diabetes diagnosis. BMI ranks second, reflecting the strong impact of obesity-related factors and CatBoost's ability to capture non-linear relationships. DiabetesPedigreeFunction and Age show comparable importance, indicating the combined influence of genetic predisposition and aging. Variables such as BloodPressure, Pregnancies, Insulin, and SkinThickness contribute moderately, offering complementary information. Overall, the findings demonstrate that CatBoost relies on clinically meaningful features while ensuring high interpretability for medical decision-support systems.

Table 4. Feature Importance of input variables.

Rank	Feature	Importance
1	Glucose	24.110013
2	BMI	18.853005
3	DiabetesPedigreeFunction	13.366652
4	Age	13.026740
5	BloodPressure	8.642035
6	Pregnancies	8.289265
7	Insulin	7.439320
8	SkinThickness	6.272971

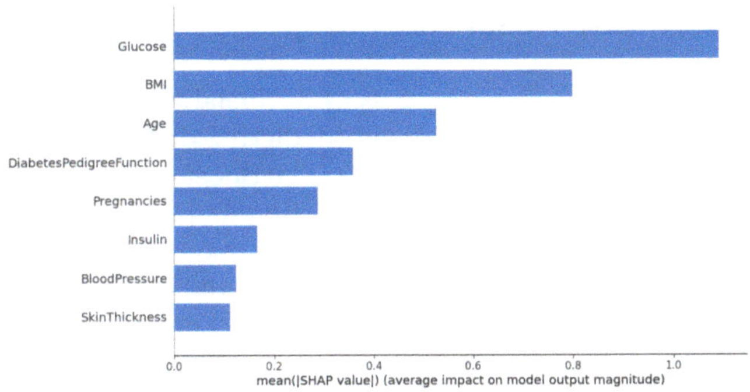

Fig. 11. Explainable AI (XAI) Feature Importance Visualization.

The SHAP beeswarm plot (see Fig. 12) illustrates the contribution of each feature to the CatBoost model's predictions for diabetes. Glucose and BMI are the most influential features, with higher values strongly increasing predicted risk. Age and DiabetesPedigreeFunction also contribute positively, reflecting the impact of age and genetic predisposition on diabetes likelihood. Variables such as Pregnancies and Insulin have moderate influence, while BloodPressure and SkinThickness contribute minimally. Overall, the plot confirms that the model captures clinically relevant risk factors, and the SHAP values provide clear interpretability, supporting the reliability and transparency of the predictions for potential clinical application.

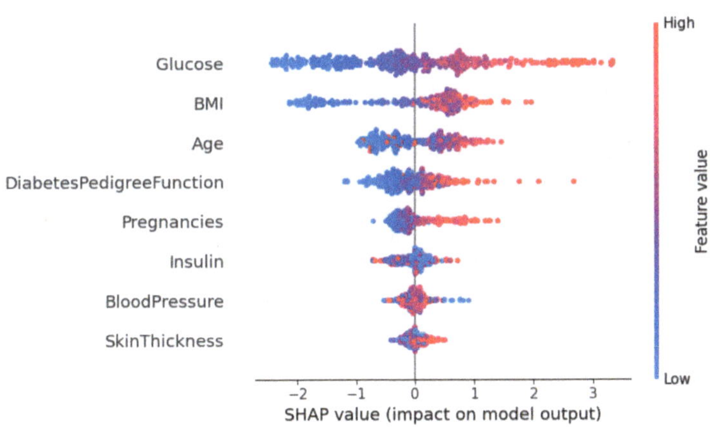

Fig. 12. Explainable AI SHPE VALUE Feature Importance Visualization.

6 Conclusion and Research Perspectives

This study demonstrates the effectiveness of Artificial Intelligence—particularly Machine Learning and Deep Learning—in predicting diabetes using simple clinical variables. Among the evaluated models, CatBoost and LightGBM achieved the best individual performance, while the Voting and Stacking ensemble methods exhibited strong stability and balanced accuracy. These findings highlight the relevance of hybrid learning approaches in enhancing diagnostic reliability. The results confirm the great potential of Machine Learning techniques to support early diagnosis and prevention of diabetes-related complications, paving the way toward more data-driven and efficient healthcare systems.

This work opens promising avenues for advancing predictive and personalized medicine, fostering data-assisted clinical decision-making. Future research will focus on:

- Integrating additional clinical and behavioral variables to enrich prediction accuracy.
- Exploring advanced deep learning architectures such as CNNs, RNNs, and Transformers for more complex pattern recognition.

References

1. Nagesh, M.R., Vijayakumar, N., Bhavani, K.: A review on diabetes mellitus–an annihilatory metabolic disorder. J. Pharm. Sci. Res. **12**(2), 232–235 (2020)
2. Tinajero, M.G., Malik, V.S.: An update on the epidemiology of type 2 diabetes: a global perspective. Endocrinol. Metab. Clin. **50**(3), 337–355 (2021)
3. Huang, Q., et al.: Global burden and risk factors of type 2 diabetes mellitus from 1990 to 2021, and forecasts to 2050. Front. Endocrinol. **16**, 1538143 (2025)
4. Zhang, J., Zhang, Z., Zhang, K., Ge, X., Sun, R., Zhai, X.: Early detection of type 2 diabetes risk: limitations of current diagnostic criteria. Front. Endocrinol. **14**, 1260623 (2023)
5. Khokhar, P.B., Pentangelo, V., Palomba, F., Gravino, C.: Towards transparent and accurate diabetes prediction using machine learning and explainable artificial intelligence. arXiv preprint arXiv:2501.18071 (2025)
6. Ahmed, A., et al.: Machine learning algorithm-based prediction of diabetes among female population using PIMA dataset. Healthcare **13**(1), 37 (2024)
7. El-Sofany, H., El-Seoud, S.A., Karam, O.H., Abd El-Latif, Y.M., Taj-Eddin, I.A.: A proposed technique using machine learning for the prediction of diabetes disease through a mobile app. Int. J. Intell. Syst. **2024**(1), 6688934 (2024)
8. Kurt, B., et al.: Prediction of gestational diabetes using deep learning and Bayesian optimization and traditional machine learning techniques. Med. Biol. Eng. Comput. **61**(7), 1649–1660 (2023)
9. Shrestha, M., et al.: A novel solution of deep learning for enhanced support vector machine for predicting the onset of type 2 diabetes. Multimedia Tools Appl. **82**(4), 6221–6241 (2023)

10. Rajagopal, A., Jha, S., Alagarsamy, R., Quek, S.G., Selvachandran, G.: A novel hybrid machine learning framework for the prediction of diabetes with context-customized regularization and prediction procedures. Math. Comput. Simul. **198**, 388–406 (2022)
11. Chang, V., Ganatra, M.A., Hall, K., Golightly, L., Xu, Q.A.: An assessment of machine learning models and algorithms for early prediction and diagnosis of diabetes using health indicators. Healthc. Anal. **2**, 100118 (2022)
12. Ahmed, N., et al.: Machine learning based diabetes prediction and development of smart web application. Int. J. Cogn. Comput. Eng. **2**, 229–241 (2021)
13. Hadji, A., Kholladi, M.K.: Disaster information extraction: evaluation of NLP techniques using JAPE rules, ontologies and machine learning approaches. In: Proceedings of International Conference on Computer Science Complex System Applications (CCSA), pp. 294–313. Springer, Cham (2025). https://doi.org/10.1007/978-3-031-90758-6_22
14. Hadji, A., Boumaza, F., Bali, D.S.: Sentiment analysis of digital currency discussions: a machine learning and ontology approaches. In: Proceedings of International Conference on Informatics Applied Mathematics (IAM24), CEUR Workshop Proceedings, vol. 3922, pp. 22–35 (2024)
15. Hadji, A., Kholladi, M.K., Borisova, N.: Enhancing spatial information extraction from Arabic text: a hybrid approach with ontology and rule-based methods. Ingénierie des Systèmes d'Information **29**(4), 1261 (2024)
16. Hadji, A., Boumaza, F., Hadji, F.: Enhancing smart grid management through machine learning-based hourly energy forecast. ITEGAM-JETIA **11**(56), 136–147 (2025)

Author Index

A
Abdessalam, Lokmane 234
Abid, Faroudja 224
Aissaoui, Rahma 44
Ali, Wided 57
Amamra, Abdenour 150
Aroui, Abdelkader 30
Asmaa, Chatta 132

B
Barkache, Abderrahman 258
Belaidi, Hadjira 150
Benbelgacem, Mohamed 13
Benbelgacem, Sonia 13
Benhamza, Karima 22
Benkhaled, Imad 166
Benmerzoug, Djamel 234
Benselim, Rania 22
Bouakkaz, Fatima 57
Boubakeur, Latreche 132
Bouhaouche, Ibtissem 114
Boumaza, Farid 30, 283
Bouramoul, Abdelkrim 234
Boussadia, Ahlam 95

C
Chaib, Mohammed 258
Chaouche, Ahmed-Chawki 77

D
Daddiouamer, Redouane 234
Demim, Fethi 150
Dilekh, Tahar 211
Djebbar, Akila 188

F
Farah, Rania 44
Fatima, Haloua 250
Ferrag, Mohamed Amine 65

G
Guerfi, Roudeina 77
Guezouli, Larbi 13, 211

H
Hadjadj, Ranya 234
Hadji, Atmane 30, 283
Hallaci, Samir 258
Halladja, Ghada 188
Hamouda, Djallel 65
Hattab, Abdessalam 211

I
Imane, Kadi 250

K
Kerdoudi, Saoueb 211
Khebizi, Ali 114
Khelifi, Manel 95
Kouahla, Zineddine 44, 65

M
Mansouri, Elhocine 1
Melkemi, Kamal Eddine 1
Merouani, Hayet Farida 188
Merzougui, Hadil 22
Messioud, Wassim 258
Michelucci, Dominique 1
Mokhtari, Bilal 1
Mukhaimar, Samaa 188
Mustapha, Bouakkaz 132

O
Ounnas, Badreddine 166

Z. Kouahla et al. (Eds.): IAM 2025, LNNS 1973, pp. 301–302, 2026.
https://doi.org/10.1007/978-3-032-26031-4

R
Rouigueb, Abdenebi 150

S
Sardou, Adel 150
Seghir, Rachid 13
Seridi, Hamid 258

Y
Yahiaoui, Achraf 258

Z
Zebiri, Rania 77
Zitouni, Farouq 166
Zouache, Djaafar 30

FSC
www.fsc.org
MIX
Papier | Fördert
gute Waldnutzung
FSC® C083411

Kolibri 360 GmbH
Ferdinand-Jühlke-Straße 7
99095 Erfurt, Deutschland
produktsicherheit@kolibri360.de